Computer Telephony Demystified

Other McGraw-Hill Books of Interest:

Computer Telephony Demystified

Putting CTI, Media Services, and IP Telephony to Work

Michael Bayer

McGraw-Hill

NEW YORK   SAN FRANCISCO   WASHINGTON, D.C.   AUCKLAND   BOGOTÁ
CARACAS   LISBON   LONDON   MADRID   MEXICO CITY   MILAN
MONTREAL   NEW DELHI   SAN JUAN   SINGAPORE
SYDNEY   TOKYO   TORONTO

Library of Congress Cataloging-in-Publication Data

Bayer, Michael (Michael Thomas)
        Computer telephony demystified : putting CTI, media services, and
    IP telephony to work / Michael Bayer.
            p.   cm.
        Includes bibliographical references and index.
        ISBN 0-07-135987-7
        1. Internet telephony.   2. TCP/IP (Computer network protocol)
    3. Telephone systems.   4. Mobile communication systems.
    5. Multimedia systems.   I. Title.

    TK5105.8865 .B39 2000
    621.385--dc21

                                                            00-043230

McGraw-Hill

A Division of The McGraw·Hill Companies

1 2 3 4 5 6 7 8 9 0   DOC/DOC   0 9 8 7 6 5 4 3 2 1 0

P/N 0-07-136328-9

ISBN 0-07-135987-7

The sponsoring editor for this book was Steve Chapman. It was set in Palatino by Michael Bayer of Computer Telephony Solutions. Printed and bound by R.R. Donnelley & Sons Company.

This book is printed on recycled, acid-free paper containing a minimum of 50% recycled, de-inked fiber.

McGraw-Hill books are available at special quantity discounts to use as premiums and sales promotions, or for use in corporate training programs. For more information, please write to the Director of Special Sales, McGraw-Hill, 2 Penn Plaza, New York, NY 10121-2298. Or contact your local bookstore.

To my parents, whose love, support, and encouragement
– which I have always taken for granted –
has been more than anyone could ever hope for.

Notices

Contents

Contents

Contents

Contents

Contents

Contents

Contents

Contents

Contents

Contents

Contents

Contents

Contents

Contents

Contents

Contents

Sidebars

Sidebars

8. Switching Fabric Implementation

Figures

1. What Is Computer Telephony?

2. CT Solutions and Benefits

Figures

Figures

Figures

8. Switching Fabric Implementation

Figures

11.CT System Configurations

Figures

13. CT Solution Examples

Tables

13.CT Solution Examples

Acknowledgments

Thank you to Mark Hume, Bob Papp, Frederic Miserey, Mike Spieser, Jean Michel Karr, John Stossel, Susan Schuman, and all the other members of the extended MTA team that got this ball rolling.

Thanks to my friends Susan King, Bob Flisik, and Lisa Heller, who planted the seeds for this book.

Thank you to Dave Anderson, Frederic Artru, Marc Fath, Ellen Feaheny, Jose Garcia, Jim Knight, Peter Kozdon, Tom Miller, Neal Oliver, Jeff Peck, Gib Ritenour, and Steve Rummel for their years of dedication to making CT Plug & Play a reality.

A big thank-you to Dave Anderson, Frederic Artru, April Bayer, Marc Fath, Lisa Heller, Laura Mack, Tim Schweitzer, Mike Spieser, and Tamara Wilkerson for all the constructive feedback and editorial input.

Last but not least, thanks to Steve Chapman for sponsoring this project and fighting all the necessary battles.

Introduction

This book is for anyone who wants to know about Computer Telephony (CT), specifically anyone who is:

- Interested in opportunities to develop or deploy CT solutions or products; or

- Interested in putting CT to work in a home or business.

This book does not assume that you have any prior knowledge of telephony. All the telephony concepts and terminology you need in order to master CT are explained and illustrated.

This book shows you how to put time and resources to better use if you:

- Manage or work in a telecommunications or information systems (IS) department

- Manage or work in a call center or help desk environment

- Develop or deploy new services for a telecommunications company, including long distance and local exchange carriers, CLECs, ISPs, and ASPs.

- Develop application, utility, or system software

Introduction

As CT continues to emerge as a mainstream technology, you'll find this book indispensable as both an introductory text and a reference guide if you work for a:

- Telephone company

- Cable company

- PBX company

- Consumer electronics company

- Computer or electronics store

- System integrator

If you do have a telephony background, you will find this book's review of telephony from a universal perspective to be invaluable. In the world of telephony technology, every company uses a slightly different vocabulary, and this complicates CT product and solution development. To address the requirement for a unifying model, this book presents a common set of concepts and terminology which reflect industry standards. These are explained in a fashion that makes them easy to relate to the models with which you are familiar.

The Time for CT Is Now!

The field of CT technology has been developing slowly but steadily for more than a decade. Early pioneers and visionaries have championed and struggled to bring about the ubiquitous integration of computers with telephony functionality.

Unfortunately, CT is unavoidably a complicated topic. The results of more than 100 years of frenetic innovation in the telephony world, and an even faster and more innovative explosion of technology in the computer world, meant that considerable effort was needed to build a framework for no-compromise CT solutions. The bulk of this framework has now been completed and this book represents the result.

Despite the challenges involved, the emergence of CT as an essential element of everyone's technology environment is inevitable. Many incremental developments in the telephony, telecommunications, and computer industries have now brought CT to the point where it has emerged as a mainstream technology.

Three key factors mark the maturity of CT technology:

- *Interoperability,* through standard CT Plug and Play protocols

- *Power,* through easy-to-use software tools

- *Access,* through low-cost multimedia communications hardware

This book provides a thorough introduction to telephony and CT technology based on the latest industry standards. These standards represent a universal abstraction, or unification, of the historically divergent telephony world. The framework presented by this book will help you understand the capabilities, limitations, and potential of existing products and equipment, and to anticipate the next generation (and future generations) of new telephony and CT products.

With the help of this book you will be able to make the right CT planning and implementation decisions now. With knowledge of the core CT framework you will be able to in anticipate future CT developments and you won't get trapped at a technological dead end.

Organization of this Book

To be a passenger on an airplane you really don't need to know anything about the physics of flying, the training and licensing of pilots, the safety record of different airlines, or even how to buy a ticket. Whether you are an airplane designer, a new pilot, an investor in airline stocks, or just a passenger, however, a little bit of insight into each of these areas can greatly enhance your enjoyment of the experience and the likelihood of your success.

Introduction

With this in mind, the goal of this book is to provide you with a complete insight into CT and its underpinnings, regardless of what your final objective might be. In fact, the more you learn about the endless possibilities for taking advantage of CT, the more you'll want to put it into practice!

This book covers all four disciplines of computer telephony: switching fabrics, CTI, media services, and administration. While many readers will be interested in how these areas relate to one another, this book has been constructed so that readers interested in just one or more of these areas can skip the sections that are not of interest. Chapter 3 serves as a common foundation for all four areas. Chapters 4, 5, and 6 apply to CTI. Chapter 7 applies to those interested in media services. Chapter 8 applies to readers interested in switching fabric implementation and Chapter 9 provides an introduction to CT administration technology.

Chapter 1: What Is Computer Telephony?

Chapter 1 describes exactly what Computer Telephony is and what it is not. It provides context for the discussion of CT and specifically relates CT technology to the broader category of CT solutions. This chapter presents the fundamental insights into why CT is now maturing, and why it is likely to become as ubiquitous as the graphical user interface (GUI) and other mainstream personal computing technologies.

Chapter 2: CT Solutions and Benefits

Chapter 2 presents a collection of eight complete CT solution scenarios in which people from different walks of life have applied CT technology to improve their working and living environments. This chapter explains the CT value chain, which reflects all of the different services and components that must be integrated to build functional CT solutions. The CT value chain provides the hierarchy around which the rest of the book is structured.

Chapter 3: Telephony Concepts

Chapter 3 presents a thorough introduction to the telephony concepts you need to know in order to take full advantage of CT technology. This chapter presents the universal abstraction of telephone system technology that is used in computer telephony technology. This simple abstraction is used throughout the book as a framework for easily describing any telephony product or capability.

Chapter 4: Telephony Device Concepts

Chapter 4 builds on the concepts presented in Chapter 3 and shows how the logical and physical elements and components of telephone system devices are modeled and identified to support the diversity of telephone system implementations.

Chapter 5: Call Processing Features and Services

Chapter 5 builds on the concepts presented in Chapters 3 and 4. It shows how the elements within telephone systems of any size interact to provide the comprehensive range of telephony features and services commonly available.

Chapter 6: CTI Concepts

Chapter 6 introduces the concepts of CTI as a means for observing and controlling telephone systems of any size. This chapter discusses how telephony features and services are accessed through a CTI interface, and presents those capabilities of a telephone system that are specific to CTI interfaces.

Chapter 7: Media Services Concepts

Chapter 7 introduces the concepts of CT media services as a means for generating, processing, and manipulating telephony media streams. This chapter presents the model used to provide access to telephone system media processing resources through media service interfaces.

Chapter 8: Switching Fabric Implementations

Chapter 8 describes the diversity of switching fabric implementations found in telephone systems. This chapter explains the role of the switching functions within a telephone network and the various transport options for implementing individual media streams over various network technologies ranging from traditional analog networks to digital telecommunications networks, wireless networks, and packet-based networks.

Chapter 9: CT Administration

Chapter 9 deals with the most often overlooked aspect of telephone systems: administration. While technologies for administering CT systems lag behind the technologies for CTI, media services, and switching, this chapter introduces the key concepts of CT administration and provides an indication of the innovations likely to emerge.

Chapter 10: Telephony Equipment and Network Services

Chapter 10 relates all the telephony concepts presented to tangible telephony product and service implementations. It describes the options that exist for assembling a telephone system ranging in size from a single telephone to a private telephone network.

Chapter 11: CT System Configurations

Chapter 11 presents the diversity of physical CT configurations that are now possible and that will become easier and easier to assemble, given the maturation of CT technology and availability of products based on standardized CT protocols.

Chapter 12: CT Software

Chapter 12 complements Chapter 11 by presenting the diversity of software configurations that are available to support application development, CT system integration, and user customization of CT solutions.

Chapter 13: CT Solution Examples

Chapter 13 revisits the CT solution scenarios presented in Chapter 2 and looks under the covers to explain how each can be implemented using the technologies explored in the book.

To pull everything together, each chapter ends in a review of the key concepts covered. While the book is intended to be read from beginning to end, it is designed to permit the use of individual chapters. References to key concepts explained in other parts of the book will make it simple for you to navigate.

This icon is used throughout the book to mark key definitions.

This icon is used to mark fundamental concepts of which to take special note.

This icon is used to mark points that clarify sources of potential confusion.

This icon is used to mark real-world examples and applications of computer telephony technology.

You will find that this book is extensively illustrated, to help make the frequently complex and abstract concepts surrounding telephony and CT easier to comprehend. Standardized graphical notations are used wherever applicable. The symbols and icons used in these graphical notations are summarized on the inside of the front and back covers for easy reference.

Sidebars and footnotes are used throughout the book to present material that is useful but nonessential. You can read these for diversion and further insight as you proceed through the book, or you can return to them later.

Exploring Further

This book is intended for users of any type or brand of telephone system, computer, or operating system. To avoid any bias in the illustrations used throughout the book, all diagrams use stylized icons to represent the various components of a system. References to particular products are made only when the product in question represents a recognized, de facto standard.

Use this book as a resource to help you identify your CT opportunities; determine the way that you want to approach CT; identify the type of solutions, products, or components you want to assemble or build; and assess your needs and preferences. Armed with the insights you'll have, you can identify the products and the strategies that best satisfy your needs.

If you wish to delve into greater technical depth, or wish to arm yourself with specific details from the applicable published standards and specifications, you will find the included CDROM to be a natural extension of this book. The CDROM is packed with specifications and information of use to application developers, product designers, and those who must specify system requirements. If you are interested in developing CT products, you may also wish to consult the bibliography at the back of this book for a list of other valuable development documentation.

The pace of new product development in the world of telephony and CT is so rapid that product-specific information is not included here; it would become obsolete far too quickly. To find out about the latest in available products, you might wish to:

- Contact your existing telephone equipment vendor(s);

- Subscribe to industry magazines;

- Monitor the efforts of groups such as the ECTF (their web site is at *http://www.ectf.org*);

- Attend industry trade shows; and

- Check the author's web site at *http://www.CTExpert.com.*

In particular, updates to the material in this book will be regularly posted to the web at *http://www.CTExpert.com*/book.

1.
What Is Computer Telephony?

Telephones and computers indisputably are the two technologies that most greatly impact every aspect of our daily lives. These technologies are central to the operation of virtually any business of any size. A vast number of organizations exist only because of these technologies. In fact, one could argue that these are the technologies that hold together the very fabric of the modern information-oriented economy. Computer Telephony, or CT, brings these two technologies together and harnesses their synergy.

The telephone network is a tremendous resource, of which everyone should be able to take full advantage. Telephony[1-1] technology, the technology that lets people use the telephone network, is extremely powerful and has great potential for empowering people in every walk of life.

For the vast majority today, however, the only means of accessing the telephone network is through a telephone set or answering machine of some sort. The one thing all these devices have in common is their

---

1-1. Telephony — Incidentally, the word "telephony" is pronounced "teh·LEF·eh·nee." It should not be pronounced "teh·leh·FOH·nee" as this generally leads to some embarrassment.

arbitrary and very limited user interface. The telephone set is a terrific device for getting sound into and out of the telephone network, but it is a very poor device for getting access to powerful technology. Even the simplest features are rarely mastered by the average telephone user. All too often you've heard someone on the telephone say, "I'll try to transfer you now—but if this doesn't work, here's the number to call."

Figure 1-1. Without CT and with CT

Personal computers, the Internet, fiber optic networks, speech recognition software, and other revolutionary information technologies are reshaping how we live and work. Organizations are anxious to put these new technologies to use. However the full potential of these new technologies can only be realized when they are integrated with the telephone system. Yet just as telephone systems have offered limited interfaces to individual users, they have traditionally offered little access to those who wish to customize them to their own specific requirements.

Computer telephony changes all of this. CT provides an alternative means of accessing the power of telephony technology. Computer technology is an empowering technology because it allows people to extend their reach, and it can take on time-consuming or non-creative tasks. Computer technology allows for tremendous customization, so that you can have a user interface and work environment that is optimized for your needs. With CT, this means you can have full access to the power of the telephony technology you need, and have it in the form that is best for you. A computer working on your behalf can take actions independently, so it effectively becomes your assistant. With CT technology, this means your computer can screen calls, handle routine requests for information without your intervention, and interact with callers in your absence. As shown in Figure 1-1, CT technology not only gives you full access to telephony technology, it actually amplifies its utility!

The following examples illustrate how computer technology has the power to amplify the power of telephony and make it more accessible.

- At a pay phone:

    - *Without CT*, you have to laboriously enter your carrier preference (if you can remember the access code), credit card information, and number you want to dial.

    - *With CT*, you simply point your personal digital assistant (PDA) or laptop computer at the infrared (IR) port on the pay phone and click on a person's name. In fact, if you'll be there a while, you can have selected calls rerouted from your office number to the pay phone, which would in turn alert you with information about each incoming call.

- At home:

    - *Without CT*, you're in the living room watching TV and you remember that you have to talk to a colleague (who is traveling on the other side of the world) before he leaves his hotel. You head for the phone in your den, look up the number for his hotel, look up the country code for the country where he is traveling, and the access code. After you have dialed all of the digits correctly (on the second or third try) and complete your short conversation, you jot down a note to yourself as a reminder to expense the telephone call when you get your next bill. By the time you get back to the living room, you've missed the rest of your TV program.

- *With CT*, you wait for the next commercial break and then, with your TV's remote control, you pop up your personal directory as a picture-in-picture on your TV set and select your colleague's name. The TV then displays a set of locations, including the hotel where he is staying. You select the hotel and indicate you want to dial the number. The appropriate number is dialed automatically (using the cheapest available carrier); your TV then acts as a speaker phone, allowing you to converse with your colleague without leaving your seat. When the call completes, you return to your TV program. Information about the call is logged to your home finance package as an expensable item.

• While telecommuting:

- *Without CT*, you appreciate the opportunity to work at home as a means of avoiding the long commute and the other inefficiencies of work at the office. Unfortunately, however, it means you miss many important telephone calls that then wind up in your voice mail box at the office. You don't want to forward the calls home because you don't want them inadvertently being directed to your family's residential voice mail system.

- *With CT*, your home personal computer has a remote connection to the local area network at your office, allowing you to monitor your telephone. If an important call comes through, your computer notifies you so you can decide whether you want it to go to voice mail or be redirected to your home telephone. Using Internet telephony, you can take the call even if your phone line is busy. In fact, the system works so well that even the call center agents in your company, who spend their whole day handling customer telephone calls, are able to work from home.

- At the office:

  - *Without CT*, you are trying to cope without the administrative support you lost through organizational downsizing. Throughout the day calls pile up in your voice mail box instead of being directed to people who could take care of callers right away. You place many telephone calls yourself, and when you need to reach someone, you spend a lot of time placing calls—only to get their voice mail systems, and you end up having to try again later.

  - *With CT*, your desktop personal computer has taken the place of your secretary. Using CT technology, it screens every call that comes in. If you're busy, you aren't disturbed. If the call is best handled by someone else, it is redirected without your intervention. When you need to get hold of someone, your personal computer will keep trying until the person is reached; only then does it connect you to the call, so you aren't tied up doing the redialing yourself.

- In the school:

  - *Without CT*, there are just three telephone lines in the whole school and they can be accessed only from the principal's office and from the staff room. Despite the fact that teachers believe students would benefit greatly from the ability to use telephones as a resource for research and other projects, there are no telephones in the library or in the classrooms because there is no way of controlling their use.

  - *With CT*, the school has an Internet telephony gateway and each of the computers in the classrooms and the library are equipped with speaker phone applications. These allow groups of children to sit around computers and place telephone calls to resource people in the community, as well as children at other schools. The school doesn't need a high-speed connection to the Internet because most calls are placed using one of the phone lines the school already has. The school's electronic bulletin board on the Internet, which lists homework assignments and lunch menus, is available not only to parents who have computers at home, but also those who want to dial in from any touchtone telephone.

- In a meeting:

  - *Without CT*, you may miss an important call that you are expecting because you must attend an equally important meeting. You have no secretary and cannot forward all of your telephone calls to the meeting room because it would be too disruptive. You must rely on voice mail to catch the call. You leave the meeting frequently to see if it has come through and then you try to return the call.

  - *With CT*, your personal computer checks the origin of each call and identifies the important call when it arrives. Rather than simply taking voice mail, it tells your caller that it will try to track you down. It then sends a notification to the wireless PDA (personal digital assistant) you are carrying. This notification tells you who called and informs you that the caller is holding. You can respond by instructing your computer to take a message, hold the call while you return to your desk, or transfer the call to a nearby telephone. The others in the meeting are not even aware that you are having this wireless dialog with your desktop computer.

The ultimate promise of information technology is that it empowers people to collaborate with one another more effectively and with greater ease. Since the invention of the telephone over a century ago (in 1876), telephony technology has evolved tremendously. Innovation was fast and furious in the earliest days after Alexander Graham Bell's invention; it took less than two years to commercialize the technology. To this day the pace has not slowed. While modern computer technology is a much more recent invention, its history has been just as fast-paced. The key to both disciplines is their ability to improve the ways people communicate and collaborate, because these activities are at the heart of all human endeavor.

This book differentiates between "CT technology" and "CT solutions." This book covers all the concepts and details you need to know about CT technologies, and then illustrates how these technologies are applied in typical CT solutions. These terms are further explained in section 1.11.

1.1. The Importance of Telephony

The telephone is the single most important communication appliance in the world today. Despite the high levels of acceptance for various other forms of information technology, including consumer electronics products and personal computers, the telephone (in all of its forms) remains the only communication device that can be considered ubiquitous.

Figure 1-2. Everyone you want to talk to is somewhere on the telephone network

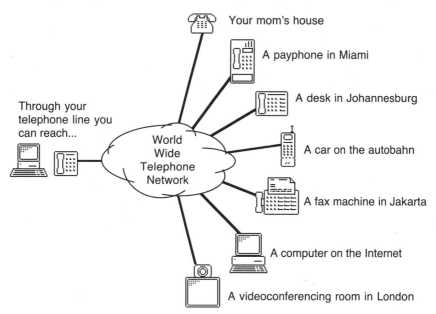

7

The telephone network is the single most important network in the world because:

- There are more "endpoints" on this network than any other (counting just the number of telephone sets).

- The telephone network requires the least effort to use because its basic format is voice. Virtually any human being on the planet is able to interact with any other through the telephone network.

- The telephone network acts as both the on- and off-ramp, and as the interconnection between, most other networks in the world. In fact, there are very few people, fax machines, computers or data networks in the world that cannot be reached through the worldwide telephone network (Figure 1-2).

1.2. The Importance of Computers

The "C" in "CT" stands for "computer." This may be a little misleading, though, because the word computer is being used here in its broadest sense. The term refers to any device that can be programmed to control or observe telephony resources. Think of it as any device using "computer technology." While this set of devices certainly includes traditional mainframes and minicomputers, as well as your own personal computer, it also includes many others. For example, PDAs are small, handheld devices that generally have much less power than traditional computers and a more limited set of uses. But as personalized information appliances, PDAs are very important when it comes to telephony integration. Similarly, a whole new generation of so-called "intelligent," programmable consumer electronics products will transform everyday devices such as VCRs, TVs, watches, games, etc., into what we consider to be "computers" for purposes of CT (Figure 1-3).

Contrary to popular wisdom, the diversity of computing devices continues to grow as computer technology becomes more pervasive and more specialized. In the early days of computer technology, a panel of experts is said to have concluded that only a small number of

Migration from the Real to the Virtual Desktop

At every stage in the evolution of the personal computer, more and more of the items that once sat on our physical desktops have migrated to the virtual desktop of the personal computer.

- The typewriter was replaced by the word processor.
- The calendar was replaced by the scheduling application.
- The desktop calculator was replaced by its software counterpart.
- The dictionary was replaced by the spell checker.
- The personal stereo system was replaced by the PC's built-in CD player.
- The "In" and "Out" boxes were replaced by e-mail.
- The Rolodex™ was made obsolete by contact management software.
- The fax machine was replaced by scanners and fax modems.

The telephone is the only item on the desktop that hasn't been greatly affected by the information technology revolution. CT is the technology that is bringing about the evolution of the telephone. Configurations for desktop CT solutions are covered in Chapters 11 and 12, where your questions about the next step in this evolution will be answered.

A significant benefit of this function migration to the personal computer's virtual desktop is that with a laptop computer or PDA you can simply pick up your entire desk and take it anywhere you want. With the next generation of information appliances, the appropriate parts of your office will be available to you in your car, your living room, or on your refrigerator door.

Figure 1-3. Diversity of applicable computer technologies

computers would ever be needed! Not so long ago it was thought that the world of personal computing revolved around one or two popular operating systems. The reality is that as innovative new devices are built for everything from automobile navigation to home entertainment, personalized computing devices become more and more diverse. Many don't even have operating systems in the traditional sense.

The motivation behind CT is *customization* and *control*. The real significance of integrating computer technology with telephony is the opportunity to tie any form of intelligent appliance, whether it be a mainframe computer, personal computer, or other personal information appliance, into the telephone system in order to create a customized, and preferably personalized, communications environment.

1.3. Communications and Collaboration Technology

Before you conclude that computer telephony deals with all aspects of human communication, it is important to put CT in the context of other areas of communications technology, or specifically, what we will refer to as *communications and collaboration* technology (C&C for short). This term refers to the superset of computing technologies and communications technologies that allow people to collaborate with one another.

Collaboration Grid

People frequently described human collaboration in terms of the grid shown below. One axis represents separation by time and the other separation by place.

When two people are not separated by either time or place, they can have a face-to-face conversation, read from and write on the same piece of paper, etc.

When two people are separated by time, they must use store-and-forward technology. If they are not separated by place, this can be as simple as leaving notes on the refrigerator door (or, in ancient times, hand paintings on cave walls). It can be as technologically sophisticated as leaving a message stored in a shared computer. If the two are separated by both place and time, then store-and-forward techniques such as postal mail, e-mail and electronic publishing come into play.

When two people are separated by place but not by time, they rely on real-time communications technology to talk and possibly share visual information at a distance.

1.3.1. Overall Vision for C&C

Already personal computers are being seen more as communications tools and media delivery devices than as the "document processors" they were just a few years ago. As information technology merges with communications technology, personal computers and other information appliances become the focal point for people collaborating with other people.

Information technology increasingly is taking the place of administrative staff who once provided much of the management of communications and collaboration activity. In business, secretaries used to place calls, screen calls, send faxes, type and send letters, track down people and information, take and prioritize messages, etc. Most people do not have someone to delegate all of these tasks to; instead they must perform them themselves or apply new technologies.

Vendors of communications technologies have been quick to respond to this demand and have generated a dizzying array of different and diverse communications products. While these new products and forms of communication have enriched our ability to collaborate with others, they have also added to the complexity of managing personal communications.

The personal computer and other personal information appliances offer a means for getting this complexity under control, and fully mastering, or putting to work, the ever-richer communications infrastructure around us.

Embedding C&C technology as a central part of personal computer use involves integrating technologies from the five disciplines that make up C&C: telephony, video and document conferencing, messaging, publishing and browsing, and networking and communications infrastructure. This is illustrated in Figure 1-4.

Figure 1-4. Personal computer and information appliances are becoming the focal point for people collaborating with other people.

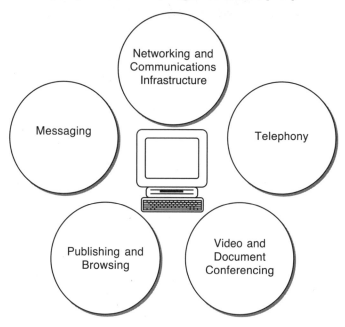

1.3.2. Five Disciplines of C&C Technology

The five disciplines, or domains, of C&C technology reflect both the key areas in which technology innovation is centered, and also the way in which people buy and use information technology. These areas can be further grouped into real-time collaboration technology, store-and-forward collaboration technology, and communications technology.

Real-time Collaboration Technology

Real-time collaboration involves technologies that overcome the barriers of distance between people who are working together at the same time from different locations.

What Is Computer Telephony?

1. Telephony

   Telephony involves bringing control of the telephone and access to telephony media streams into the computer's realm.

2. Video and Document Conferencing

   Video and document conferencing complements Telephony by supplementing it with shared-visual workspaces. For two or more people engaged in a discussion or project the computer not only sets up the call, but can also present visual information and even permit the real-time manipulation of that shared electronic material.

Store-and-Forward Collaboration Technology

Where real-time collaboration overcomes barriers of distance between people, store-and-forward collaboration involves technologies which allow people to collaborate while working at different times, independent of location.

3. Messaging

   Messaging involves the delivery of electronic mail, fax mail, voice mail, and any other form of information that is sent by one person to one or more specified people, to be delivered at some later time.

4. Publishing and Browsing

   Publishing and browsing technologies complement messaging technologies. While messaging is a directed broadcast of information to a particular address, publishing and browsing technologies allow information to be made available to a designated community of people who are granted access. The Internet's World Wide Web represents the best example of information publishing as a store-and-forward collaboration mechanism.

Communications Technology

The four functional areas described above each rely on network and communications "plumbing"—the fifth area—in order to move the real-time and non-real-time data from one place to another.

5. Network and Communications Infrastructure
Network and communications infrastructure includes hardware (such as wiring, transceivers, bridges, routers, gateways, etc.) as well as software (protocols, protocol stack implementations, drivers, etc.).

1.3.3. Bringing It All Together

The real benefit of integrating C&C technology into the fabric of everyday personal computing technology goes far beyond providing access to the various areas through a single device.

A truly complete C&C solution for a particular individual or organization draws on many select components from multiple areas to satisfy a specific set of identified needs. Such a solution derives distinct benefit from the integration that is achieved *between* the different areas.

To provide a seamless, integrated collaboration experience for each computer user, great products and technologies in all five areas take full advantage of other computing technologies and facilities from multimedia to scripting.

The potential of personal computing devices to apply intelligence, with context, and to make full use of accessible information, will take human collaboration to another level. It will make it manageable, if not effortless.

To deliver on the vision of computing devices as the focal point for C&C, computing devices must embrace the telephone. Only then can the personal computer, and other forms of information technology, fulfill their potential for making all forms of collaboration easy and accessible. While all of the areas within the C&C framework are critical in delivering on this vision, telephony is in many ways the most fundamental.

Most people identify the telephone as their most important C&C tool. If the ultimate vision for empowered human collaboration is to be achieved, CT, or the coming together of computer technology and telephony, is on the critical path.

1.4. Telephone Systems

 A *telephone system* provides end-to-end telephone services and/or access to an external telephone network. Technically, a telephone system can be any subset of the world-wide (public and private) telephone network ranging from a long distance carrier's network, to a corporate PBX, to an individual telephone. For practical purposes, when we refer to a particular telephone system, we're interested in the collection of components that make up the telephony solution for a particular customer, or system owner.

Figure 1-5. Telephone systems

The system owner's area of responsibility defines for him or her the size and scope of a particular telephone system; that is, the boundary between the functional elements inside the telephone system and the networks external to the system. For example, the office manager for a branch office or a small business might define her telephone system in terms of all the telephone equipment at a single location with all services being external. On the other hand, an individual employee might view a single desktop telephone as his telephone system because it is the only component over which he has control. Meanwhile, the executive responsible for the entire company's information systems might view the telephone equipment at all corporate sites, and the private network connecting them, as the enterprise telephone system.

Telephone systems of any complexity are made up of components drawn from four functional areas:

- Switching Fabric

- Call Processing

- Media Processing

- Administration

Figure 1-6. Telephone System Functional Areas

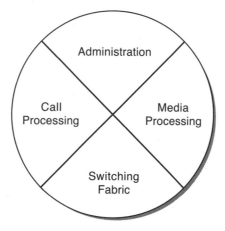

An important insight into telephone system technology is that the interfaces between the components of a telephone system are determined by the features it offers independent of its size, scope, and complexity.

Switching Fabric

A telephone system's *switching fabric* is the set of functional components that are responsible for moving media streams and signaling data between the endpoints of a given telephone network. Switching is the core of any telephone system. Without a switching fabric there can be no telephone system.

Call Processing

Call processing is responsible for managing the switching fabric and determining how commands from various sources (including telephone keypads and call control application software) should be carried out. Call processing is responsible for implementing all the functions we associate with telephony ranging from making and dropping calls to holding, picking, parking, forwarding, routing, and other supplementary services and features. Call processing is the "brain" of a telephone system and may be very simple or very sophisticated.

Media Processing

Media processing is responsible for terminating and processing media streams delivered by the switching fabric. Media processing includes tone and pulse detection and generation, recording and playback of media streams, manipulation of modulated and digital data streams (such as fax and video), and functions such as text to speech and automatic speech recognition. As with call processing, media services functionality encompasses all of the relevant resources that perform these functions as well as software that manipulates these resources and all the interfaces and layers in between.

Administration

Administration is the fourth, equally significant aspect of a telephone system. Administrative functionality includes support for system configuration (moves/adds/changes and customizing operation), fault monitoring, accounting and logging functions, performance management, and security. While the simplest of telephone systems may be "factory-configured," requiring no owner customization and have no reporting functions, the vast majority of telephone systems allow for extensive administrative control over the operation of the switching fabric, call processing, and media services.

1.5. Computer Telephony

Computer telephony refers to all of the areas where off-the-shelf computer technologies are being used to implement the telephone system functionality described in section 1.4. This includes using off-the-shelf computer backplanes and buses, operating systems, application software, protocol stacks, and other reusable hardware and software modules to construct telephone system components.

Computer telephony represents a movement away from traditional legacy telephony implementations that were based on mechanical and electronic technologies developed by vendors themselves. Arguably most modern proprietary telephone system equipment make use of the same electronic components and engineering practices found in computers, but the use of widely available, general purpose computer technologies is a recent development.

Computer telephony represents a shift from monolithic telephone systems to highly modular telephone systems as illustrated in Figure 1-7. Using off-the-shelf computer technology and development tools allows vendors to implement (or re-implement) administration, call processing, media processing, and even switching fabric functionality as independent software and/or hardware modules. Fixed functionality call processing and media processing implementations are replaced by CTI and media services applications and servers that allow for customization and extension. Likewise, administration and switching fabric implementations can be developed and installed as independent modules using mainstream computer technology.

Figure 1-7. Evolution of Computer Telephony

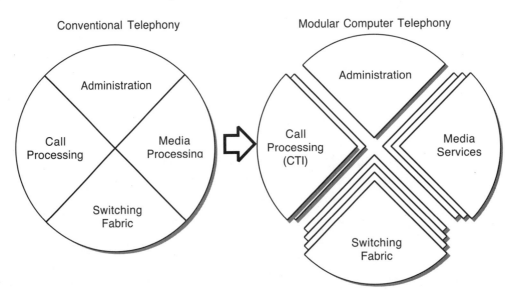

The real promise of computer telephony is that customers can build and augment their telephone systems using modular components obtained from a wide variety of disparate vendors. Ultimately system owners should be able to build telephone systems that are uniquely suited to their requirements.

Interoperability between modular components is not a direct benefit of CT technology itself. However, CT permits the definition and implementation of interoperability specifications that determine the boundaries between modular products and drives the economics that allow new vendors to enter the world of modular (sometimes called "open") computer telephony.

1.6. Switching Fabric

The foundation of any telephone systems is its *switching fabric* implementation. A telephony switching fabric consists of the cabling, networking technology, controllers, and protocols used to communicate the voice (and other media) plus control information associated with telephone calls.

Telephony switching fabric implementations consist of two key aspects:

- Telephony switching network

  The telephony switching network is a logical network made up of all the telephone system components that transmit, receive, or interconnect media and/or signaling information and the components that manage the switching fabric.

- Transmission networks

  Transmission networks are the underlying communication networks and links that provide the connectivity between components of the telephony switching network.

The switching fabric implementations in conventional telephone systems typically involve proprietary media stream switching hardware with connections for analog (POTS) and digital (T-1, ISDN, proprietary) transmission facilities. The closed nature of these implementations limits the number of transmission networks that are directly supported and limits the flexibility or functionality provided by the telephony switching network.

In the field of switching fabric development, computer telephony involves using off-the-shelf computer bus technologies as well as software protocol stacks as the basis for a new generation of transmission networks, media stream interconnection components, open media transport, signaling, and control protocols. For example, IP Telephony and other packet-based telephony switching fabrics deliver voice over IP networking infrastructures.

1.7. CTI

While computer telephony refers to the whole field of new generation telephone system implementations that draw on off-the-shelf computer technology, the term *CTI* has evolved to refer to just the call processing portion of computer telephony[1-2]. Specifically, CTI refers to the technologies used to build systems where one or more software applications running on computer platforms are integrated with the call processing functionality of a telephone system. In a telephone system that is based entirely on computer telephony technology, CTI and call processing are effectively synonymous.

CTI: Bridging Computing and Telephony

Until recently, the world of telephony technology and the world of computer technology have remained largely isolated from one another. CTI refers to the ability to combine the products, services, and systems from these different technology areas, but it does not refer to a superset of computer and telephony technologies. CTI technology is the bridge between these technology areas (Figure 1-8). CTI permits the development of solutions that further enhance traditional applications of these individual technologies.

Defining CTI

The definition of CTI has remained vague for some time, but with the maturation of CTI technology a very concise definition of this term has emerged.

1-2. CT vs. CTI — The terms CT and CTI are often confused as the distinction is not literal. The term CTI came to refer to the use of computer technology to access the call processing functionality of a telephone system because historically telephone systems were considered closed systems to which external computer equipment was integrated. As CTI products are arguably the most functionally rich computer telephony products, this term continues to be used to distinguish call processing software and hardware from other computer telephony technologies.

Figure 1-8. CTI brings together the worlds of computer and telephony technology

Computer Telephony Integration
Solutions

Computer
Solutions

Telephony
Solutions

Computer
Technology

CTI
Technology

Telephony
Technology

CTI involves three aspects:

- Call Control

    Call control is the ability to observe and control telephone
    calls, switching features and status, and call routing facilities,
    and to use switching resources that include tone generators
    and detectors.

- Telephone Control

    Telephone control is the ability to observe and control physical
    telephone devices as computer peripherals.

- Media Binding

    Media binding involves binding telephone calls to other media
    services such as voice processing, fax processing,
    videoconferencing, and telecommunications.

CTI technology achieves functionality in each of these areas by interacting with *telephony resources* that exist within CTI-accessible telephone systems. A given CTI product may provide functionality in one, two, or all three of these areas.

1.7.1. Call Control

The most essential part of CTI is the aspect that deals with a computer observing and controlling telephone calls and the telephone system features that interact with calls. *Observing* means tracking all call processing activity that takes place, and being aware of any changes to feature settings. *Controlling* refers to issuing instructions for the telephone system to obey with respect to telephone calls, features, and associated telephony resources. Call control functionality supported through a CTI interface includes both the telephony functionality available to users of a telephone system through telephone sets, as well as functionality that applies only to computer observation and control.

1.7.2. Telephone Control

Telephones are the most ubiquitous appliance in the world. Today they outnumber all computer keyboards, mice, printers, and other computer peripherals combined. They come in myriad different forms, from sophisticated office sets, to home sets, to pay phones, to small wireless telephones. As such, they represent an exciting new category of computer peripheral. This aspect of CTI involves the ability, of an appropriately enabled computer, to observe the activity of a telephone set (which buttons are being pressed, what the display says, whether the message indicator is flashing, etc.), and to control the telephone set (updating the display, simulating the press of buttons, turning on lamps, etc.).

1.7.3. Media Binding

The third aspect of CTI is media binding. This also tends to be the aspect of CTI that is subject to the most confusion. *Media binding* in CTI refers to the ability of a computer to get access to the media stream associated with a particular telephone call (that can be specifically observed and controlled through call control). Computer technologies that manipulate the media stream once it is accessed are part of media services (see section 1.8) and not part of CTI. For example, many computers support fax software that can interpret a fax document and display it on the screen, or transform an electronic document into a form that can be sent using a fax modem. This capability is not a CTI capability. In this example, CTI comes into the picture when the time comes to place (or answer) a call that is to be used to send (or receive) a fax. When the call is established, media binding functionality is used to "bind" or associate the call with available fax modem functionality so that the computer's fax software can take control of the fax modem and send (or receive) the fax.[1-3] Likewise, given the availability of appropriate media binding resources in the telephony products and the corresponding software in the computer, sound streams from the call can be recorded by the computer, sound can be played down the phone line, speech recognition and text-to-speech functionality can be applied, and digital data can be exchanged.

1.8. Media Services

A telephone system must be capable of generating and terminating the media streams that are delivered through the switching fabric. At a minimum, the telephone system requires media service resources to detect and generate tones. The ability to record and playback audio data is also required to support functionality such as auto-attendant and voicemail.

1-3. Fax media services — A complete example for fax media binding, illustrating how a fax is received from a CTI perspective, is presented in Chapter 6, section 6.11.

Implementation of *media services* involves telephony resources that consume, generate, or modify media streams. These *media resources* may be implemented within stand-alone media access devices or built into other types of devices.

Telephony media services include:

- Tone Detection and Generation
- Recording and Playback
- Text-to-Speech
- Speech Recognition
- Modulated Data (Modem/Fax)
- Digital Data (Compressed Video, etc.)
- Call Binding

By implementing open interfaces to support media services, vendors not only streamline their own development but allow system owners to customize the telephone system by adding their own computer telephony media resources and applications.

1.9. Telephony Administration

Telephone system administration is one of the most important aspects of a telephone system as it is the administrative features that are used to put a system into operation and maintain it. Unfortunately, this aspect is often the most overlooked by those building and buying telephone systems.

Telephony administration encompasses the following:

- System configuration
- System customization
- Moves / Adds / Changes (or MAC)
- Fault monitoring

- Accounting
- Performance management
- Security

The Future of Computer Telephony

Computer telephony technology is here to stay, but the term "computer telephony" is much like the terms "color photography", "HiFi audio", and "digital computer".

When most photography was black and white, it was essential to differentiate the new color technology by referring to this new branch of photography as "color photography". After thirty or forty years all mainstream photography used the color technology and black and white photography was relegated to being a specialized art form. At this point, the term "color photography" was no longer used as the term "photography" implied color.

The same evolution will take place in the world of telephony. Eventually all mainstream telephone systems will be based on off-the-shelf computer technology and we will no longer need to say "computer telephony" because this will be assumed. Certain vendors may continue to implement "proprietary telephony" technology or "monolithic telephony" technology to cater to specialized needs, but they will call attention to this to differentiate their products from the ubiquitous computer telephony technology.

1.10. CT Everywhere

The field of computer telephony is now reaching its maturity. This is marked by three key milestones:

- Interoperability – through standard CT Plug & Play protocols
- Power – through easy-to-use software tools
- Access – through low-cost multimedia communications hardware

The software industry's evolution to better graphical user interfaces, combined with the efforts of early CT developers, has led to many powerful applications and software tools that can be applied in the construction of CT solutions. Innovation in multimedia communications hardware products has brought about a new generation of products that make connecting computers to networks and telephone lines easy and inexpensive.

Despite the obvious potential for CT technology, despite the work by the pioneers of the CT industry over more than a decade, and despite the tremendous excitement surrounding each new hardware or software technology that promised ubiquitous CT, the use of CT technology has only now begun to grow dramatically. Overall, the principal barrier to dramatic growth of the CT market has been the lack of interoperable products. Without this interoperability, developers must specialize their products for very precise configurations, and customers cannot rely on anything they purchase and integrate themselves (as is the normal practice for mainstream computer products). There are many ways to differentiate CT products—but interoperability should not be one of them.

Of the three milestones in CT that have been achieved, interoperability is the most important.

1.10.1. Interoperability

 CT interoperability for hardware and software is best defined in terms of two operational requirements:

1. Customers can assemble and upgrade their CT systems in any fashion using CT hardware and software products from any combination of vendors. For example, someone can use the same CT software on the computer at home (connected to an analog residential telephone) and the computer at the office (where there is a digital feature phone desk set). The software takes full advantage of all the capabilities in each location. The customer can even substitute a new CT server at the office without having to make any changes to the telephone system or application software.

2. Users of mobile computing devices (laptop computers, personal digital assistants, etc.) can take their devices, and all their software, from one work location to another and still take full advantage of the telephony capabilities in each location, regardless of the combinations of products involved. For example, someone can use the CT software on her laptop with a pay phone at the airport, the hotel room phone, and the phone at a client's site, as well as her cellular phone.

The existing interoperability standards for telephony have been slow to emerge because in order to satisfy the implications of these two requirements, they had to simultaneously address all the following needs:

- Compatibility with existing products
    Perhaps the biggest challenge of all, interoperability specifications must minimize the efforts required of vendors, developers, and customers to migrate their earlier products to take advantage of the specifications. This means that specifications must anticipate future products and maximize the reusability of existing components.

- Plug-together, out-of-the-box interoperability
    The true measure of success is that CT interoperability standards allow customers to simply buy the CT products that best fit their needs, plug them together, and have them work—the first time.

- Deterministic and robust operation
    People expect flawless operation from their telephone systems, often holding them to a much higher standard than computer systems. No vendor or customer is willing to trade interoperability for reliability. Satisfying this need and the previous one simultaneously is among the most significant challenges. Developers must adopt *normalized behaviors* so that software can be programmed to anticipate well-understood telephone system behavior.

What Is Computer Telephony?

- Platform independence

  The increasing diversity of computer technology—the blurring of the very meaning of the word "platform"—means that complete interoperability specifications must not make assumptions about the type of computing technology being used.

- Support for arbitrary system configurations

  The virtually unlimited range of uses for telephony means that interoperability specifications must be applicable to arbitrary CT system configurations that reflect both existing usage patterns and those that cannot be foreseen.

- Appeal to the entire computer and telephony industry

  The community concerned with CT technology represents a superset of the entire communications and computer industries. This includes telephone system and telephone device vendors, software application vendors, server and client platform vendors, and those who comment, analyze, and provide consulting to all of these groups.

Fortunately, every major telephony and computer vendor has invested heavily in contributing towards interoperability goals. Each industry group that has tackled these challenges has moved the industry closer to fully satisfying the requirements; each group has built upon and added to previous efforts. These industry efforts have included the following:

- Telephone Manager (Apple and partners)
- CallPath (IBM)
- CSTA (ECMA)
- TAPI (Microsoft and partners)
- TSAPI (AT&T, Novell, and partners)
- CTI Encyclopedia (Versit and partners)
- The ECTF Framework (ECTF)

The ECTF Framework consists of an architectural model for computer telephony along with corresponding interfaces and protocols. The ECTF framework captures the work that has proceeded it and continues to expand as it incorporate architectures, models, interfaces, and protocols that provide a foundation for modular CT products.

This book provides a complete road map to the state of the art in CT, based on the latest interoperability specifications to emerge from the industry. It is based on the de facto industry standards for:

- Telephony terminology

- Telephony operational model, features, and services

- Normalized operations

- Protocols

- Configurations

- Programmatic interfaces (APIs[1-4])

1.10.2. Three Phases of CT Evolution

Consistent with the iterative development of interoperability specifications for telephony, and the maturing of technology in the overall information technology industry, there have been three distinct phases in the evolution of CT technology:

1. Custom systems

2. APIs for everybody

3. Protocols for systems, APIs for applications

1-4. API — API literally stands for Application Programming Interface. In general usage, it has come to refer to any programmatic interface that allows two independent software components to interact. It is not limited to application software programming. To avoid confusion, this book uses the term programmatic interface to refer to generic interfaces between software components. The term API is used here to reflect conventional usage.

Each phase represents a paradigm shift in the approaches used to develop CT solutions. Each shift was based on changing economics, technology, and priorities in the telephony, computer hardware, and software industries. The products and practices associated with each phase will continue to exist and interoperate for some time to come. The economics and opportunities associated with developing products for the third phase, however, will mean that more and more industry investment and activity will migrate to this approach.

This book deals with all three phases and is intended to help anyone who is considering the alternatives to choose the best approach for meeting their own needs.

First Phase: Custom Systems

The first phase in the evolution of CT involved custom development. Anyone who wanted to integrate a telephone system and a computer system had to work directly with the telephone system vendor to obtain that vendor's proprietary (and often product-specific) CT interface. These CT interfaces ranged from specifications for how a computer could be interfaced to the telephone system in question, to special APIs for specific operating systems. This is illustrated in Figure 1-9.

The economics of this arrangement were very poor. In most cases, customers themselves (or systems integrators working on behalf of customers) were developing vast quantities of highly specialized software that was hard-coded to a very specific telephone system. Rarely could this software be reused by others who had the same telephone system. Furthermore, should customers ever need to upgrade their telephone systems, they generally needed to rewrite their CT software.[1-5]

1-5. Conspiracy? — Those who like to believe in conspiracy theories have suggested that the lack of CT interoperability was a direct result of telephone system vendors seeking to "lock in" their customers. A better explanation is simple supply-and-demand economics.

Figure 1-9. The first phase: custom systems

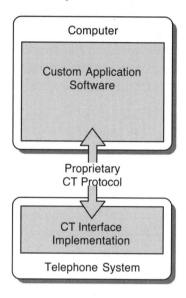

In instances where telephone vendors had invested in some type of custom API, it was often on the wrong operating system for many customers. The vendor's API still required custom software specialized to each specific telephone system. In addition, the telephone system vendor had to keep track of all the popular operating systems. Vendor's had to split their development resources between maintaining existing custom API implementations (as new versions of supported operating systems were released), and trying to adapt ("port") the API for new operating systems as they appeared.

Both of these arrangements severely limited the potential size of the CT industry for a very long time.

Second Phase: APIs for Everybody

The second phase of CT evolved from the first phase, but it represented a tremendous leap over earlier approaches. It involved APIs that were independent of the telephone system.

The phase-two approach was based on the prevailing philosophy that only a small number of pervasive operating systems needed support. Each development platform would therefore provide application developers and telephone system vendors with APIs.[1-6] Each group then could develop the necessary software components to build a complete solution without being dependent upon one another. This is illustrated in Figure 1-10.

Figure 1-10. The second phase: API layering

The intent of the new approach was to minimize the work necessary for software developers writing CT applications. Thanks to a single, stable API for a given platform, application software could, in theory, run independently of any specific telephone system or product. This allows software developers who sell shrink-wrap software (not just customers and systems integrators) to develop CT software. The result

---

1-6. Telephony APIs — The first mainstream telephony API was the Telephone Manager, developed by Apple in the late 1980s. IBM, Microsoft, Novell, Sun, and others were quick to follow over the next few years.

should have been more applications, which in turn should have meant a more attractive incentive for telephone system vendors to develop the necessary CT interfaces for their products.

This approach was not perfect, however. While it addressed the needs of software developers, it did little for telephone system vendors relative to the first phase. Telephone system vendors still had to write code for all of the popular operating systems, and they still had to rewrite those pieces of code every time a particular operating system was revised. From their perspective, at a technical level, the only real difference between the first and second phases was the fact that they no longer had to take responsibility for the design of the API.[1-7]

Another challenge associated with the second phase is that solutions built using this approach tend to be less robust or reliable than desired. The reliability that people take for granted in the world of telephony is in large part the result of international standards that define, in precise terms, the protocols for the ways systems interact. APIs, unlike these protocol definitions, are open to much greater levels of selective interpretation. Applications that use a particular standard API might not work with certain telephone system implementations supporting that same API. This is in part because the behaviors of the telephone system software are not normalized. As it was for phase one, in these cases, the applications must still be specifically modified to work with a particular telephone system. (Generally, however, there is less of this work to do.)

Third Phase: CT Protocols for Systems—APIs for Applications

While the second phase was characterized by its focus on the needs of application developers, the third phase in the evolution of CT is characterized by its focus on the needs of telephone system vendors and customers. It involves improving on phase two by addressing the reliability issues and eliminating the key bottleneck, that is, the effort

---

1-7. API design — Despite the fact that all CT API developers worked with telephony vendors when designing their APIs, some telephony vendors did not consider this loss of design control to be a positive thing.

required by telephone system vendors to implement and maintain all the different APIs for all the different operating systems. Both of these objectives are accomplished through the addition of standardized CT protocols to the APIs developed during phase two.

Figure 1-11. The third phase: CT protocols

In the third phase, telephone system vendors are responsible only for implementing software that runs internally on their own products. The only interaction with application software is through standardized CT protocols. This is illustrated in Figure 1-11.

Now telephone system vendors no longer need to know with how many operating systems they support or which operating systems their customers are using as the standards are the same regardless. This also allows their telephony products to be used by devices that don't even have operating systems (or CT APIs in the conventional sense). Freeing up their resources from platform-specific development projects allows telephone system vendors to focus on providing richer functionality and more robust operation.

The benefit for customers and application developers is that the normalized protocols provide a much higher level of reliability and robustness using the existing operating system-based APIs. Application developers don't have to "special-case" specific telephone systems (unless they want to take advantage of special and unique features)[1-8] and customers then can use applications without having to worry about compatibility.

1.10.3. CT Plug & Play

The term *CT Plug & Play* refers to the ability to take two CT products out of the box, hook them together, and have them work together without having to install special "driver," "mapper," or other proprietary software in either.

CT Plug & Play[1-9] is made possible by the CT protocols of the third phase. The use of CT protocols allows one device to negotiate automatically with other devices to determine their CT capabilities. No effort is required beyond instructing one to connect to the other.

In the paradigm of the second phase, telephone system vendors had to develop special pieces of software[1-10] for a computer to work with the CT interfaces on their products. If these pieces of software are not installed in the appropriate devices, or do not exist, the products simply will not work with one another.

CT will truly be universal when all CT products are CT Plug & Play.

1-8. Vendor specific extensions — Vendor specific extensions are defined in Chapter 5. They refer to unique capabilities supported by a telephone system beyond the standard superset.

1-9. Plug and play — It should be noted that CT Plug & Play is different from Microsoft's Plug and Play. A device that supports CT Plug & Play will work with any other CT Plug & Play device. A device that supports Microsoft Plug and Play will work with a PC that is running the correct version of the Windows operating system and has the correct driver installed.

1-10. Telephony vendor software — The names of these pieces of telephony vendor software differ, depending on the platform. They include telephone drivers, telephony service providers, telephone tools, and mappers.

1.11. CT Technology, Products, and Solutions

In this book we differentiate among the terms "CT technology," "CT products," and "CT solutions."

The term *CT technology* refers to the specific technologies that provide the basis for linking computing and telephony technology. The term *CT products* refers to products that implement CT technology and make it accessible to CT solutions. Finally, the term *CT solutions* refers to the application of CT technology (by taking advantage of CT products) in a specific product or working system that combines any number of other communication and collaboration technologies.

CT products are always necessary either to support, or be part of, a working CT solution. Not all CT solutions are CT products, however; a CT solution does not necessarily make CT technology accessible to other CT solutions.

Computer-based voice processing products provide an excellent example to illustrate these distinctions:

- Computer-based voice processing systems are CT solutions because internally they incorporate CT technology in the form of computer-controlled call processing and media services.

- A voice processing system may take advantage of additional CT technology provided by another CT product in order to simplify its own design and/or to get access to a richer telephony feature set.

- A voice processing system is also considered a CT product if it makes its CT functionality (either its own technology or that of CT products it is using) available to other CT solutions.

1.11.1. CT Solution Categories

CT solutions often are grouped into a number of basic solution categories. While these categories generally overlap in arbitrary ways and incorporate other categories, they do make it easier to refer to individual CT solutions as being used for a particular purpose.

The solution categories referred to in this book are briefly listed below. These explanations are further expanded throughout the book:

- Call accounting

  Call accounting applications involve using software to track information about individual calls (who was called, when the call was placed, the length of the call, etc.) in order to track telephone usage, recover costs, bill for services, reconcile telephone bills, and more. Call accounting applications are generally easy to justify in environments where telephone use is intensive and time is billed. These products tend to pay for themselves quickly.

- Auto-dialing

  Auto-dialing refers to automating the placing of telephone calls. Auto-dialing allows a computer user to simply indicate a desire to talk to someone; the CTI software takes complete care of the process. This can include looking up the right person, finding an appropriate telephone number, determining the best way to call the person (factoring in the current location, time zone, etc.), dialing the number and associated billing information (calling card or credit card), retrying as necessary, and indicating its progress throughout the process. While arguably among the most basic of all CTI solutions, auto-dialing is the most universal in terms of appeal and has the potential for tremendous time savings.[1-11]

- Screen-based telephony

  Screen-based telephony refers to the use of application software that presents a telephone user interface on a computer screen and acts as an alternative to the user interface provided by a physical telephone set. This virtual telephone is usually (but not always) more functional than the telephone set that it supersedes. More important, though, is that it can be customized to meet the specific needs and usage style of a particular individual, who might need more or less functionality than is otherwise available through the telephone set. The functionality of screen-based telephony encompasses auto-dialing and adds support for other telephony features and services. A fully featured *screen-based telephone* application, *SBT* for short, provides access to every telephony function available on a given telephone system.

- Screen pop

  Screen pop solutions involve having a computer system immediately present (or "pop") information pertinent to an incoming telephone call onto an individual's computer screen. This allows the person to prepare for the call before it has been answered, or to choose not to answer the call based on the information presented.

- Programmed telephony

  Programmed telephony refers to the broad range of CT solutions that involve unattended computer interaction with telephone calls. In contrast to screen-based telephony which involves creating an alternative telephony user interface for a human actively managing a call, programmed telephony delegates to a computer the responsibility for interacting with telephone calls. In fact, most applications of programmed telephony can be thought of as creating a user experience for the person at the other end of the phone call. Programmed telephony involves CT software that allows rules for individual call disposition to be established (or *programmed*).

1-11. Auto-dialing time savings — Dialing a telephone number manually (including the time to look up the number if necessary and provide any applicable billing information) can take from 15 seconds to a few minutes. In contrast, the time needed to instruct a CTI application to call a person might be 1–10 seconds. Given the thousands of telephone calls made by computer users each year, the potential time savings are staggering.

Programmed telephony applications can be as simple as an application that rejects all calls from a certain list of telephone numbers, or as complex as interactive voice response systems that take messages, locate people and information, and redirect calls.

- Voice processing

    Voice processing refers to a subset of programmed telephony solutions that involve using media services to interact with human callers in some fashion. Any application that interacts with callers by playing messages, recording messages, working with speech information, and detecting or generating tones is referred to as a *voice processing solution*.

- Call routing

    Call routing solutions belong to a class of programmed telephony solutions that automate the delivery of calls to selected individuals. Calls can be routed based on associated information provided by the telephone system, or on an actual interaction with a caller using media services in some fashion.

- Call screening

    Call screening solutions involve using CTI technology to filter calls and handle them differently, depending upon who is calling or what they are calling about. Call screening solutions may involve screen-based telephony, programmed telephony software, or a combination of the two. *Attended call screening* is just a screen pop solution in which a screen-based telephony software user actively decides to accept or reject a call before the call is actually answered, based on information from the telephone system that indicates who is calling. On the other hand, *unattended call screening* involves routing software that is set up to redirect calls automatically on some basis. For example, an application could act as a private secretary on behalf of a user. It could answer each call, capture information about the purpose of the call, pass only urgent calls on to the user, and take messages in all other cases.

- Auto-attendant

    An *auto-attendant* solution uses voice processing, in lieu of a human operator or attendant, to interact with callers and direct their calls to the desired person.

- Information retrieval

    Information retrieval solutions are voice processing solutions that allow callers to track down and retrieve information without having to interact with a human clerk. Examples include access to prerecorded messages, access to information stored in databases (such as retrieving a bank balance or order status), and retrieval of documents that can be returned via fax or electronic mail.

- Fax-back

    Fax-back is an information retrieval solution specifically involving the retrieval of faxable information. All information requested is provided to the caller in the form of one or more fax transmissions to a selected fax number.

- Personal agent

    A *personal agent* is a piece of programmed telephony software that can act as an autonomous agent for a given computer user. Typically it utilizes voice processing to interact with callers on a user's behalf. The personal agent may provide call screening, if the user is present, or will handle calls independently, if he or she is away. A personal agent might provide any or all of the programmed telephony capabilities described above in the process of handling a call.

- Call center

    A *call center* is a team of two or more (potentially many hundred) individuals in a particular location handling calls on a dedicated basis. The individuals are referred to as *call center agents*. A call center may be just inbound (such as a hotel reservation call center) or just outbound (such as a telemarketing organization), or it may be a blended call center with a pool of call center agents who handle both kinds of work.

- Distributed call center
    A *distributed call center* involves all of the functionality and activity of a call center, but the call center agents themselves may be working from two or more locations. A distributed call center might involve a single pool of call center agents working two different time zones, overflowing calls from one to the other at peak times. A distributed call center could involve having each call center agent work from his or her own home, so that no central work location even exists.

1.12. Conclusion

In this chapter we explored the motivation and context for computer telephony and identified the functional areas of telephone systems that are being implemented using computer telephony technology (call processing, media services, switching, and administration). We delved into the definition of CTI as the subset of computer telephony dealing with call processing and identified three areas encompassed by CTI (call control, telephone control, and media access). We also reviewed the evolution of CT in the context of interoperability and learned to differentiate CT technology, CT products, and CT solutions.

In the next chapter we look at a number of ways in which CT technology can be applied to build customized CT solutions in a wide variety of circumstances.

2.
CT Solutions
and Benefits

CT brings to the world of telephony the kind of revolution that personal computers brought to the world of computing: the ability to get more value out of technology by empowering more people to participate in the development of complete solutions.

Opportunities to apply CT technology and techniques are as infinitely varied and numerous as the people and organizations that use telephones. In fact, the primary benefit of CT technology is that it shifts the view of telephone systems to solutions that can be tailored precisely to meet the specific needs of market groups, specific customers, and most importantly, of specific individuals.

CT makes customization of telephony systems possible because it:

- Opens up access to telephone systems using the well-established flexibility, programmability, and customizability of off-the-shelf computer technology;

- Allows direct integration with an organization's business systems or an individual's work flow tools; and

- Creates many new opportunities for vendors, developers, and individuals with special skills, knowledge of needs, and awareness of individual preferences to add customized functionality to a telephone system.

While the existing diversity of telephony products is staggering, this diversity actually reflects the fact that a huge demand exists for telephony solutions that address an almost infinite variety of needs. It is really an indication of the much larger opportunity that exists when telephony features and services can be more easily brought to bear on the needs of an individual or organization.

2.1. Who Benefits from CT? (The CT Value Chain)

The benefit of CT technology comes from each individual user or organization's ability to select various components and services and to integrate them to form a unique and optimal solution. This ability to obtain different parts of a solution from different sources, and to have someone assemble it in a tightly integrated fashion, is at the heart of what makes CT valuable, and what has traditionally made it challenging.

By definition, a complete CT solution involves the integration of hardware, software, and services from different vendors. This collection of participants in the final CT system is known as the *CT value chain*. Each tier in the value chain represents an important part of the final solution; CT creates new opportunities for the companies and individuals that provide products and services at each layer. This concept is illustrated in Figure 2-1. A summary of the tiers in the CT value chain appears in Table 2-1 (at the end of the chapter).

Figure 2-1. CT value chain

2.1.1. Telephone Service Providers

Telephone service providers offer services in the form of access (wired or wireless) to the worldwide telephone network. They also provide optional telephony features to their customers (subscribers) through their portion of the telephone network.

The average customer traditionally uses only a small fraction of the features and services available. This is in large part due to the fact that customers have had no good way of accessing much of this functionality, given the simple telephone set's poor user interface for handling many of these capabilities. (Chapter 3 explains the fundamental concepts behind telephone networks; Chapter 5 explains the common features and services; and Chapter 10 describes the forms in which these services are offered by telephone companies.)

CT represents a big opportunity both for traditional telephone companies and for new telephone service providers entering the business because of industry deregulation worldwide. These new service providers include wireless network operators, competitive local exchange carriers (CLECs), internet service providers (ISPs), and application service providers (ASPs). In order to sell more telephony features and services, these companies have to make services more useful and easier to access. At the same time, these companies don't have the expertise or might be legally unable to develop the other pieces of a complete solution. However, the concept of the CT value chain means that they can work with others, who do have the appropriate expertise or products, to build the solutions their customers need. Not every telephone company will actively pursue CT solutions as a means of differentiating or adding value to their offerings, but they all will benefit from them.

2.1.2. Telephone Equipment Vendors

Telephone equipment vendors make the products that connect to telephone networks, as well as to the telephone equipment of other vendors. Telephone equipment vendors include companies that develop, manufacture, and sell:

- Consumer telephone sets and systems

- Business telephone sets and systems

- Cellular and other wireless telephone sets

- Pay telephones

- Computer peripherals, add-ons, and subsystems for interfacing with telephone networks

- Special-purpose telephony equipment

Telephone equipment vendors have a great deal to gain from CT. In the past, they have had to compete with one another by staying abreast of the latest advances in telephony technology, differentiating through industrial design, and trying to build telephone systems with the broadest possible appeal. Like the telephone companies, however, their challenge has always been to find ways to let more customers actually take advantage of the features they have built into their products. In most environments, few individuals ever use the full functionality of their telephone system because it isn't tailored to their needs and it isn't easy to use. (Chapter 5 explains the features and services implemented by telephone system vendors, Chapter 10 describes the forms in which these are implemented in tangible telephony equipment, and Chapter 6 and Chapter 7 reveal how these vendors open their products to expose CTI and media services interfaces.)

By supporting CT, telephone equipment vendors are better able to meet the individual needs of customers by taking advantage of the other players in the CT value chain.

CTI allows much easier access to the functionality of a telephone system, this means that potential customers will actually place greater value on systems that offers a rich set of telephony features and services because they can assemble a CTI solution to take advantage of them. CTI effectively unleashes the latent power of telephone systems, and the result is a bigger and healthier market for telephony equipment in all categories.

Similarly, support for media services interfaces makes telephony products more attractive to customers because they can be viewed as foundations for customized interactive telephony solutions and save the vendor from having to build one-size-fits-all media processing solutions.

Telephone equipment vendors that embrace modular computer telephony specifications are able to focus on the specific technologies where they have leadership and leverage the work of others wherever possible.This is illustrated in Figure 2-2.

Figure 2-2. Before and After: Modular CT technology for telephony vendors

Before:

After:

2.1.3. Computer Hardware Vendors

Computer vendors play a critical role in the CT value chain because they build the platforms on which all of the software components in a CT system run. Computer vendors include companies that develop, manufacture, and sell personal computers, server machines, and multi-user computers, as well as consumer electronics information appliance products and PDAs. Through CT, they are able to add value to their products by:

- Supporting slots, ports, and connectors that allow for connection to standard LANs and CT peripherals;

- Including an operating system with CT support;

- Building in CT functionality; and

- Bundling CT applications.

All of these efforts to bring value to a computing platform through CT rely on all the other parts of the CT value chain. For a customer actually to take advantage of a given computer product in a CT solution, the products and services represented by the other tiers are essential. The computer vendor is relying on telephone network providers and telephone equipment vendors to provide services and products that can be accessed through standard CT protocols, or with appropriate software. They also rely on the vendors above them in the chain to create the CT software components that will appeal to customers. (Chapter 11 describes how CT systems are configured by assembling telephony and computer hardware components.)

2.1.4. Operating System Vendors

Operating system (OS) vendors develop the system software that runs most mainstream computer systems. Operating system software provides access to the facilities available on a given computer system, allows applications to be launched, and provides a means for applications to make use of system-wide capabilities.

In most cases, the operating system and not the computing hardware represents the environment targeted by the application software developer. As described in Chapter 1, most operating system vendors have developed programming (or programmatic) interfaces that allow application software running on a given platform to take advantage of appropriately configured and connected telephony functionality. (These interfaces are further discussed in Chapter 12.)

Operating system vendors are much like computer hardware vendors in that they must rely on telephone network providers and telephone equipment vendors to support their proprietary APIs, or to support standard CT protocols that can be used with these APIs. They also rely on application developers to use their telephony architectures, tools, and interfaces to build products that make a given operating system useful to customers building a CT solution.

2.1.5. Telephony Software Developers

Telephony software developers are a special category of developers that produce telephony-specific applications. These are applications that take direct advantage of telephony features and services; their primary reason for existence is to extend the functionality of a telephone system or product in some way. The applications produced by these developers are highly specialized towards telephony. However, as the primary telephony application installed on a given computer, even the most casual user of a CT system will require one. (Chapter 12 describes the software configuration of CT systems.)

One example of a typical CTI application presents a user interface with all of the functionality (and more) available through the buttons, lamps, and other components on a telephone set.[2-1] Another example of CTI software is a so-called "firewall" server program that allows multiple computers to safely share a single CTI interface on a piece of telephony equipment.

Telephony software developers exist only because of CT technology. As CT becomes ubiquitous, the task of writing CT software will be easier and the market will be larger. CT developers are able to take advantage of the layers provided by all of the others in the CT value chain. They are able to thrive only because of the opportunities that result from the access made possible through CT interfaces, and because of the modularity that results from the division of responsibilities implicit in the CT value chain.

2-1. Screen-based telephone application — An application that presents an alternative user interface to the telephone is referred to as a screen-based telephone or SBT. This is discussed further in Chapter 12.

2.1.6. Mainstream Application Developers

Mainstream application developers also can play an important role in CT solutions. These developers are the software vendors responsible for writing the popular applications that justify most computer purchases today. These application programs represent the mission-critical applications that everyone relies on day to day to manage their businesses (or their personal lives). A few examples include:

- Contact managers

- Messaging software

- World Wide Web browsers

- Database applications

- Electronic forms applications

- Accounting applications

- Spreadsheet applications

- Calendar applications

- Time and billing applications

- Office management software

The focus of the work performed by mainstream application developers is not on CT. These developers specialize in solving other problems, yet in many cases their applications can be used with dramatic results as part of a CT system. In fact, CTI's ultimate value for most computer users is integrating their telephone systems with these mainstream applications, which they use for the majority of their daily work. (Operating systems vary in the level of support provided for applications in this category. See Chapter 12 for a complete discussion of CTI software configurations.)

2.1.7. CT System Integrators

CT system integrators are a special breed of developer; they assemble CT systems on behalf of their clients. This involves assembling and integrating the necessary components from each layer in the value chain, and providing whatever configuration and customization of application software is necessary.

Figure 2-3. Modularity from an integrator's point of view

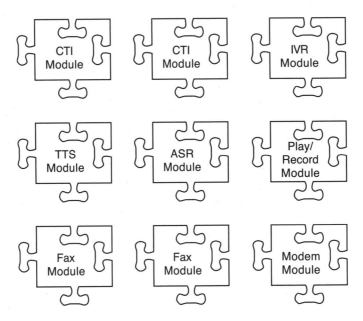

Like telephony software developers, CT system integrators credit their existence to the maturation of CT technology. They rely on the modularity of CT product implementations (see Figure 2-3) to be able to apply their talent and expertise to the needs of many clients cost-effectively, while customizing each CT system to specific individual requirements.

2.1.8. CT System Customers

Regardless of an organization's size (from a home business to a global enterprise), or its objectives (for-profit or not-for-profit), a CT system brings a multitude of very significant and measurable benefits. These include (but certainly are not limited to):

- Saving money

- Making more money

- Increasing customer service and customer satisfaction

- Increasing internal efficiency

- Projecting a more professional image

- Increasing employee morale

Customers acquire a CT solution either by having a CT system integrator do the work, or by assembling and integrating the CT components themselves. Even after a CT system integrator has completed an installation, the customer will be able to continue to build on and extend the CT system as needs change.

The flexibility offered by a CT-based approach to telephone system design is one of the things that makes it very attractive to customers. Modular CT technology allows new components, new software, new services, etc., to be added to a solution without having to replace the entire system. Another key benefit of a CT solution is the opportunity it affords to make changes very quickly in reaction to fast-changing business needs and requirements.

The scenarios in the remainder of this chapter further illustrate the range of benefits that customers can realize by applying CT technology.

Figure 2-4. Before and After: Scalability of modular telephone systems

2.1.9. CT System Users

Individuals who use CT technology are among the biggest beneficiaries because of the potential for customization at the individual level. Beyond specific solutions to optimize work flow from an organizational perspective, modular CT technology allows an individual user to further customize their personal working environment. By obtaining complementary pieces of software needed to support a particular work style and activities, and customizing the interface and operation, an individual can shape the CT solution to his or her needs and preferences.

This ability to address simultaneously the needs of the organization and the individual user through CT technology is an exciting and invaluable dimension to these solutions. The scenarios later in this chapter illustrate this dual benefit.

2.1.10. Callers (Customers, Colleagues, and Friends)

A fact often overlooked is that a CT system serves not only its owner, but also—just as importantly—those who interact with the owner through the telephone. Whether the CT system in question is in a home, small business, or enterprise, callers will benefit from the greater responsiveness, efficiency, and professionalism made possible through a CT solution.

For example, when CT technology is in place at a particular business, customers are much less likely to be asked to repeat their request to three different people or spend hours trying to track down a particular piece of information. Individuals empowered by the enhanced functionality of a CT solution will present to their callers a better, more professional interaction. Callers also can benefit significantly from CT solutions such as automated attendants, interactive response software, and telephony agents that help them get what they need in the absence of a person with whom to speak.

2.2. CT Solutions

This chapter introduces typical CT solutions for a wide variety of environments. The scenarios presented, and the CT technology used, illustrate how certain CT features, functions, and services may be applied. These examples will stimulate ideas for how you can apply CT technology in your own environment.

These scenarios will also provide context as you read this book and become familiar with CT technology and the underlying telephony functionality it exploits.

The CT solution scenarios in this chapter are:

1. Screen-Based Telephony

2. Mobile CTI

3. Power Dialing

4. Personal Telephone System

5. Personal Telephone Agent

6. Interactive Voice Response (IVR)

7. Help Desk

8. Call Center

9. Internet Telephone System

10. Ecommerce Business

Each scenario is presented in terms of the challenges faced by a hypothetical individual or organization before applying CT technology (if applicable), and how the implementation of a CT solution has transformed day-to-day activities. In Chapter 13, armed with knowledge of the technologies and functionality provided by each layer of the CT value chain, you will see the actual implementation of the CT system for each scenario.

2.3. Screen-Based Telephony

Andrew, an account executive for a large public relations agency, realized that he and his assistant were not getting the full benefit of their company's telephone system. Despite the personal computers and advanced digital phones on every desk, they were still doing business the same old way. Andrew was an intensive telephone user and spent most of the day talking to editors and his clients. The fact that the telephone system wasn't easy to use was getting in the way. The final straw came when he overheard his assistant saying, "I'll try to transfer you now, but in case this doesn't work, please call back and ask for extension 40220." Everything changed when he put a CT solution to work.

Andrew connected both his phone and his assistant's phone to their personal computers and installed screen-based telephone application software. This alone was a dramatic improvement. He then could start taking advantage of certain advanced features of the telephone system that had previously been to difficult to actually use. Transferring and conferencing calls became a snap and always worked; callers were never cut off or left waiting to be connected. In addition, built-in telephone system features of which he'd been unaware, such as call parking, became useful tools in his day-to-day work.

Every voice mail is now easier to access. His screen-based telephone software alerts him when there are messages waiting; all he has to do to retrieve them is click on the application's voice mail button. The application dials the voice mail system and logs in automatically. By clicking on buttons on his computer screen, he can listen to each of his messages, save and delete messages, rewind and fast-forward individual messages, and even change his greeting—something he had never figured out how to do by himself.

Another benefit Andrew noticed was reduced training time for new employees. When his assistant went on vacation, her replacement was able to use the software after running through a quick tutorial on her first morning.

However, in many ways the real productivity gains have come from hooking the company directory database and his personal contact manager application (an off-the-shelf software product) to the screen-based telephone software. The result greatly streamlines the placing of calls.

To place calls before installing the CT solution, he had to look up the number of the person he was calling using his contact manager, and then manually enter the phone number on his phone. Looking up the telephone number was the easy part: he had a function key macro to bring up his contact manager software, and he could find entries just by typing the first few letters of the last name. The annoying part of the process was having to dial the number itself. Depending on what area code he was calling, he had to dial 9 or 8 before the number. (He could never remember the rules, so he had a cheat sheet beside the phone.) If his contact was out of the office, he would hang up and repeat the process using the person's cellular phone number, home number, or ultimately the pager number or the person's assistant.

After setting up his CT solution, all Andrew has to do now to place a telephone call after looking up the person's name is click on the appropriate telephone icon beside the name or press the Enter key on his keyboard. The screen-based telephone software does the rest. It even puts his telephone set into speaker-phone mode so that he doesn't even have to move his hands away from the keyboard. With his new software in place, Andrew can call any contact using an average of five keystrokes. From the time he decides to call someone to the time he hears the phone ringing is less than five seconds, and his hands don't even have to leave the keyboard (or mouse).

In fact, Andrew can dial any number on his computer with just a single gesture of the mouse. Any telephone number that appears in a piece of electronic mail, a word processing document, a spreadsheet, or an Internet page can be dialed just by selecting it and dragging it to the screen-based telephone application. The application figures out the type of number (internal, local, or long-distance) and dials the number appropriately. Andrew is now looking forward to installing similar software on his personal digital assistant so he can perform many of the same functions when he is away from his desk.

In summary, Andrew's CT solution has allowed him to be much more effective in his job. His clients and press contacts are treated with a high level of responsiveness and sense of professionalism, and from Andrew's perspective the system is more efficient and reliable. He now is making far better use of the agency's substantial original investment in a telephone system and personal computers, despite the small incremental cost of the solution and the short time it took to get it working. Perhaps most importantly, work is a lot more fun for Andrew and his assistant.

2.4. Mobile CTI

Betty is a sales representative for a sportswear company that specializes in short-run custom manufacturing. She spends all of her time traveling and meeting with prospective customers. Her office literally is her notebook computer and the nearest phone. She does all of her work from hotel rooms, airport lounges, and taxicabs. When she's lucky enough to be at home, she works from her spare bedroom. Her telephone system changes with her location and include pay phones when she's on the run, a cellular phone when she's in motion on the ground, the phone at her airplane seat when she's in the air, and whatever is installed in a given hotel room at night.

While Betty enjoyed the pace of her job, loved the travel, and relished the challenge of working with new people every day, this array of different telephones and ways of dialing numbers was the one sore point. In fact, she often delayed returning phone calls for this reason, despite the importance of the telephone to her job. The sales process involved meeting with prospective clients, leaving samples, working out the purchase details over the telephone, faxing or e-mailing a proposal, and getting a confirmation of the sale. When her prospects wanted to reach her with questions, requests for additional information, or were ready to place an order, they left her electronic mail or paged her. She would call back from wherever she was, answer their questions, take down their requirements, and then follow up with information or a proposal using the customer's preference of fax or e-mail.

Every time she placed a call, sent a fax, or connected to her electronic mail service she had to enter up to 36 digits. For example, to return a call to a customer in San Jose from her hotel in Boston she had to dial the following:

- 9 to get an outside line

- 1-800-MYT-ELCO to call her long-distance telephone company

- 408-555-1234 to call her customer's number

- 666-777-8888-9999 to enter her billing number

If she made a mistake at any point in the sequence, she had to start over. Though she was very good at punching in all of these numbers, the minute or so that it took still distracted her from thinking about what she wanted to say to her customer.

If this weren't bad enough, the numbers to dial also differed depending on where she was and what kind of phone system she was using. This meant that she always had to think about (or actually look up) the right way to dial a call from her current location. For example, in Toronto she had to remember that to call Hamilton (area code 905) to the west was a long-distance call, but dialing east to Pickering (also area code 905) was a local call.

Needless to say, Betty was very excited when her company had a system integrator develop a sales-force automation solution that included a CTI component. The new system dramatically simplified every aspect of using communications technology in her job, and let her really focus on the time she spent with her customers.

With the new system, a typical sequence of events when Betty arrives in her hotel room might be as follows:

- She attaches the modem in her notebook computer to the hotel phone line and tells the software what city she is in and what hotel she is calling from (if it is one where she has stayed before). If it is a new hotel, she enters the information for placing local and long distance calls and saves it for the next time.

- She instructs her computer to retrieve her outstanding electronic mail while she unpacks. The dialing software knows that her Internet service provider has a local number, so it dials that one rather than placing a long-distance call.

- If a new piece of electronic mail from a customer requests a phone call, she simply drags the phone number from the electronic mail to the screen-based telephone application she is using. Placing the call is fast, effortless, and error-free regardless of where she is. She lets her computer dial the number and then lifts the handset to wait for the customer to answer.

- Whenever she calls a customer, a special sales-force automation application developed by the system integrator pops up on the screen. It is aware of who she is calling because it monitors use of the screen-based telephone application and provides context-based assistance. Before the call is even completed, this software has retrieved all of the information about the customer she is calling, based on the phone number called.

- If during the course of her telephone conversation a customer asks for information about a particular product (such as the price), Betty just presses the "product info" button in the sales-force automation application window and enters the product code. Because the application already knows which customer is on the phone, it can produce the correct pricing information for that particular customer. If the customer wants a hard copy of the information, she selects the "send fax" option; her computer generates an appropriate cover letter to go with the product information and places the resulting fax in queue, from which it will be sent as soon as the phone line is free. The system knows the fax number already because it is in the customer database.

- When a customer wants to place an order, Betty presses the "place order" button. An electronic form is presented with most of the fields already completed, based on the customer context. She completes the rest of the form according to the customer's request. The form then is automatically sent to the customer by electronic mail or fax for confirmation.

- When the customer wants to schedule a meeting, Betty presses the "set up meeting" button and is presented with her calendar. As with the other automated functions, once she agrees on a meeting time, a confirmation letter is automatically generated and sent via fax or e-mail.

The software works in any place she can connect her notebook computer to the phone line. This includes her hotel rooms, the frequent traveler lounges at the airport, her cellular phone, and the phone in the airplane. Betty has already customized the solution to use a commercially available screen-based telephone application that has a smaller window. (The one provided by the system integrator supported many features for use with the company's office system, but because she never visits the office, these features were never active.) Betty is now looking forward to being able to use the infrared port on her notebook computer to control pay phones, and to use simultaneous voice and data (SVD) technology to be even more responsive to her customers.

In summary, Betty's new CTI solution transformed her job. It took the technological drudgery and communications nightmare out of her job and actually put her notebook computer to useful work. Dialing not only is fast and easy, but calls are always placed correctly the first time and always through the least expensive carrier. Overall, Betty enjoys every aspect of her job and her customers see much greater professionalism, consistency, and responsiveness.

2.5. Power Dialing

Chuck works in Accounts Receivable for a division of a Fortune 500 company. Until recently, his week had a very specific, predictable schedule. Four days a week he processed payments and updated customer accounting information in the company's AR software database. Every Thursday evening he would run the past-due accounts report, and on Friday he would spend the day pursuing customers whose accounts were past due. Tracking down outstanding payments was easy once a customer was on the phone: it was usually just a clerical error or oversight that needed to be addressed. Reaching the customer was what took most of the time, and leaving voice mail never got results. The routine went as follows:

- Starting at the top of the report, look up the number for the customer and dial it.

- If the number is busy or if it goes to voice mail, hang up and remember to try this customer a little later.

- If the call goes through to a live customer, pull up the appropriate customer information on the computer terminal and resolve the issue.

- Repeat for the next customer in the report.

- When the bottom of the report is reached, start at the top and try all of the people who weren't reached on the preceding pass.

Fridays were very long and tedious days for Chuck, and he didn't look forward to them. All this changed, however, when a friend of his in the IS department connected the accounting computer to a new CT server and wrote a simple program (based on the logic above) that automated the whole process.

Using the new solution, the computer does all of the dialing for Chuck automatically. In fact, it uses a telephony capability known as *predictive dialing*, which allows the telephone system to place the call automatically on Chuck's behalf and to hang up automatically if it detects a voice mail system or a busy line. Chuck is only involved in the calls that successfully reach a person. When a customer is reached, the appropriate customer record is presented on Chuck's terminal and Chuck's speaker phone is activated.

The new CT solution makes the process of collecting overdue accounts simple for Chuck. In fact, because the new software does virtually all of the work, it frees Chuck to work on his other AR responsibilities between calls. This means that collections activity can run all week long. (Extra logic was added to make sure that customers were not called more than once per week.) Chuck activates the dialing software while he is doing other work and deals with customers as the system successfully finds them.

Chuck is now looking forward to the new accounting system, which is being developed to run on a network of personal computers. With this system his dialing software will become even more powerful.

In summary, this simple CT solution has both improved Chuck's job and speeded the collections process, saving the company money. It was relatively simple for someone in the IS department to implement, and it took little time for Chuck to learn how to use it.

2.6. Personal Telephone System

The Morgans are a typical family with a not-so-typical phone system and home computer—but it wasn't always that way. The Morgan home computer is their all-in-one communications appliance. The kids use the family computer for their school work and they particularly enjoy surfing the Web. Dad uses it to manage the household finances and to connect to his company's network for his electronic mail. During the day, though, it's down to business: Debbie Morgan has a home business and everything revolves around the computer.

It wasn't long before she discovered that Debbie couldn't take telephony for granted. The telephone quickly became central to her business. She started out using her home telephone line, but it became awkward to keep track of which calls on the monthly bill were business calls and which were not. Privacy became an issue when business calls came during the evening. She would take the calls in her office, but occasionally the kids would pick up the phone in the kitchen and disrupt her call. Furthermore, with only a single line, she was missing calls from important clients when she was using the telephone line for sending faxes.

That's when Debbie invested a little effort in setting up a CT solution. In assessing her needs, she concluded that while the home definitely needed a second phone line so she could use one for faxes and another for voice calls, the cost of a third line could not be justified because there was virtually never a time when all three lines would be busy. She opted for a personal PBX (a personal telephone system the size of a computer hard drive) and a second phone line. She rewired the existing phones in the house so that both telephone lines came directly from the telephone company into the personal telephone system. She then connected her fax modem, her office telephone, and the lines leading back to the kitchen and bedroom extension phones into the new telephone system. With this system she got the benefits of powerful CT features, along with the ability to share two lines among all of the users in the household.

Debbie's customers are located all over the country, but with her screen-based telephone application, they're all just a mouse click away. When Debbie wants to follow up with a client, she just retrieves the right page in her client database and clicks on a dial button. Much more exciting than automatic dialing, however, is the support her new system has for presenting information about a caller when the phone rings—even before she answers. When a client calls, the appropriate page of her client database is displayed and she can decide whether she wants to talk to the person or have her computer take a message. If she answers the call, all the information she needs is already on the screen. In fact, the computer actually acts as a speaker phone, so her hands stay free during a call and she doesn't have to invest in another, separate piece of equipment.

The time and duration of every call is tracked, so it's a snap to sort out which calls are personal and which are business. Client billing is automatic because, with a little bit of scripting, Debbie was able to tie the call-logging application into her time and billing software!

The Morgan computer even acts as the household's full-time answering machine and fax machine. It sorts out which calls are faxes and which are voice calls and presents both in an on-screen "mailbox." Callers on Debbie's business line are presented with an appropriate greeting; callers on the home line can choose the family member for whom they wish to leave a message. Faxes sent to either line are detected automatically, so that no telephones ring and the fax goes directly to the computer.

To summarize, Debbie Morgan's home business in many ways is made possible by the combination of her personal computer and her personal telephone system. The combination delivers an easy-to-use set of capabilities that are seamlessly integrated, allowing Debbie to focus on working with her clients. Debbie saved a lot of money by using this type of CT solution rather than buying extra phone lines, a dedicated fax machine, a speaker phone, and separate answering machines. The system also saves Debbie a great deal of time in many areas. She is always working to a tight deadline, so knowing who is calling before answering the phone avoids having to spend valuable time with a nonpaying client. In addition, the logging feature means saving the time that otherwise would be spent sorting through phone bills, and ensuring that every call is billed back to an appropriate client if possible.

2.7. Personal Telephone Agent

Edmund is a very busy person. He is a freelance photographer who is always on the move. While he is at work his time is split between his office, his studio, and his darkroom. The rest of his time is spent traveling to outdoor locations. Edmund doesn't have a secretary or other administrative support; his personal computer does everything from tracking his schedule to automating his accounting and billing.

The one thing his personal computer wasn't doing for him was handling phone calls. If he wasn't in his office (which was the majority of the time), an answering machine answered the calls. From time to time he would check his messages either by returning to his office or calling into the answering machine from his cellular phone. After a few occasions when he missed out on some business because he didn't get back to an important client quickly enough, he decided to invest in a CT solution.

Edmund spent some of his spare time setting up a personal agent. Once his system was fully customized, it was able to autonomously handle all of his calls. If he is away from the office or doesn't want to be disturbed, the personal agent answers all calls automatically. The personal agent interacts with each caller to find out who they are and the reason for their call. If a call is indicated as being urgent, the personal agent asks the caller to wait while it searches for Edmund. The software then tracks down Edmund as follows:

- An intercom system connects the office to the studio and darkroom. The personal agent first tries to find Edmund by announcing the caller's name and the subject of the call on the intercom speakers.

- If Edmund doesn't respond, it then tries calling him on his cellular phone, and then at his home number.

- If all of these attempts fail, it takes a message from the caller and sends an urgent pager message to Edmund, indicating the nature of the message that has been taken.

Not only does Edmund have the ability to call into his personal agent to take urgent calls that his agent has tracked him down for, he also can call in to retrieve important information from his personal computer. For example, if he forgets the time and location of his next meeting, his personal agent can read this information to him over the phone, and if he forgets an important document, his personal agent can fax it to him.

In summary, Edmund's personal telephone agent is an example of a very powerful CT solution that allows anyone to turn his or her personal computer into a full-fledged assistant. In Edmund's case, it gave his photography business a significant edge over his competitors. His business now projects a more professional image, and he is accessible 24 hours a day without having to change his lifestyle or the way he runs his business. This CT solution also saved Edmund money. Instead of hiring a receptionist or answering service, he put his computer to work. An extra benefit is that he now has 24-hour remote access to his computer from any telephone, anywhere, at any time.

2.8. Interactive Voice Response (IVR)

Frances is the vice-principal of a primary school. At the last PTA meeting, a number of parents presented an interesting idea. They wanted to be able to call the school to find out about each evening's homework, special events, snow closures, etc. There were only two problems with this request. The first was that the school secretary already was very busy. The second was that the secretary didn't work past 5:00 PM and the parents needed to be able to call in the evenings.

Frances decided to put the school's administrative personal computer to work at nights. With the help of one of the school's computer whiz kids, she set up the computer to answer the school's phone number at night. At some point each day, each teacher drops by Frances's office and records a message for the parents of the kids in his or her class. These messages include descriptions of the evening's homework and hints the parents can use to help their kids. Frances records any special announcements and then leaves the computer running when she goes home for the day. Parents can call the school at any time, the computer answers, and they can punch in their child's grade on their touchtone phone and hear the appropriate prerecorded announcement from the child's teacher. They can also listen to prerecorded school-wide bulletins, special notices, award announcements, etc. In the event of an emergency, such as a snow closure, Frances is able to dial into the system and leave a special message for parents.

The new system is such a success that Frances already is planning expansion. She wants to add two more telephone lines and make the information accessible over the Internet for those families who have Internet access from home.

This scenario demonstrates that CT is not just for business, but for any organization that wants to take full advantage of technology to provide a better service to their community.

2.9. Help Desk

Gunther works for the information systems group at a university. The information systems group is responsible for all of the computers and networks in the university and its dormitories. Gunther is a member of the help desk team that provides technical assistance when people have problems. Because everyone in the university community uses a computer and these tools are so essential to their academic pursuits, a priority has been placed on making sure that all members of the academic community have all the necessary support to keep their systems running. The university also is extensively networked. Every office, lab, and dorm room has an Ethernet connection to the university's computer network and a telephone connected to the university's telephone network.

The help desk is a telephone number that is answered 24 hours a day by a team of highly trained computer technicians, like Gunther, who can help users troubleshoot their problems.

When the help desk was first established, it was almost immediately overwhelmed with calls. Gunther and the other help desk team members identified the following problems:

- Despite the fact that each technician had a different area of specialization, calls would be delivered to the different technicians at random. Often it was not until a minute or more into the call that the technician could determine that the problem might be better handled by someone else. At that point the call could be transferred (and the background information repeated), or the technician could take a stab at the caller's problem.

- After a technician had explained how to a fix a problem, callers preferred to stay on the line while they implemented the solution; they didn't want to risk having to explain the problem (and the fix that was recommended) to a different person if the fix didn't work. The result was that calls that should have taken only two or five minutes actually were tying up a technician and a phone line for as much as half an hour.

- Another reason callers wanted to keep their technician on the line was that waiting times were so long that they didn't want to have to call back again if their problem wasn't solved.

The team decided to let Gunther implement a CT solution to eliminate these problems. Here's what he did:

- Gunther began by building a simple database, using an off-the-shelf database product, that would track all the information relevant to every problem report. Every technician's computer has access to this multi-user database, and the appropriate record is presented whenever one of them takes a new call. All the information is available to them even if they didn't handle the call that generated the record.

- Gunther set up a voice processing system that greeted each caller to the help desk and asked basic questions about the problem. It could associate the call with an existing request from the caller, or could create a new record in the database.

- Gunther set up call-routing software that used the information about each call in the database to queue and then deliver it to the technician who last handled the call (in the case of a repeat call), or to the technician with the correct expertise based on the nature of the problem (if it's a new call or the original technician is not on duty).

- Gunther also set up a simple Web site[2-2] on the university's intranet. The site provides both answers to the most frequently asked questions (to eliminate the need to even talk to a technician if possible), and also an escalation mechanism if the needed advice cannot be found. The escalation mechanism is an alternative to the voice processing system for inserting a request into the call queue. If a particular computer user's problem is such that the web site can still be accessed, he or she can create a new problem report or indicate an existing one, and request a call to his or her location. When the user's turn comes up, the system automatically calls and connects the call to a technician. In this way, calls are queued on a first-come, first-served basis, but they don't have to tie up phone lines. Web site users also can monitor their position in the queue so they know roughly how long they'll have to wait.

2-2. Web site — A Web site is a collection of Web pages on an Internet World Wide Web server, or on an equivalent intranet server. A Web page is an HTML document accessible to Web browser applications. These applications translate the document and present it on the computer screen in the form of a hypertext display. In this scenario, the HTML document viewed by each client is created dynamically and is specific to a particular request.

In the future, Gunther would like to add support for screen sharing so that a technician can use the network connection to look at a user's screen, speeding the diagnosis and the fix.

By implementing this system, Gunther was able to make the help desk a much more effective resource in the university community. It saved the university from having to hire many more technicians and provided much more efficient service to callers.

2.10. Call Center

In the case of Henrietta's virtual travel agency, there is no life before CT. Her business is one that is possible only because of CT technology.

Henrietta manages a small travel agency with six agents. Competing with the big chains is tough, but Henrietta has CT technology on her side. CT gives her agency the capabilities of a big business with the personal touch of a small business, and the overhead of cyber-business.

Henrietta's CT solution delivers unprecedented customer satisfaction, streamlines work, and creates a more pleasant work environment for her employees because they all work from their own homes as a distributed call center. The hub of her business is a server and networking equipment in her basement.

Her server hosts an auto-attendant and a customer database, which her agents use to track clients and activity. Each of her travel agents has a computer with a dial-up connection to her server and a telephone for business use. The different agents work at different times of the day, depending upon when demand is highest. When customers call (either to the local number or Henrietta's 800 number), their calls are automatically forwarded to the agent with whom they last spoke (if he or she is working). All the information about a given customer is presented on the computer screen before the agent even picks up the phone. If the appropriate agent is unavailable, the caller can be redirected to another agent or can leave a message. The agents all have pagers, so if they are not working when a call comes in they are notified as to how the call was handled.

Calls that come in overnight (typically emergency calls from stranded customers) are handled by a designated on-call agent. These events are rare, so the agent doesn't have to sit by the phone; he or she merely stays home that night. When such a call arrives, the server will send the call first to the designated agent and, if this does not produce a response, a message is taken and the agent is paged repeatedly until he or she does respond. The emergency service is rarely used, but it is a significant differentiator for Henrietta and allows her to handle corporate clients who would otherwise go to big chains that can afford to staff a call center 24 hours a day.

Another benefit of the CT system is that it saves money by streamlining outbound calls. Agents just click on the phone numbers for clients, hotels, airlines, etc., and the call is placed instantly. Misdialed numbers are a thing of the past, and calls are always dialed the most cost-effective way available. When Henrietta's agents reach an airline or other travel service provider that has its own multi-step voice processing system (phone maze), their screen-based telephone software is preprogrammed to navigate to the desired service with the click of a mouse.

In the future Henrietta will be growing her business by adding a Web site to her server. By using an Internet Telephony Gateway she'll be able to reach a new market, cyber-customers, and save charges to her 800 number by feeding calls from the Internet into her network.

Henrietta's business is indistinguishable from the advanced distributed call center she has built. CT technology makes her employees very happy because they can work at home; improves on customer service and satisfaction because of the personalized treatment they get; and also significantly reduces the overhead of the business while it keeps variable costs to an absolute minimum. Callers are automatically directed to the agent who can help them the most effectively, so calls are shorter and agents can handle more customers.

2.11. IP Telephone System

The telephone system used by Ian's small business, like Henrietta's call center, was only made possible by the latest advances in computer telephony technology.

Ian is the founder of a small Silicon Valley start-up company. While the company is only six months old, it has grown from a single employee to twenty-five and has moved from Ian's garage to temporary office space Ian was able to secure from a company that had outgrown its facilities. Ian projects that the company will double in size every six to twelve months and that he will need facilities for over five hundred employees within three years. This means that the company will continue to move and acquire new office space on a regular basis. Ian expects to begin splitting his workforce among multiple offices and even home offices as the company grows.

While finding office space is a challenge, an even bigger problem is picking the right telephone system. Ian needs the advanced features of a modern PBX and knows that at a minimum he'll need to deploy CTI software to support the company's customer service and technical support personnel. He needs a telephone system that grows as fast as the company and can span all of the different office facilities that the company will be using.

Ian's solution was to opt for an IP-based telephone system that is hosted by the same network service provider that provides Internet services to the company and will be providing the network connectivity between offices in the future. By delivering all telephony services through the company's Internet connection, the system can scale and adapt to whatever size and distribution of company resources emerges in the future. The telephone system can be accessed over the Internet so employees can access it from their home offices and even from hotel rooms and client sites.

The core of the telephone system is software running on a server that is managed by the service provider but delivers all of the functionality traditionally associated with a customer owned PBX. Ian's staff can configure it from any location with Internet access and the company can add or remove users instantly. Most importantly for Ian, this telephone system solution provides complete flexibility to grow, move, and change at will while only paying for the functionality and number of users required at any given time. Ian doesn't have to make a capital investment in a PBX that would either be too big or be outgrown in a short time. The service provider operates the service and provides all of the power protection, backup systems, and security necessary to guarantee robust and reliable operation.

A key feature of the system is that it consists primarily of software running on mainstream operating systems and exposing mainstream CTI capabilities. This makes it easy for Ian to integrate off-the-shelf software that manages call flows, matches callers to technical support agents based on customer history, and supports software-based dialing applications on every workstation in the company.

2.12. Ecommerce Business

Jane is the webmistress for a popular online shopping web site specializing in cooking equipment and gourmet food. Like many "dot-com" companies, the founders began with the idea that they could easily compete with brick-and-mortar stores because all they needed was a web server and an Internet connection. The theory was that they could operate with very low overhead because they could reach a world-wide audience with a single electronic presence on the Internet. They hired Jane to build and manage the computer systems that would track inventory, manage credit card billing, and present the ecommerce store on the world wide web. They hired an operations team to take care of warehousing inventory and for picking, boxing, and shipping each order.

Jane's web and information system team did a superb job in building the back-office systems. They purchased an SQL database server upon which they built a complete suite of inventory, billing, customer account, and order tracking systems. They purchased web servers and developed web applications that provide a compelling shopping experience for web site visitors and deliver orders to the back-end systems. They also set up an e-mail server to handle customer inquiries and to distribute order confirmations.

Unfortunately, Jane's company, along with many others like it, began having difficulties when the busy Christmas season began. They spent millions of dollars advertising their web site URL to attract Christmas shoppers, and the web shoppers came. Their web servers were able to handle all of the traffic without any difficulty or delays. However, Jane began noticing that a very large number of their shoppers were abandoning their online shopping carts before actually making a purchase. Despite this, the demand for products was strong and the operations team was hard-pressed to ship every order within 24 hours so some orders suffered an extra day's delay. Then the phones started ringing. The web site listed the company's main telephone number in the section that dealt with background on the company. Customers, wondering about the status of their orders, found this number and dialed it. The result was an avalanche of calls that the company's receptionist was ill-equipped to handle. She transferred the calls to available staff members as fast as she could. Unfortunately, the staff members could only take down information on the order in question and promise a return phone call as they had no way of checking on the customer's order.

Meanwhile the company's e-mail server was also being inundated with e-mails from customers inquiring not only about the status of their orders, but also requesting clarification on the description of various products and with questions on applicable taxes, shipping rates, and methods of payment. Evidently the reason for the large number of abandoned shopping carts was the fact that customers had no way to pose their questions to a person with access to additional information. As the call volume grew, more e-mails began arriving complaining about the busy signal at the company's headquarters.

Jane's company survived the Christmas season, but learned a difficult lesson. While the world wide web offers a new way for customers to interact with a company, it does not eliminate the need for the personal contact that the telephone is able to provide.

The company responded by shifting their focus from being a web-centric business driven by web-based transaction data to a customer-centric business driven by customer data in a customer relationship management (CRM) system.

A customer contact department was staffed and Jane was responsible for re-architecting the company's systems to facilitate handling customer inquiries through mail, web browser, and telephone. Customers opting to call the company can choose between an automated interactive voice response system and talking with a live customer contact staff member. Customers that have questions during their online shopping experience can request help from a live staff member through a web-based chat or a telephone call.

By taking advantage of IP telephony technology and other web-centric development strategies, Jane's new customer-centric system allows for easy migration to future customer support technologies such as video and will allow customer contact staff to work from home offices.

2.13. Conclusions

In this chapter we have seen the tremendous opportunities and benefits that exist for those who invest in CT technology. This applies both to customers and to those in the CT value chain who provide the individual components that make up working CT systems.

This chapter has demonstrated the importance of interoperability and modularity in the design of CT components. The people described in each scenario were able to build and use their CT solutions because they were able to go to different sources for the pieces that make up the final solution. No single vendor could have provided the complete solution. Each involved the customer integrating multiple components and customizing the solution to meet specific needs.

These scenarios have become possible as a result of the maturation of CT technology and the associated recognition of the CT value chain. Given how very compelling these scenarios are, market forces (the economics of interoperability) will encourage vendors to build CT Plug and Play products so that solutions like the ones described can be realized with even greater ease.

The steps in building or customizing a complete CT solution from individual components, or for building an individual component that is to be used as part of a CT solution, include the following:

1. Familiarize yourself with all of the technology, functionality, and value that is added at each layer.

2. Decide which layers you can or want to do yourself, and which ones you'd be better off relying on someone else to provide.

3. Decide what functionality you need to provide in your CT component, or at each layer in your CT system.

4. Armed with the resulting knowledge, proceed with the implementation of your system or component.

CT Solutions and Benefits

Regardless of where you fit into the value chain, it's a good idea to be aware of what takes place above and below. Then, and only then, can you be sure that you are taking full advantage of the opportunity that CT represents to you or your organization. This book provides a thorough insight into the technologies and considerations associated with each layer. As shown in Table 2-1, we will be working our way through the value chain from the bottom up. (This book is designed, however, so that you can go directly to the chapter corresponding to the layer you are most interested in if you wish. References will guide you back to key concepts you might have skipped.)

Table 2-1. CT value chain

Layer	Role	See Chapter
Telephone Network Providers	Provide telephone network access and services	3, 4, 5, 8, 10
Telephone Equipment Vendors	Provide equipment for connecting to telephone network	3, 4, 5, 10
	Implement CTI interfaces	6
	Implement media services interfaces	7
	Implement new switching fabric technologies	8
	Implement administrative services	9
Computer Hardware Vendors	Provide hardware platforms for CT software	11
Operating System Vendors	Provide software platform for CT software	12
Telephony Software Vendors	Provide indispensable CT-specific software	12
Mainstream Application Vendors	Provide mainstream/mission-critical application software	12
CT System Integrators	Assemble and integrate components for CT solutions	10,11,12,13
CT System Customer	Build and/or customize CT solutions	13
Individuals	Use and customize CT solutions	13
Callers	Benefit from CT solutions	

3.
Telephony Concepts

More than a century of innovation and creative product development have created an extraordinarily diverse range of telephony products and technologies. This diversity, like the diversity seen in computer technology, has benefited customers with a wide array of technology choices. But unlike the computer industry, which always has been reined-in by the need for highly functional interoperability and standardization of component interfaces, the telephony industry traditionally has not had to address the need for integrating with the products of other vendors. (Their traditional worry has been interoperability among telephone networks.) As a result, vendors have developed their own unique feature names, terminology, and user frameworks.

Despite the tremendous diversity in how telephony equipment vendors actually implement the products and interfaces that allow telephone calls to be placed, the basic concepts are, in fact, universal.

The functional diversity, lack of consistency, rapid innovation, and competitiveness found in the telephony industry have been both a blessing and curse for those challenged with making CT technology ubiquitous.

In recent years various industry organizations such as the ITU, ECMA, Versit, and ECTF have undertaken to build a common framework with standardized terminology so that any telephony product or feature can be described universally. This framework and related interoperability specifications not only provide a consistent overall architecture for everyone working with computer telephony technology, but also supports the rich diversity of features and capabilities found in proprietary approaches. A fundamental aspect of these efforts has been in developing a consistent body of nomenclature to replace the arbitrary and proprietary terminology that obscures the many fundamental similarities between implementations. In this way developers, users, and everyone else in the CT value chain can base their plans, designs, and implementations on a common framework.

This book explains telephony technology using this vendor-independent framework. With these concepts, you'll be able to interpret and compare the telephony features of any set of products.

This chapter presents the universal telephony concepts by explaining the various types of telephony resources that can be found in a telephone system and the entities they manipulate. These concepts can be applied to any telephony product by modeling the mechanical, electronic, and software components[3-1] that make up its implementation as a telephony resource set. (Telephony products are discussed and modeled in Chapter 10.)

3-1. Human system components — To be complete, it should be noted that some telephone systems (especially antique ones) include humans to implement one or more of the telephony resources in addition to mechanical, electrical, and software components. This is referred to as "putting humans in the system." In most cases humans are considered users of the system and are therefore outside it.

3.1. Telephony Resource Framework

A *telephone system* is a collection of interconnected or related *telephony resources*. Any subsystem, or portion, of a telephone system is itself a telephone system and represents a particular *telephony resource set*. A telephone system may be as simple as an individual telephone or as vast and complex as the worldwide telephone network.

A complete set of telephony resources can be visualized as shown in Figure 3-1. Each block in this diagram represents an abstraction of a distinct area of telephony functionality. Most telephone systems have only a subset of the telephony resources shown.

Figure 3-1. Telephony resource framework

3.1.1. Switching Resources

Switching resources represent the switching fabric in a given telephone system. Switching refers to the ability to connect telephone calls and is therefore the most fundamental function within a system. Within the switching resources are a finite number of channels that are used to convey information from one place to another. The switching resource in a given telephony resource set is the only required resource. In a basic telephone it could be as simple as a mechanical switch that connects the phone to a telephone call. In modeling the worldwide telephone network, this resource represents the ability to connect any two telephones in the world.

3.1.2. Call Processing

Call processing is the "brain" of a telephony system. It manages all the information about how the telephony resources should behave with respect to one another, it accepts commands from the other resources, and it manipulates the resources in order to carry out these commands.

3.1.3. Devices

Devices are the resources in a telephone system between which the switching resource is able to provide connections. They are referred to as *endpoints* and, as we shall see, devices are essential to the way that we model telephony functions. While there is no upper limit to the number of devices a particular telephone system may have, a functional system has no fewer than two devices that may or may not be connected together by the switching resource at a given instant.

There are eight basic types of devices:[3-2]

- Station devices (telephone sets)
- Network interface devices
- Pick group devices
- Hunt group devices
- Park devices

3-2. Device types — As with all aspects of the abstractions presented in this book, any implementation may invent unique functionality and represent that functionality by extending this model with new types of resources.

- ACD group devices

- ACD devices

- Media access devices

3.1.4. Dynamic Objects

Dynamic objects are abstractions that represent dynamic relationships between resources in a telephone system. Along with devices, these objects are central to the conceptual model for telephony. There are three types of dynamic objects:

- Calls

- Connections

- Agents

Calls and connections are explained in detail in this chapter. Agents are explained in Chapter 4.

3.1.5. Interfaces

The last category of telephony resources are *interfaces,* which convey command and status information to and from the telephone system. These interfaces include:

- CTI interface
    The *CTI interface* allows computer technology to interface with call processing in order to send commands and receive status information. This interface is also referred to as an *open application interface* or *OAI* by some vendors. (The nature and use of this resource is the principal topic of Chapter 6.)

- Operations, Administration, and Maintenance (OA&M) Interface
    The *OA&M interface* provides a configuration mechanism for instructing call processing and the other resources how to behave when put into operation. It also may provide

diagnostic or status information and allow telephony resources to be taken in and out of service. (This interface will be covered in Chapter 9.)

- Accounting interface
  The *accounting interface* provides a mechanism for tracking the usage of resources in order to appropriately account or bill for their use. (Accounting interfaces are another administrative function that will be covered in Chapter 9.)

- Media service interfaces
  The *media service interfaces* provide external access to media data such as voice, video, fax, or other media stream types using media resources and media access devices. (These concepts are the principal topic of Chapter 7.)

3.2. Switching

Switching refers to the function that establishes and clears calls and manipulates the connections associated with those calls. Switching is therefore the most fundamental and essential activity that takes place in a telephone system. The telephony resources that implement switching in a system are referred to as *switching resources*.

Switching resources manage calls by:

1. Collecting and conveying control information associated with a call, and

2. Conveying the media stream associated with a connection by allocating and deallocating channels.

The operations that are performed by switching resources are referred to as *switching operations*.

3.2.1. Telephone Calls

At the heart of a telephone system's functionality is the ability to make and manage telephone calls, so our exploration of telephony concepts logically begins with the concept of calls.

In section 3.3 we'll learn about *devices* and *connections* which round out the three most important concepts in the world of telephony. As we shall see, no matter how complex the telephone system, its basic operation can be described in terms of these three concepts.

Two Cans and a String

The simplest way of thinking about a telephone call is to visualize it as a pipe that delivers streams of voice information. Perhaps the simplest of all telephone calls is the one that can be established by two people using tin cans and a string (Figure 3-2).

Figure 3-2. Tin can telephones

In the case of the tin can telephone system, calls are established when the string is pulled taught between the cans. Once established, the sound of one person's voice is captured by one tin can, conveyed across the string, and delivered to the ear of the person using the second tin can.

Media Streams

A telephone call can be thought of as a pipe that delivers streams of voice information through a telephone network, to and from a device such as a telephone (Figure 3-3).

These voice streams are referred to as *media streams*. In general, this media information is a stream of data that may be voice or some other type of arbitrary data.

Figure 3-3. Voice on a simple telephone call

Control Information

As we saw in the case of the tin can telephones, there is more to telephone calls than just their media streams. In addition to the media information associated with a telephone call, *control information* must be associated with a call. Sometimes referred to as *signaling*, this information is used by the telephone system to set up, manage, and clear the telephone call. This is illustrated in Figure 3-4. In the case of the tin cans, this signaling was in the form of stretching the string prior to speaking.

Figure 3-4. Control information on a telephone call

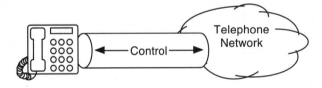

Control information is just as important as the media stream itself. Without that associated information the system wouldn't, for example, know how to connect the call.

3.2.2. Media Stream Channels

 Media stream channels are paths of communication that can be established to convey a media stream. There are a finite number of media stream channels in any implementation of switching resources, and a finite number of media stream channels that can be associated with connections for any given device. A media stream channel is able to carry a single media stream in each direction simultaneously.

When a channel is in use it is referred to as *allocated*, and when it is retired from use it is said to have been *deallocated*.

3.2.3. Switching Concepts

Switching involves the allocation and interconnection of media stream channels. To illustrate the rationale behind switching, imagine that you wanted the ability to communicate with a number of different people using the tin can telephone technology illustrated in section 3.2.1. You would need a separate telephone (or tin can) for every person you called because each telephone would have to be connected directly to a corresponding telephone owned by each of the other people.

Not only would such a system be quite impractical, it also would be very inefficient because you can use only one telephone at a time. An example of this is shown in Figure 3-5. A seven-endpoint fully connected network (where every endpoint has a direct channel to every other) contains 21 direct media stream channels, yet only three of these can be active at a given time (assuming that each endpoint can only use one channel at a time). All of the capacity represented by the other dedicated media stream channels and telephones would go unused most of the time.

In contrast, a switched network of the same size has only seven media stream channels, and all of these can be utilized simultaneously (assuming the switching resource supports multi-point calls). This is why the notion of switching was developed.

Switching allows a limited number of media stream channels, lines,[3-3] or *transmission facilities* to be used as efficiently as possible by sharing them. This is referred to as *line consolidation*. Telephone networks are systems of these transmission facilities, interconnected with switching

3-3. Lines — The terms line and telephone line can be misleading because many types of telephone lines are capable of supporting multiple channels. While the term line tends to imply a piece of wire, it actually refers to any type of link. The concepts shown here apply equally to wireless networks and fiber-optic networks. Chapter 8 explores different types of telephone lines.

Figure 3-5. Networks

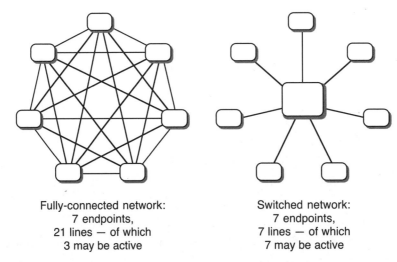

Fully-connected network:
7 endpoints,
21 lines — of which
3 may be active

Switched network:
7 endpoints,
7 lines — of which
7 may be active

resources that ensure the calls always reach their destinations and a minimum number of media stream channels are needed to connect any two points.

This is particularly important as the size of the network increases. Figure 3-6 shows the interconnection of two small networks. Only two additional media stream channels are needed to connect the two networks into a larger network, by directly connecting the switching resources. This ability to scale is theoretically unbounded; for example, the endpoints shown in the diagram could themselves actually represent different networks.

In the particular example shown, most calls established by the members of each small network are *internal* to their networks, and the need for more than two active calls between the networks is extremely rare. This provides insight into the limitation of switched networks. The ability to establish calls between two points may be limited by the capacity of the transmission facilities between two sets of switching resources somewhere in the network. In practice, the allocation of transmission facilities based on usage statistics represents a very important science in the business of operating a switched network. If the estimate for capacity requirements is incorrect for a specific span

Figure 3-6. Joining two switched networks

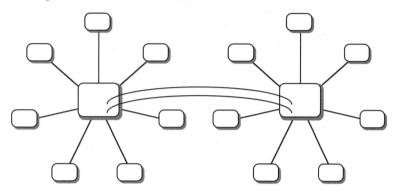

between two adjacent nodes, the network may have underutilized capacity, may consume more switching resources than needed to reroute calls around overloaded spans, or may simply fail to complete calls if no paths exist.

3.2.4. Telephony Switching Fabric

The switching resources in a given telephone system include the following components:

- Transmission facilities for both media stream channels and signaling between all of the physical components in a system;
- Mechanisms for allocating and deallocating the media stream channels associated with each transmission facility;
- Mechanisms for encoding and delivering signaling information using the appropriate transmission facility;
- Facilities for interconnecting allocated media stream channels; and
- A switching control function that manages the switching resources.

Working together, these elements are known as a *telephony switching fabric*, or just *switching fabric*. Switching fabric is the collection of networking and communication resources that allow the telephone system to operate.

Figure 3-7. Telephony switching fabric

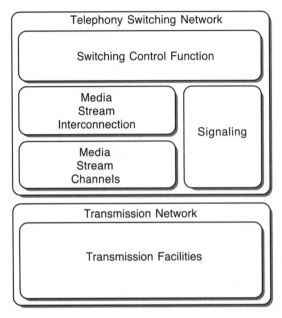

It is of great significance that the switching network established between the endpoints in a given telephone system is distinct from the supporting transmission network which provides the physical and logical connectivity between endpoints. So while the implementation of a particular telephony switching network provides the ability to interconnect, or switch, media stream channels, the transmission network being used to transport the signaling and media stream data may or may not be based on data switching[3-4] technology.

There is tremendous diversity in switching fabric implementations. Limitations of certain switching fabric implementations determine the CTI, media services, and administration features that can be made available in a given system. However, switching is otherwise independent of these other CT technologies. Switching fabric implementations are the subject of Chapter 8.

3-4. Switching — Unless stated otherwise, references to switching in this book are to telephony switching networks and operations and not to the use of switching in local area and telecommunications networks.

Telephony and Telecommunications

The terms telephony and telecommunications are frequently confused. Traditionally these words have been used to differentiate between voice telephone networks used for telephony, and other networks (including those based on telephone networks) that are used to move data between computer systems (telecommunications). Over time, the transmission facilities of the telephone network have evolved into a powerful digital communications infrastructure in which voice is but one type of data. This has led to some confusion because one can build a telephone network on top of a telecommunications network that supports telephony and one can build a telephony-specific network that supports telecommunications. Telephony has more stringent requirements, however, and its functionality is generally a superset of telecommunications functionality; you can always layer telecommunications on top of a telephony infrastructure, but you can't always do the reverse.

3.2.5. Telephone Networks: Inside the Cloud

In Figures 3-3, and 3-4, a cloud was used to represent the telephone network in which a call existed. These clouds[3-5] represent all the telephony resources being used to establish a call.

Telephony Resources in a Telephone Network

A telephone network is made up of a "web" of interconnected sets of telephony resources. To establish a call between two devices, one or more sets of telephony resources are used to patch together a complete

3-5. Clouds — The cloud symbol has become the standard graphical representation of a switching fabric or network. It is a convenient abstraction because one need not be concerned with how large or small the network is, or what portion of the resources it represents are consumed in taking care of a given call. If the network is operational, all the endpoints on a call are appropriately connected. Without this ability to work with a simple abstraction of the overall network, even the simplest operations would be quite complicated to explain. It has been said that telephone companies "are in the cloud business."

end-to-end call across the network. Figure 3-8 shows a call that has been established across a telephone network between the devices labeled D1 and D2. The network "cloud" is used to abstract the network; the call between the two devices then is easily abstracted.

Figure 3-8. Call between D1 and D2

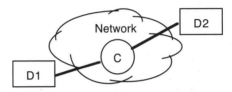

Figure 3-9 shows the same scenario, but instead of abstracting the network as a cloud, we see the actual system of individual sets of telephony resources (represented as hexagons[3-6]) that make up the network. The call between devices D1 and D2 is actually a series of calls. Each participating set of telephony resources establishes a different segment of the overall path that the call takes through the network. Each segment of the complete call is actually an individual call made up of connections with two devices, just as in the overall abstraction involving the entire network.

3.3. Fundamental Objects

In the telephony resource set model, the call processing and switching resources carry out the work of the telephone system by acting on three fundamental types of objects: calls, connections and devices. (A fourth type of object, agents, is related to a specialized type of device, the ACD Group, and is covered in Chapter 4.) As we shall see, these objects are fundamental because they reflect virtually everything that is of interest to the user of a given telephone system.

3-6. System boundaries — The shaded hexagon is used throughout this book to symbolize the boundaries of a particular telephone system.

Figure 3-9. The network of telephony resources inside the cloud

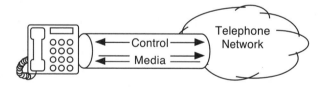

3.3.1. Calls

The abstraction of telephone calls (which are referred to as simply *calls* from here on) consists of a media stream and associated signaling information, as shown in Figure 3-10. Keep in mind that this is just an abstraction; in Chapter 8 we'll see how this abstraction relates to different implementations.

Figure 3-10. Telephone call abstraction

Call refers to both a media stream that is established between endpoints in a telephone network and all associated control information.

In understanding telephone systems we are concerned with both the control information and with managing the media streams.

3.3.2. Devices

Another important element in the abstraction of a call are the endpoints of the media stream.

An endpoint to which a telephone network is able to connect calls is a *device*. In the preceding illustrations, a simple telephone is shown as the device at one end of a call. (Devices are the subject of Chapter 4 where we will look at specific types of devices and further explore the device abstraction[3-7].)

A functional telephone call involves two or more devices as endpoints to the media stream associated with the telephone call. At certain points in the life of a call, however, such as when the call is being originated or cleared, it may have only one device or no devices at all.

3.3.3. Connections

The relationship between a particular device and a particular call is referred to as a *connection*. If two devices are involved in a call, then in the abstract representation of that call there are two connections, and each connection corresponds to one of the devices[3-8].

As we shall see in Chapters 5 and 6, connections are the most used objects within this abstraction because they allow efficient manipulation of both a call and a device simultaneously.

3-7. Device configurations — Devices are further subdivided into elements that are arranged according to a device configuration for the individual device. There are two types of device elements, physical elements and logical elements. Unless otherwise stated, all references to devices are to logical device elements. Logical device elements and device configurations are explained in sections 4.4 and 4.5. Physical device elements are explained in section 4.1.

3-8. Call appearances — Devices have call appearances that correspond to connections. A device with multiple call appearances can be associated with multiple calls simultaneously. Call appearances are described in section 4.4.1.

Voice Connections

By default, references to connections represent media streams that are compatible with the *voice network*. This means that they carry voice or modulated data (see the sidebar "Modulated Data"). A *voice call* is therefore a call made up of *voice connections*. A rule of thumb is that if a piece of analog telephony equipment can be added as a device on the call, the call is considered a voice call.[3-9] A voice connection allocates, at most, a single media stream channel.

Digital Data Connections

Some switching implementations are able to create connections that are associated with digital data media streams. *Digital data media streams* support data traveling at much higher rates than are possible with modulated data because they take advantage of the digital switching capability in a *digital data network*. These *digital data calls* (calls made up of digital data connections) are treated specially by the switching implementation, however, because they cannot interoperate directly with voice calls (calls made up of voice connections) and only a subset of switching functionality applies to them. A digital data connection may be associated with any number of media stream channels.

Quality of Service

The nature of a call or connection (i.e., whether it is voice or digital data) and attributes relating to digital data rates, the number of media stream channels used, and transmission characteristics, are referred to as *quality of service* or *QoS*. The quality of service applicable to a new call or connection is limited by what a given switching implementation supports. The quality of service associated with an existing call or connection determines the switching services that may be applied to it. (Quality of service will be discussed in greater detail in Chapter 8.)

3-9. Voice call — Technically speaking, a voice call is one that carries media streams requiring no more than 3.1 kHz of bandwidth.

Modulated Data

Telecommunications (the transmission of digital data by computers) is accomplished either by connecting a computer directly to the digital data portion of the telephone network, or by transmitting the data over the voice network using modulation technology.

Modulation involves transmitting a carrier signal, that is, a tone of a particular frequency, and then varying it according to the rules of a particular modulation protocol. Modulation protocols dictate the use of techniques that vary the carrier signal amplitude, frequency, phase, or other properties, often in combination. Simplex data transmission occurs when there is only one carrier signal on the media stream, so data transmission can take place only in one direction. The term half-duplex means that the devices at either end of a simplex media stream are able to take turns transmitting and receiving. Full-duplex communication takes place when two different carrier frequencies are used on the same media stream so that data can be both transmitted and received simultaneously. Demodulation involves decoding the modulated carrier signal into the original digital data stream.

A modulator-demodulator, better known as a modem, is a piece of hardware or software that is able to perform the modulation and demodulation of signals, given full access to a media stream. Hardware-based modems are typically packaged along with all of the electronics needed to establish and gain access to the media stream on a call. Software modems, on the other hand, depend on independent hardware and software mechanisms to deliver the required isochronous media stream data.

The modulation protocols and other related standards that allow modems from various vendors to interoperate are established by the ITU-T and are assigned names beginning with "V." Commonly used V-series standards include V.32, V.32bis, and V.34, which define full-duplex 9600 bps, 14400 bps, and 28800 bps protocols, respectively. The V.42 standard defines error detection used with V.32, and the V.42 bis standard defines the compression algorithm for use with V.42. The use of compression allows data transmission rates that are much higher than the raw data rates of the modulation protocols. The V.17 standard defines the modulation protocol used for the exchange of documents between Group 3 fax products.

3.3.4. Directional Streams

By default, all connections and their associated media stream channels are *bidirectional*; that is, there is a media stream associated with the call that is flowing to the device, and another media stream flowing away from the device.

Some switching implementations are able to create unidirectional connections. In these *unidirectional connections* there is only one media stream, and it flows either toward or away from the device. Figure 3-11 illustrates the unidirectional streams towards and away from a station device.

Figure 3-11. Directional connections

3.3.5. Symmetric and Asymmetric Communication

All connections and associated media stream channels in a particular call are also symmetric. *Symmetric connections* involve media streams that deliver data at the same rate in both directions. In an ordinary telephone call, this means that the same amount of sound information is being sent in both directions.

Asymmetric communication, where the data rate in one direction is different from the other, is abstracted as two separate calls consisting of unidirectional connections. In Figure 3-12, the media stream associated with one call carries high-speed data while the other carries low-speed data in the opposite direction.

Figure 3-12. Asymmetric communication

3.3.6. Point-to-Point and Multi-Point Calls

Most calls are point-to-point calls. This means that the call involves associating two devices with two connections. However, most switching implementations support calls with three or more connections. These are referred to as *multi-point calls*.

A multi-point call is generally referred to as a *conference call*, although the term *bridged call* is also used when the implementation is hard-wired in some way.

In the basic case of a multi-point call consisting of bidirectional voice connections, all of the media streams coming from the associated devices are combined and the result is sent back out to all the devices. (Some implementations may use signal processing capabilities to eliminate echo effects and send a slightly different media stream to each device.). All devices "hear" everything on the call.

Multi-point calls are applicable to voice calls only. Audio is a unique data type in that in its analog representation it may simply be mixed with other audio sources and the result is still a useful audio stream. The same cannot be done with digital data streams or modulated data streams. Adding these data streams together would simply corrupt the data. Synthesizing multi-point calls for digital data and large numbers of voice calls is described below.

Of course, every connection in a multi-point call need not be bidirectional. A multi-point call can be made of any combination of bidirectional and unidirectional connections as long as the media streams are all voice. Three special cases of unidirectional connection usage in a multi-point call are discussed below.

Silent Participation: Tapped Calls

One special case of a multi-point call is referred to as *silent participation* and is illustrated in Figure 3-13. In this case, a unidirectional connection is added to a call such that the new device can hear everything on the call but cannot be heard by the others participating in the call. This also is commonly referred to as a *tapped call*. Silent participation is very useful in situations where media resources are to be connected to a call for the purpose of recording audio or doing speech recognition, for example.

Figure 3-13. Three-point call with silent participant

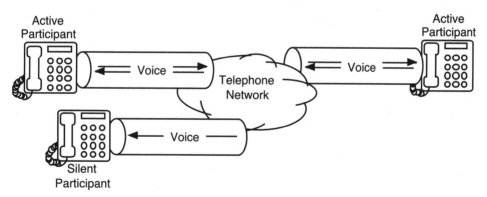

Announcement

Another special case of multi-point calls is the *announcement*. As shown in Figure 3-14, this involves the addition of a unidirectional connection to the call that delivers a media stream from the added device. This is used primarily for adding a media access resource that plays an announcement.

Figure 3-14. Three-point call with announcement

Broadcast

A third special case of multi-point calls is the *broadcast*. In this case, all of the connections in a call are unidirectional as depicted in Figure 3-15. Exactly one of the connections is unidirectional away from its device and all the others are unidirectional toward the devices. This allows one media stream to be delivered to many devices.

Figure 3-15. Three-point call with broadcaster and listeners

States and State Diagrams

The term state refers to the mode or condition of some entity. An entity's state indicates settings of its attributes and its current circumstances, thus determining what it can and cannot do. A given entity can be in only one state at any given instant.

A state diagram, or state graph, shows the states of which an entity is capable, as well as the transitions between states that are permitted. Once an entity is in a given state, it can only assume a new state if the state transition model represented in the state diagram indicates that the change is allowed.

In the example below, an entity is permitted to transition between three states identified as one, two, and three. According to the graph, the entity can transition from state three to state one, from state one to state two, and may go back and forth between states two and three. It may also transition from state two back to state two.

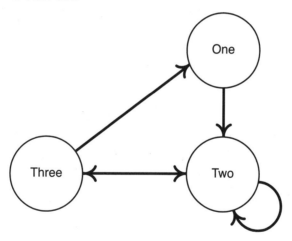

States greatly simplify the modeling of systems because, once an entity is in a known state, there is only a finite set of states to which it can transition.

3.4. Connection States

 Connection states are among the most important concepts in telephony. Each and every connection has an individual state that determines the condition of the connection and what can and cannot be done with that connection. The life cycle of a device's participation in a call (or call progress) is represented by the sequence of states through which the corresponding connection transitions. A state transition indicates that some service was performed on the connection by the switching implementation. The switching services that may be applied to a connection at any time are determined by its connection state.

Figure 3-16. Connection state diagram

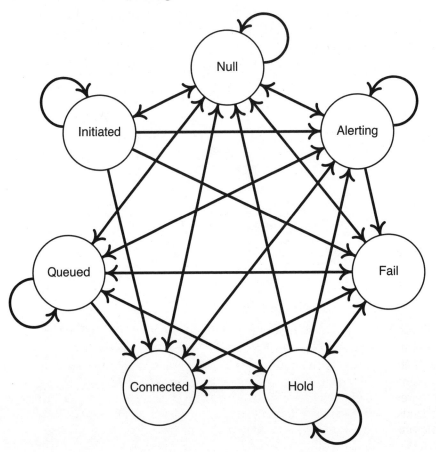

3.4.1. The Seven Connection State Model

There are seven connection states (described below), and they are permitted to transition as shown in the state diagram in Figure 3-16.

Initiated

While a device is requesting service or dialing a digit sequence to initiate a call, the corresponding connection is in the *initiated* state. The *initiated* state typically is entered when the device goes off-hook[3-10] and, if supported, when the user is prompted[3-11] to take the device off-hook. The media stream received by the device, if any, is initially dial tone followed by silence after the first digit or command is issued.

Connected

After a call has been created and the telephone system is establishing connections with other devices, the connections where media stream channels are allocated and the associated media stream(s) are flowing, are in the *connected* state.

Null

A connection is said to be in the *null* state if it no longer exists. A nonexistent connection is synonymous with a connection in the *null* state. Occasionally a connection is said to "transition through the *null* state." This is a simple way of saying that the original association between a device and call was replaced with a new one.

3-10. Off-hook — In general usage, the term off-hook refers to an action to be taken on a device, typically lifting a telephone handset, signifying that the device is requesting service. The term comes from the days when a telephone's microphone hung from a hook. The use of the term off-hook as it applies to the operation of analog telephone sets is explained in Chapter 8.

3-11. Prompting — The term prompting refers to an indication, typically audible, a telephone set makes to indicate that it should be taken off-hook in order to progress out of the initiated state. Prompting is distinct from ringing, which is an indication made by a device that a connection is in the ringing mode of the alerting state. Prompting is used when some service is initiated for a device but is unable to go off-hook without manual intervention.

Alerting

While an attempt is being made to connect a call to a device, the connection representing the association between the call and device is in the *alerting* state. There are actually three *modes* of the *alerting* state that determine what type of action the device can take:

- *Entering distribution* mode

- *Offered* mode

- *Ringing* mode

These modes will be discussed in Chapter 5.

The fact that a connection is in the *alerting* state is independent of whether the corresponding device is indicating an incoming call in some way (by ringing, for example).

Typically for voice calls, all of the other active (e.g., *connected*) connections to the call will hear ringback,[3-12] while a connection in the *alerting* state is associated with the call.

Fail

The *fail* state indicates that call progress was stalled for some reason and an attempt to associate a device and a call (or keep them associated) failed. The most common example of this state is attempting to connect to a device that is busy.

In most cases, all of the other active (e.g., *connected*) connections to the call will hear a busy tone, or another appropriate failure tone, while a connection in the *fail* state is associated with the call.[3-13]

3-12. Ringback — Ringback is the "ringing" sound heard after you have placed a call and are waiting for it to be answered. It is not the actual sound of a phone ringing, but rather a sound the switching implementation generates to supply the caller with feedback on call progress.

3-13. Blocked — One case where the fail state may not be associated with an audible tone is the case where a bridged connection is blocked from a call. This is described in section 4.4.3.

Hold

When a connection is in the *hold* state, it continues to associate a particular device with a call (signaling information continues), but the transmission of associated media streams is suspended. Depending on the implementation, the media stream channel(s) associated with the connection's media stream(s) may or may not be deallocated while the connection is in the *hold* state. A channel that is not deallocated is said to be *reserved*.

Hold should not be confused with mute. Mute is a telephone set feature which will be described later in this chapter. Mute deals with turning off a speaker or microphone and is not related to the transmission of media streams or the allocation of media stream channels.

Queued

A connection is in the *queued* state when call progress is suspended pending subsequent application of certain switching services. Like the *hold* state, connections in the *queued* state do not have active media streams and the associated channel may or may not be deallocated.

3.5. Graphical Notation

The abstraction of calls, devices, and connections is illustrated graphically using a notation based on simple symbols. This graphical notation allows the depiction of both a specific situation in a telephone system at a moment in time, or a whole scenario involving a sequence of activities. Much the way that standard musical notation allows composers, arrangers, and musicians to easily describe and interpret a piece of music, this graphical notation allows material of an equally complicated nature—the activity inside a telephone system—to be easily and precisely communicated. This notation is used in standards documents, product documentation, and telephony reference materials to explain how certain telephone system functions behave. It is also used by those developing CT solutions in order to plan and troubleshoot their implementations.

Calls are represented as circles as shown in Figure 3-17. In order to make reference to a specific call, the calls are labeled with the letter "C" followed by a number.

Figure 3-17. Symbol for calls

Devices are represented as rectangles, as shown in Figure 3-18. They generally are labeled with the letter "D" followed by a number, so that different devices can be explicitly referenced.

Figure 3-18. Symbol for devices

Connections represent the relationship between a call and a device. They are represented graphically as a line between a device and call, as shown in Figure 3-19. Connections do not require explicit labels because they can be uniquely identified by making reference to the labels for the device and the call (in that order) that they associate. Figure 3-19 shows two devices, D1 and D2, connected to call C1 with connections D1C1 and D2C1 respectively.

Figure 3-19. Symbolic representation of connections

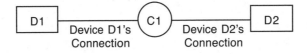

Figures 3-20, 3-21, and 3-22 illustrate the use of this notation in a number of typical examples. In the first example, call C1 has two connections D1C1 and D3C1. The second example shows a multi-device call in which D1, D2, and D3 are all participating in call C2 using connections D1C2, D2C2, and D3C2 respectively. The last example shows call C3 in the process of being set up or cleared, as it is associated only with connection D4C3.

Figure 3-20. Two-device call

Figure 3-21. Three-way call

Figure 3-22. Single device in a call

3.5.1. Representing Directional Connections

Figure 3-23 shows how unidirectional connections are represented in graphical notation. By default, connections are bidirectional. If a particular connection is unidirectional, this is indicated by placing an arrowhead in the appropriate direction.

Figure 3-23. Directional connections

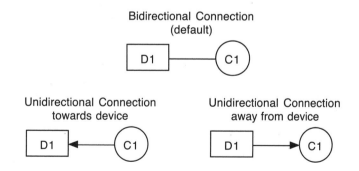

Asymmetric communication, where the data rate in one direction is different from the other, is abstracted as two separate calls consisting of unidirectional connections. In Figure 3-24, call C1 could be a high-speed data call, while call C2 could be a low-speed data call.

Figure 3-24. Asymmetric communication

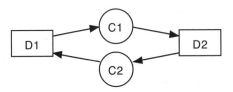

If all the connections associated with a call are bidirectional, as shown in Figure 3-25, each device in the call receives a media stream representing a mix of the media streams from all of the other participants. Examples of calls involving unidirectional connections are shown in Figures 3-26 and 3-27.

Figure 3-25. Point-to-point and multi-point calls

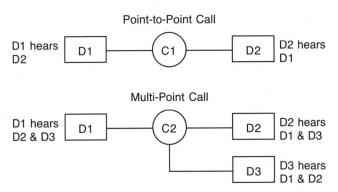

In a scenario involving silent participation (as shown in Figure 3-26), the connection to the device which is only listening (D3C3) has an arrowhead to indicate that it is unidirectional towards the listening device (D3).

Figure 3-27 illustrates two scenarios involving unidirectional streams away from a device. In the first example, device D3 is delivering an announcement to an active call ("Please insert another 25¢ to continue

110

Figure 3-26. Multi-point with unidirectional stream towards device

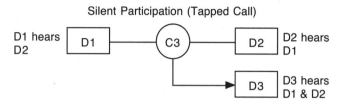

this call" for example). All the other connections in the call are bidirectional and all hear the announcement. The second scenario involves broadcasting where all of the connections are unidirectional. One device, D3 in this case, is delivering a speech while the other devices listen so D3C4 is unidirectional from the device. D1 and D2 are not able to provide any media stream data to the call so their connections (D1C4 and D2C4) are unidirectional towards their devices.

Figure 3-27. Multi-point with unidirectional stream away from device

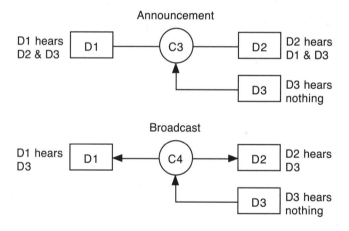

3.5.2. Connection State Representation

In graphical notation, connection states are represented by placing symbols over the line representing the appropriate connection. The symbols for the states are:

- 'a' representing *alerting*

- 'c' representing *connected*

- 'f' representing *fail*

- 'h' representing *hold*

- 'i' representing *initiated*

- 'n' representing *null*

- 'q' representing *queued*

If a particular connection is in the *null* state, meaning that it doesn't actually exist, the line representing the connection is generally omitted from the diagram altogether.

Figure 3-28 shows an example of how these symbols are used in graphical notation.

Figure 3-28. Connection state representation

In this example, call C1 has two connections: D1C1 and D2C1. Connection D1C1 is in the *connected* state, indicating that it is active and that media streams are flowing. Connection D1C2 is in the *fail* state. A likely reason for the situation depicted in the example is that device D1 tried to place a call to device D2 but D2 was busy, so the attempt to connect to D2 failed. In this case, Device D1 probably would be receiving busy tone from the call.

3.6. Call Control Services

Call processing carries out operations on the calls and connections in a telephone system. In terms of the basic telephony abstraction, these *call control services* involve one, some, or all of the following five actions:

1. Creating new calls

2. Disposing of calls

3. Adding connections to a call

4. Removing connections from a call

5. Manipulating connection states

Each call control service changes the relationships between particular devices and calls by acting on connections that relate them.

A given call control service therefore can be described in terms of the states of the connections to which it can be applied, the state transitions that result, and whether it creates or disposes of calls or connections. This is generally presented in terms of "before" and "after" the call control service has taken place.[3-14] A call control service is represented in Figure 3-29.

Figure 3-29. Switching service representation in terms of "before" and "after"

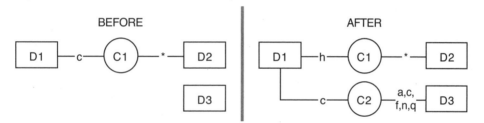

In this example, the call control service shown is called "Consultation Call" being applied to the connection D1C1. Before the service, the connection D1C1 is in the *connected* state. After the service, D1C1 is in the *hold* state; a new call, C2, has been created; and the connection D1C2 is in the *connected* state.

3-14. Call control services — Switching operations with well defined state transitions and other behavior are referred to as call control services. Refer to Chapter 5 for details of specific call control services. Between the states shown as "before" and "after" for a given service, connections may transition through intermediate states. Each complete sequence of event transitions is referred to as a flow. The concepts of flows and normalized flows are explained in Chapter 6.

If a connection might be in one of a number of different states before or after a call control service, this is shown in the notation either by providing a list of the symbols (as shown in the preceding example), or by using one of the following special symbols. The meaning of one of these symbols depends on whether it is on the "before" or "after" side of the diagram.

- '!' represents *unspecified*

  - This symbol is equivalent to "a,c,f,h,i,n,q"

  - Before a service, it indicates that any connection state is applicable.

  - After a service, it indicates that any connection state may result and the original state (if the connection existed before the service) has no bearing on the final state.

- '#' represents *unspecified non-null*

  - This is equivalent to "a,c,f,h,i,q"

  - Before a service, it indicates that any connection state other than *null* is applicable.

  - After a service, it indicates that any connection state other than *null* may result and the original state (if the connection existed before the service) has no bearing on the final state.

- '*' represents *unspecified/unaffected*

  - Before a service, it indicates that any connection state is applicable and that the connection state will be unaffected by the service.

  - After a service, it indicates that the connection state was not affected by the service and will be the same as it was prior to the service.

An example using these symbols is shown in Figure 3-30. In this example, the call control service *Clear Connection*[3-15] is applied to the connection D2C1. Before the service, there are three connections to

Figure 3-30. Representation of the Clear Connection[3-15] call control service (applied to D2C1)

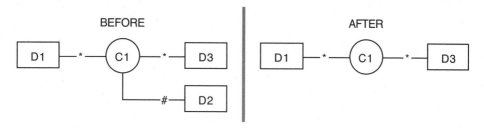

the call (D1C1, D2C1, D3C1) and the only restriction is that D2C1 be in any state other than *null*. After the service, there are only two connections (D1C2, D3C2). The connection states for D1C2 and D3C2 were not affected by the service.

The set of call control services supported by a particular product represents a large portion of its telephony feature set. See Chapter 5 for descriptions of all the common call control services and other telephony features.

3.7. Media Resources

Media resources are telephony resources that interact with the media streams associated with calls.

In many cases, media resources are simply features of a particular device, or of the switching resources, that are transparent to the operation of that device or those resources. In other cases, media resources are accessed through independent *media access devices* that may be associated with calls using appropriate call processing services. Examples of media access devices are *voice response units* (*VRUs*), voice mail systems, and generic *media servers*.

3-15. Clear Connection example — Note that this is actually just one case of the clear connection operation, where D2 is not a shared–bridged device (see section 4.4.3) and both D1C1 and D3C1 are non-null.

3.7.1. DTMF (Touchtone) Detectors and Generators

In existing telephone systems, the most common type of media resources are those that generate and detect DTMF tones.

DTMF generators typically are built into most station devices so that people can generate touchtones by pressing dial pad buttons. In some telephone systems, DTMF tones are generated only to communicate a request to the other endpoints in the call. In other systems, the DTMF tones are also used to send commands (such as the number of a device to be called) to the call processing resources. In this case, general-purpose *DTMF detectors* are required by the telephone system to interpret the commands issued by various devices.

3.7.2. Pulse Detectors and Generators

Before the development of the DTMF scheme for encoding digits, *pulse* sequences were used to communicate commands. A pulse is a very short break (less than 0.1 seconds) in the media stream. A sequence of pulses represents the corresponding number, and each number is separated by a pause of at least 0.7 seconds. Pulses are typically generated by a rotary dial on a telephone station. *Pulse detectors* and *pulse generators* are telephony resources responsible for working with pulse encoded digits.

3.7.3. Telephony Tone Detectors and Generators

While various mechanisms exist for providing feedback from a telephone system to the person using a particular station device, the only lowest common denominator is the media stream itself.

As a result, telephone systems working with the voice network use two sets of tones that can be generated and detected easily, one in order to provide feedback and the other to indicate that modulated data is to be used. These are referred to as *advisory tones* and *telecom tones* respectively.

DTMF

DTMF stands for Dual Tone Multi Frequency. It refers to a standard mechanism for encoding the 16 digits that can appear on a telephone dial pad as a combination of tones that are easily generated and detected in the media stream of a voice call. DTMF tones are also frequently referred to as touchtones.

As the name implies, DTMF tones are formed by unique combinations of two precisely defined tones. The scheme uses a set of four "high" tones and a set of four "low" tones. Using a combination of one low and one high tone (for 16 combinations) makes the detection of digits much more reliable; it is unlikely that some other source of media information on the call (like a person speaking) would accidentally generate one of these precise combinations of tones.

Generating these tones is also very easy to implement through a push-button interface, as shown below. Each column is associated with one of the four high-frequency tones and each of the rows is associated with one of the four low-frequency tones. Pressing a particular button connects tone generators for the appropriate row and column to the call in order to make the appropriate dual-tone or "touchtone."

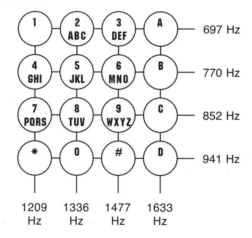

The first three columns represent the twelve DTMF tones in widespread use. The last column represents four additional DTMF tones labeled "A" through "D" which are generally referred to as military or autovon tones. These additional tones should not be confused with the alphabetic labeling found on most telephone dial pads.

The following are the tones that are most commonly encountered in a telephone network:

- Dial Tone

  Dial tone is an advisory tone indicating that call processing has created a new call on behalf of a given device and a new command can be sent. It is typically associated with the *initiated* connection state.

- Billing

  A *billing tone* is an advisory tone indicating that call processing is expecting to receive billing information, typically a credit card or calling card number. This tone is also referred to by some as *bong tone*.

- Busy

  Busy tone is an advisory tone indicating that there is no device currently available to which to present the call. Call progress for the call in question has stalled, so this tone is associated with a connection state of *fail*.

- Reorder

  Reorder tone is an advisory tone indicating that a call has become blocked in a telephone system because a necessary network interface device or other switching facility is unavailable. The cause may be a misdialed number. This tone is also referred to by some as *fast-busy*. Call progress for the call in question has stalled, so this tone is associated with a connection state of *fail*.

- Special Information (SIT) Tones

  Special information tones, or *SIT tones,* are sequences of three precisely defined tones used to indicate that call progress has stalled for some specific reason. SIT tones usually precede a prerecorded message describing the problem. They are associated with a connection state of *fail*. The four SIT tones currently in use are:

Vacant code – the number dialed is not assigned.

Intercept – all calls to the number dialed are being intercepted (typically because the number has changed).

No circuit – no circuits are available for the call.

Reorder – the call cannot be placed; the number may have been misdialed.

- Ringback

  Ringback is an advisory tone indicating that call processing is attempting to connect the call to another device. Ringback is the "ringing" sound that a caller hears after placing a call and while waiting for it to be answered. It is associated with the *ringing* mode of the *alerting* connection state.

- Beep

  A *beep* is the tone generated by a voice answering machine, voice mail system, or other media access device that records from a voice media stream. The beep is a prompt to the person calling, indicating that the recording has begun and that he or she should begin speaking (e.g., "…please leave your message after the beep…").

- Record Warning

  A *record warning* tone is a short 1400 Hz tone used to indicate that a conversation is being recorded. It is required by law in some places and typically is used in all situations (such as 911 services and security dispatch) where there is a high degree of accountability or the need to collect evidence of a conversation through a call.

- Fax CNG

  Fax CNG, or fax calling tone, is a tone generated by a fax machine or fax modem that wishes to initiate data modulation for fax transmission on a given voice call. It is a telecom tone.

- Modem CNG

  Modem CNG, or modem calling tone, is a tone generated by a modem that wishes to initiate data modulation on a given voice call. It is a telecom tone.

- Carrier

  The detection of *carrier* refers to the presence of modulated data transmission on a given voice call. It is a telecom tone.

- Silence

  Silence is the absence of any tones, voice, or modulated data in the media stream associated with a voice connection. It is quite useful to have a silence detection capability to determine that a modem transmission has completed, that a caller may have hung up, or that a message may be played.

The actual frequencies and cadences corresponding to a tone of a particular meaning may vary widely from country to country or between implementations of telephony products. This makes detection of these tones in every case very difficult, if not impossible to guarantee. There is sufficient standardization, however, that makes tone detection for calls within a particular country or system reliable.

3.7.4. Media Services

Media service interfaces provide external access to the contents of media streams using the specific capabilities individual media resources. A particular media service interface, along with the media resources that can be used to access the media streams of a given call and any required media access devices, is known as a *media service instance*.

Some media resources capture information from a media stream. These are able to convert the media stream into a media data format for use with the media service instance. Other media resources are able to convert media data specified through the media service interface into a media stream for the call. Still others are able to do both simultaneously.

The number of possible media service types is virtually unbounded, but the most popular involve media resources that work with raw sound, speech, modulated data, and digital data.

Live Sound Capture (Isochronous)

A *live sound capture* media service is able to capture the audio content from a media stream and deliver it to a component external to the telephone system such as a digitizer, tape or CD recorder, or even a speaker. The physical media interface involved might be digital or analog.

Live Sound Transmit (Isochronous)

A *live sound transmit* media service is able to obtain an isochronous stream of raw sound from an external source such as a computer audio output, a tape or CD player, radio, or even a microphone and transmit it through the telephone system. The physical media interface involved might be digital or analog.

Sound Record

A *sound record* media service is able to capture sound from the media stream and store it for future use. In this case, the media service interface is used simply to start and stop the recording and specify where and how the sound is to be stored.

Sound record is different from sound capture in that the telephone system itself is doing the recording, and the sound data never leaves the telephone system.

Sound Playback

A *sound playback* media service is able to play previously recorded sounds to the media stream. In this case, the media service interface simply is used to start and stop the recording and specify what sound is to be played.

Sound playback is different from sound transmit in that the telephone system itself is providing the prerecorded sound.

Text-to-Speech

A *text-to-speech, TTS,* or *speech synthesis,* media service is able to transform text into a stream of speech-like sounds generated by a synthetic, electronic voice. The media service interface is used to specify the text to speak and the attributes (male/female voice, accent, prosody, volume, speed, etc.) of the speech desired.

Text-to-speech is very useful because it allows arbitrary or dynamic text information to be spoken over the phone automatically. The alternative, prerecording all of the necessary information or, at a minimum, all of the necessary words that make up the information, is generally much more complicated and expensive.

Concatenated Speech

Concatenated speech is a media service comparable to speech synthesis, but it uses strings of whole prerecorded words or syllables rather than synthesizing each syllable. Concatenated speech generally provides much higher quality than text-to-speech, but is limited to a certain vocabulary of prerecorded words or sounds.

Speaker Recognition

Speaker recognition media services identify the person speaking in the media stream based on voice energy characteristics unique to each individual.

Speech Recognition

Speech recognition services convert human speech in a media stream to text. The principal attributes of a speech recognition implementation are:

- Speaker-dependent/independent

  Some speech recognition implementations must be trained to understand the speech patterns of a particular individual. These are called *speaker-dependent* systems. Other implementations have been extensively trained and can understand virtually any speaker of a given dialect. These are called *speaker-independent* implementations.

- Continuous/Discrete

  Continuous speech recognition implementations are those that can automatically identify word breaks, allowing speakers to talk continuously (normally). *Discrete* speech recognition implementations cannot identify word breaks. A person speaking to such a system must place distinct pauses between words.

- Vocabulary

  Most speech recognition implementations rely on a particular set of speech grammar rules, which limits their vocabulary to a particular set of words. Implementations vary in the size of the vocabulary they can support and whether they are limited to a predetermined vocabulary and grammar.

The media service interface is used to specify, as necessary, the speaker and/or the vocabulary and grammar, and to deliver the text corresponding to the recognized speech.

Fax Printer

Fax printer media services refer to the fax receive-and-print functionality available in a fax machine. If a telephone system connects a fax printer media service to a call on which the presence of a fax CNG tone is indicating an attempt to transmit a fax, the fax will be received and printed on the appropriate device.

Fax Scanner

The *fax scanner* media service refers to the fax scan-and-send functionality available in a fax machine. It is the complement to the fax printer media service. If a telephone system connects a fax scanner media service, the fax scanner media service will attempt to establish a modulated fax data connection with another fax-capable device on the call, and then will transmit any sheets of paper fed into the fax machine's paper scanner.

Fax Modem

Fax modem media services provide fax data modulation for sending and receiving fax transmissions. (See the sidebar "Modulated Data" on page 98.) The media service interface is used to send and receive the compressed image data and fax transmission control information.

Data Modem

Data modem media services provide data modulation for establishing bidirectional modem communication. (See the sidebar "Modulated Data" on page 98.) The media service interface is used to configure the modem service and to send and receive asynchronous data.

Digital Data

A *digital data* media service provides access to the raw stream of digital data associated with a digital data media stream. The media service interface is used to convey the data.

Video Phone

A *video phone* media service is analogous to the fax scanner and fax printer media services, but applies to media streams containing video data. When attached to a media stream, this media service displays video on the video screen associated with the appropriate device, and captures and transmits video from a camera associated with the device.

Media services concepts are discussed in greater detail in Chapter 7.

3.8. Equipment and Network Options

The telephony resources described in this chapter are used to model the implementation and operation of any telephone, telephone system, or telephone network including those implemented on networking technologies such as IP. The remainder of this book will expand on these basic concepts. However, before we delve into more concepts dealing with call control, media services, switching fabrics, and administration, we'll look at how the telephony resource framework is used to model networks and equipment.

3.8.1. Public, Private, and Virtual Private Networks

Telephone networks are more frequently categorized in terms of the ownership or allocation of their constituent telephony resources and transmission facilities rather than in terms of the technology that they use. In this sense, there are three types of telephone networks:

- Public networks

  The term *Public Switched Telephone Network* (*PSTN*) refers to the publicly accessible worldwide telephone network that ultimately interconnects with virtually every other telephone network. The public network is made up of many individual public networks around the world. Any network within the worldwide network (France Telecom's network, for example) is itself a public network. Some portions of the public network are owned by commercial telephone companies and others by government-run enterprises. Collectively these service providers are referred to as *common carriers*. In all cases, the operators of the public network are intensely regulated by regulatory bodies in each country. In general, this regulatory effort is aimed at ensuring universal access to service and fair or, if appropriate, competitive rates.

- Private networks

  Private networks are sets of telephony resources that are privately owned or leased from a common carrier and connected together with dedicated transmission facilities. A private network can be as simple as a home's internal phone system. It can be as complex as a worldwide network made up of switching resources in hundreds of locations that interconnect with the public network in many different countries. Private networks may be operated for the exclusive use of their owners, or may be operated by private carriers who sell access to their networks, or some combination of the two.

- Virtual private networks (VPN)

  A *virtual private network* is a capability provided by a common carrier that offers service equivalent to a private network but uses the shared facilities of the public network. The result generally is a more cost-effective network because unused capacity is not wasted.

3.8.2. Multiple Carriers in the Public Network

In many parts of the world, multiple common carriers compete for customers. In these environments, customers may choose which carriers operating in the public telephone network they wish to have carry their calls. The ways that the networks of competing carriers can be accessed varies from one location to another and from one carrier to another.

Depending on the type of carrier and the context, any combination of the following arrangements are possible:

- Direct access

    In most cases a private telephone system is connected physically to a single specific carrier, referred to as the *local exchange carrier* (*LEC*) or *dialtone provider*. In the case of a wireless telephone, the actual carrier used is determined dynamically. A private telephone system with more sophisticated switching resources can be connected directly to two or more carriers (Figure 3-31). In this case, the telephone system decides which carrier to use for a given call. In the case of a *carrier bypass* arrangement, one of these carriers provides local service and one provides long-distance service. It is called carrier bypass because access to the long-distance carrier bypasses the local carrier's network. This also may be called a *foreign exchange* (*FX*) facility if the second network is actually a local network in a different location.

Figure 3-31. Directly connected carriers

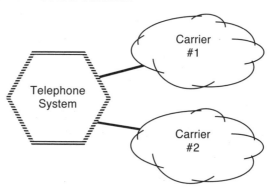

- Default carrier

    A local exchange carrier may be used to access alternative carriers where a default carrier is established for specific types of calls (Figure 3-32). For example, one could specify that all long-distance calls are to be routed through a specific long-distance *inter-exchange carrier,* or *IXC.*

Figure 3-32. Default carriers

- Dialed carrier

  With each outbound call, the local exchange carrier is informed which alternate carrier should be used. This typically is accomplished by dialing special sequences when establishing calls (Figure 3-33).

Figure 3-33. Dial selectable carriers

3.8.3. Telephone Equipment

Telephony products can be generally categorized as being:

- *telephone switches,*

- *telephone station equipment* (*telephone sets*), or

- *add-ons* or *peripherals* to one or the other.

Telephone equipment is differentiated based on where it is installed:

- Customer Premise Equipment

  Customer premises equipment, or *CPE* for short, refers to the telephony products that you purchase or lease and install at your own locations in order to assemble any private telephone system.

- Carrier Switching Equipment

  Carrier switching equipment consists of the telephone switches and transmission equipment used by a carrier to implement its network. This equipment is further categorized as *Central office equipment* if it is located in a Central Office (also know as a *CO, End-Office* or *Local Office*).

- Co-Located Equipment

  Co-located equipment is equipment that is not owned by the carrier but is co-located with carrier equipment in a carrier facility such as a central office.

In this section we'll look at how telephone systems and the products that constitute them, can be modeled. In Chapter 10 we'll pull everything together by actually looking at the implementations of various types of products.

Carrier Switching Equipment and Telephone Company Services

If you are assembling a new telephone system for your home or business, or are assessing your existing system, one of the first steps is to determine what services you want to obtain from your local exchange carrier(s), that is, your local telephone company or companies, and any other carriers with which you wish to have connections.

The functionality of your telephone system (whether it is just a single telephone or a collection of switches forming a private network) will be a combination of the features and services built into your telephony equipment, and any telephony features and services that you *subscribe to* from the telephone company (or companies) that your equipment can access.

As described in section 3.8.2, a LEC, generally provides the connection(s) between your telephone system and the PSTN. In other words, the LEC owns the switch or switches that extend service to your telephone equipment, for local calling and typically for access to IXCs. You may, however, connect directly to one or more IXCs through what is referred to as carrier bypass. The notion of bypass applies only in the context of competing carriers. In countries where a single telephone company provides all of these services, this option does not apply.

A switch owned by the LEC to which your lines or trunks connect is known as a *central office switch*, (*CO switch* for short), or a *public exchange*. The combination of line interfaces and telephony features and services available for subscription on a given CO switch represents the *service offerings*.

Customer Premises Equipment (CPE) Options

The trunks or lines that you already have and those that are available to you from various telephone companies will, in part, dictate what types of telephony products are applicable to your needs. Depending on the requirements, your telephone system might be as small as a single telephone to be connected to a subscriber line from your local telephone company or cellular carrier, or it may be as complex as a global private network.

3.8.4. Modeling Telephone Switches

 A *telephone switch*, or just *switch* for short, is a telephony product characterized as an implementation of telephony resources that, at a minimum, includes call processing and switching resources, and at least two station or network interface type devices.

A generic telephone switch is illustrated in Figure 3-34.

Figure 3-34. Generic telephone switch

A switch may consist of any number and combination of telephony resources built around the core functionality of call processing and switching resources (which may be minimal in some cases).

The switch makes connections to other telephony products through *lines* that represent the cabling, wireless transceivers, or other means of connection. Lines fall into two categories, *extension* (or *subscriber*) lines and *trunks*.

The point where a line is connected to a switch is often referred to as a *port*. In addition to the connections a switch can make using lines, the switch may connect with other special peripherals through *interface ports*.

Extension or Subscriber Lines

Extensions are lines that correspond to a switch's station devices. From the perspective of the switch, each extension line corresponds to the logical element of one station device. When the switch is part of the public network, these lines are referred to as *subscriber lines, subscriber loops,* or *CO lines.*

If a given switch is able to detect and control one or more physical telephones attached to a particular line, the corresponding station device either includes a physical element, or is associated with other devices containing physical elements through a device configuration.

In this case, the switch keeps track both of the station device with the single logical element corresponding to the line, and of all the physical element station devices associated with the port.

Additional telephony resources making up an operational switch—including any CTI, OA&M, accounting, and media interfaces—may be either built-in or implemented as add-on peripherals. Depending on the implementation, peripherals may be attached through the switch's internal connections, or may be attached using line ports. In these cases, the switch keeps track of the other device types and resources associated with the port.

Trunks

Trunks are lines that correspond to network interface devices. They are used to connect to one or more external networks by attaching to switches that are part of the external network.

 A trunk associated with one switch can be connected to either a trunk or a subscriber line of another switch. In other words, the network interface device of the first switch may correspond either to a station device or a network interface device of the second switch.

Interface Ports

Interface ports are the means for physically connecting computers or other peripherals to a switch's CTI interface, OA&M interface, accounting interface, or media interfaces (in switch implementations that do not use line ports for this purpose).

3.8.5. Modeling Telephone Station Equipment

Telephone station equipment, including telephone station peripherals, are telephony products that are connected to the extension/subscriber lines of a switch. They are fundamentally implementations of station devices, along with associated switching, call processing, media services, and interface (CTI, OA&M, accounting, and media) resources. A particular piece of station equipment may be modeled one of three ways, depending upon the context:

1. If it is modeled independently of any connection to a telephone system, it is viewed simply as a physical device element made up of the appropriate physical element components.

2. If it is connected to a switch, the telephone station may be modeled from the switch's perspective as a physical element associated with one or more of the logical station devices in the switch (as described in section 3.8.4 above). This view of a telephone station is illustrated in Figure 3-35.

Figure 3-35. Generic telephone station

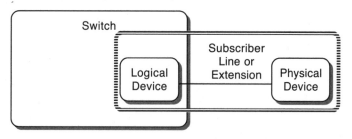

3. If it is receiving telephone service from a switch but is viewed as being a stand-alone telephone system, the station equipment is modeled as a complete set of telephony resources. This includes a station device, or device configuration, that corresponds to the station's physical element and associated logical device element(s), along with associated telephony resources that include switching, call processing, media services, and any CTI, OA&M, and media interfaces. In most cases, call processing in the telephone station accepts commands from the physical element interface and/or the CTI interface, and translates them into appropriate commands for the switch using the telephone station's network interface device as a proxy. This view of a telephone station is illustrated in Figure 3-36.

Figure 3-36. Generic telephone station modeled from station's perspective

Refer to Chapter 4, section 4.1 for a complete description of physical element components.

3.8.6. Modeling Telephone Networks

Each individual telephony product is a telephone system, that is, a set of telephony resources. Each telephone system generally is a subsystem of a larger telephone system, and virtually all telephone systems are ultimately connected to, and thus are part of, the world-wide telephone network (the PSTN). As a result, one can model any portion of the PSTN or any private network in the same way that an individual switch or telephone station can be modeled.

In Figure 3-34 a generic telephone network, a LEC for example, is being modeled as a set of telephony resources. Though the network is actually made up of multiple switches, it can still be modeled as a single set of resources because the network operates as a single entity. The network interface devices in the network represent its connections to various IXCs and the station devices correspond to subscriber lines.

Figure 3-37. A LEC's telephone network

3.9. Review

In this chapter we have seen that a *telephone system* may range in scope from an individual telephone to a vast telephone network. Telephone systems represent a collection of *telephony resources* that also may be referred to as a *telephony resource set*. Types of telephony resources include *call processing, switching, interfaces, devices,* and *dynamic objects*.

Calls are dynamic objects in a telephone system that represent a *media stream* and associated control information traveling between two or more points.

Devices are the endpoints that can be associated with calls. Devices are resources responsible for consuming and generating the media streams and control information associated with calls.

Connections define the relationship between a particular call and a particular device. The most important attribute of a connection is its *connection state*, which may be *null, initiated, alerting, connected, hold, queued,* or *fail*.

Media stream channels are allocated for connections as needed and are responsible for conveying the associated media streams.

Switching resources are responsible for carrying out services that create, clear, and manipulate the states of connections, and for allocating and deallocating media stream channels as needed. These *switching services* can be illustrated through a simple graphical notation that shows the devices, calls, connections, and connection state transitions involved.

Call processing is primarily responsible for *routing* or directing calls through the telephone system by managing the switching resources. It takes action based on feature settings, timers, default rules, and commands received from devices and through telephone system interfaces. Certain types of devices exist within a telephone system in order to provide specialized routing functionality.

Now that an abstraction of the basic telephony resources has been defined along with the graphical notation for describing them, we can delve into more of the details surrounding device modeling.

4.
Telephony Devices

Devices are the endpoints of telephone call media streams. They are a principal resource in the telephony resource framework because the rest of the telephony resources are in place to serve them.

So far we have seen how switching operations revolve around associating calls with devices through connections and manipulating the states of these connections. The switching fabric—the implementation of the switching resources used in a given telephone system—maps these connection state manipulations into actions. Media stream channels in appropriate transmission facilities are allocated and deallocated in order to establish calls between devices.

In this chapter we'll look more closely at the variety of different types of telephone system devices and how they are modeled and used. Telephone system devices have logical elements and associated components and attributes that determine their functionality and the operations that can be performed with them. Telephone stations also have physical elements and a variety of related physical components and attributes.

4.1. Telephone Stations

Station devices are the telephony resources that correspond to the tangible telephones and telephone lines with which we are all familiar. Station devices take an almost unlimited number of different forms. There are literally hundreds of telephone vendors around the world, each of which manufactures many different models of telephones. Each telephone is designed to appeal to the particular needs and preferences of a different type of telephone user.

Despite the fact that at the most simple level all these telephone devices provide the same basic functionality, there is an endless number of variations in the form that the physical user interface to a telephone system can take. Research has demonstrated, and the marketplace has validated, that people have very diverse and particular preferences when it comes to the form that their interface to telephony services should take.[4-1]

It is important to note that from the perspective of telephony concepts, every type of telephone is a station device regardless of its functionality, interface to the telephone network, or other properties. This includes POTS (plain old telephone service) telephones ("500" and "2500"[4-2] telephone sets) and multi-line telephones. It includes wireless phones such as cordless and cellular phones, and coin and card pay phones. It includes novelty phones from "football phones" through "shoe phones." It includes full-featured digital phones and attendant consoles, and hybrid devices such as fax phones and video phones.

4-1. Diversity of telephony user interfaces — The fact that user preferences are so diverse when it comes to a personalized interface to telephony functionality (traditionally the telephone) plays a very important role in both the motivation for CTI and the architecture of CTI implementations.

4-2. 500 and 2500 telephone sets — AT&T assigned the model number 500 to the old rotary-dial deskset telephones. When touchtone was developed, the touchtone model was numbered 2500. Since then, the term 2500 set has come to mean any telephone that is functionally equivalent to the original model 2500 telephone set.

4.1.1. Physical and Logical Device Elements

The portion of a device that is involved with call control, that is, the creation, management, and clearing of connections, is referred to as the *logical element* of a device. Most types of devices have only a logical device element.

Figure 4-1. Physical and logical elements

Station devices typically have a tangible portion—a physical telephone set—that is referred to as a *physical element*. To interact with calls, a physical device element may have a corresponding logical device element and/or may be associated with the logical device elements of other devices (Figure 4-1). (The concepts relating to logical elements and the relationships between physical and logical elements, referred to as device configurations, are described in sections 4.4 and 4.5.) While the logical element of a station device may be associated with both digital data and voice calls, the physical element only interacts with voice calls. In addition, a physical device element may only be associated with a single connection in the *connected* state at any given time, although any number of associated connections may be in other connection states.

Figure 4-2. Physical device element components

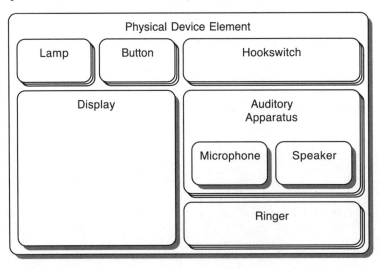

A physical device element consists of many different *components* that make up the physical user interface of the device. These components may be directly accessible to the user, or may be internal to the device (i.e., virtual). Types of physical element components include:

- Auditory apparatus
- Hookswitch
- Ringer
- Button
- Lamp
- Display

As shown in Figure 4-2, the physical element of a particular device may or may not have a display, and may have zero or more lamps, buttons, hookswitches, ringers, and auditory apparatuses. Auditory apparatuses are always associated with a particular hookswitch that governs whether they are active or not. Lamps are optionally associated with buttons.

The way that these physical device components can be combined to form different products is described in Chapter 10.

4.1.2. Auditory Apparatus

An *auditory apparatus* component is one instance of a source and destination of speech-quality media streams.

Though typically an auditory apparatus consists of both a microphone and a speaker, it need only have one or the other. An auditory apparatus usually is one of the following types:

- Handset
    A *handset* is the most common type of auditory apparatus. It is a straight or elbow-shaped device that is held in a person's hand, with a microphone at one end and a speaker at the other.

- Headset
    A *headset* is an auditory apparatus that is worn on the head. It has both a microphone and speaker. The microphone may be on a boom of some sort or built into the ear piece.

- Speaker phone
    A *speaker phone* auditory apparatus is the combination of a microphone and a speaker built into the base of a telephone set, or arranged in some other fashion to allow for untethered operation.

- Speaker-only phone
    A *speaker-only phone* is an auditory apparatus that is like a speaker phone but has only a speaker and no microphone.

A physical element may have any number of auditory apparatuses. Every auditory apparatus is associated with exactly one hookswitch[4-3] (although a single hookswitch may be associated with more than one auditory apparatus).

4-3. Hookswitch devices — Because auditory apparatuses are governed by a hookswitch, they are called hookswitch devices by some.

Microphone

A *microphone* allows sound to be converted to a media stream for use in the telephone system. Microphones have two properties:

- Mute

  Microphone mute determines whether or not the microphone is active.

- Gain

  Microphone gain determines what amplification of the captured sound is applied.

Speaker

A *speaker* allows a voice media stream from the telephone system to be converted to sound. Speakers, like microphones, have two properties:

- Mute

  Speaker mute determines whether or not the speaker is active.

- Volume

  Speaker volume determines what amplification is applied to the sound.

4.1.3. Hookswitch

A *hookswitch* is a component that determines what auditory apparatus or set of auditory apparatuses are in use. A hookswitch can be either *on-hook* (inactive) or *off-hook* (active). If a hookswitch is off-hook, every associated auditory apparatus is actively consuming and/or generating sound. If a hookswitch goes on-hook, every associated auditory apparatus is made inactive.

A physical element may have any number of hookswitches and each operates independently of all the others (Figure 4-3). A hookswitch may be a spring-loaded switch on the telephone, it may be a single locking switch that is pressed to set it off-hook and pressed again for on-hook, or it may be a pair of switches where one is used to go off-hook and another is used to go on-hook. It also may be an internal (or virtual) switch.

Figure 4-3. Hookswitches

4.1.4. Ringer

The ringer is the component in the physical element that notifies people when the telephone system is attempting to connect a call to a logical device element,[4-4] with the intent that it be answered by the physical device in question (i.e., a connection in the *ringing* mode of the *alerting* state is present) or when the telephone system is prompting.[4-5]

4-4. Multiple logical device configurations — The relationship between physical and logical device elements is described in section 4.5. Every operational physical device element is associated with at least one logical device element, but it may be associated with two or more.

4-5. Prompting — The prompting feature is described in Chapter 5. It refers to a feature where the telephone system is unable to make a physical device element go off-hook automatically and must signal for a person to do so.

The ringer has a bell or buzzer to provide an audible indication that the telephone set is ringing, but it also may utilize lamps or the display as *ringing indicators* that provide a visual indication that the telephone set is ringing (Figure 4-4). The ringing indicators may be built into the telephone set or may be connected to it remotely. For example, a telephone may have ringing indicators in the form of a large flashing light and bell mounted outdoors so that it can be seen and heard at a distance.

Figure 4-4. Ringing indicators

Ringers have the following attributes:

- Volume
  A ringer's *volume* is the sound level at which the ringer's audible indicator will ring.

- Mode
  A ringer's mode indicates what it is doing. It may be either *ringing* or *not ringing*.

- Pattern

  Pattern refers to the cadence, frequency, and other properties of the sound generated by a ringer when it is ringing. The pattern is set by the telephone system to indicate the type of call that is being presented, or to indicate that a particular number was called. If it is being used to indicate the type of call, it is one of the following:

  unspecified – meaning call type ringing is not supported
  internal – the call is from inside the telephone system
  external – the call is from outside the telephone system
  priority – the call has been marked as high priority
  callback – the callback feature[4-6] has been used to place the call
  maintenance – the call is maintenance-related (e.g., a test)
  attendant – the call is from an attendant
  transferred – the call is being transferred to the device
  prompting – the telephone system is prompting[4-9] the device

- Count

  A ringer's *count* is the number of ring cycles[4-7] that have been completed since the ringer began ringing. It is zero if the ringer's mode is *not ringing*.

A physical device element may have more than one ringer, but only one may be actively ringing at a time.

4-6. Call back feature — The call back feature is explained in Chapter 5, section 5.14.3. It involves a call from a device that was previously busy or unavailable.

4-7. Ring cycles — A ring cycle is the time between the playing of each ringer pattern while a device is ringing. The individual ringer pattern may be a simple on-off pattern or a more complex sequence. In any case, this attribute does not reflect the number or length of cycles within the pattern.

4.1.5. Dial Pad Buttons and Function Buttons

A *button* is a component that can be pressed to request a particular function or feature of the telephone system (Figure 4-5).Typically a button component is either a physical button that can be pushed and released, or an internal (virtual) switch that is pressed or triggered indirectly.

Figure 4-5. Buttons

Buttons may have the following attributes:

- Label

    A button's *label* is the name by which it is called. Typically the label is printed on the surface of a physical button.

- Associated Number

    Speed-dial buttons and others that depend on remembering a particular telephone number have an *associated number*.

- Active/Inactive

    At any given time a particular button may be active or inactive. An active button can be pressed and an inactive one cannot.

Most physical element implementations have at least 12 buttons, corresponding to the buttons on a standard dial pad. Depending on the context, pressing these buttons either communicates part of a device address, indicates a particular service to be carried out, or requests that a particular tone be generated.

The most typical example of a function button is a speed-dial[4-8] button that autodials a particular number.

4.1.6. Lamps

Lamps are components that provide simple visual feedback from the telephone system. Lamps often are implemented as simple light sources on the physical element (Figure 4-6) but they can also take other forms. For example, a lamp might be implemented as a special symbol that can be made to appear and disappear on the physical element.

Figure 4-6. Lamps

4-8. Repertory button — Speed-dial buttons are sometimes referred to as repertory buttons or rep-buttons or just auto-dial buttons.

Lamps may have the following attributes:

- Label

    A lamp's *label* is the name by which it is called. Typically the label is printed on or beside the lamp. If a lamp is associated with a particular button, its label could be the same as the button's.

- Associated Button

    A lamp may or may not be associated with a single *associated button*.

- Brightness

    A lamp's *brightness* indicates the intensity of the lamp when it is on[4-9] (e.g., steady, winking, fluttering, or broken fluttering). A lamp's brightness may be one of the following:

    normal – the lamp is at its normal intensity

    dim – the lamp is dimmer than normal

    bright – the lamp is brighter than normal

- Color

    A lamp's *color* is a fixed attribute that describes the hue of the lamp when it is on[3-20]. The color may be any value but is typically one of:

    no color – meaning the lamp is just a white lamp or liquid crystal display (LCD) indicator

    red

    yellow

    green

    blue

4-9. "On" — Technically speaking, "on" in this context refers to the "active phase of the lamp's duty cycle."

- Mode

    A lamp's *mode* indicates what it is doing. A lamp's mode may be one of the following:

    off – the lamp is off
    steady – the lamp is on
    wink – the lamp is flashing slowly
    flutter – the lamp is flashing quickly
    broken flutter – a combination of wink and flutter[4-10]

A physical element may have any number of lamps and any number of these lamps may be associated with a particular button.

4.1.7. Message Waiting Indicator

A physical device element may or may not have a *message waiting indicator* of some sort. If present, this indicator could be implemented in a fashion similar to a lamp, could be an audible chirping mechanism, or could be implemented in some other fashion. Message waiting indicators have only two simple attributes. They are either on or off, and they are either visible or not.

4.1.8. Display

A physical device element's *display* is a grid of alphanumeric characters that allows the telephone system to communicate text-based information (Figure 4-7).

A display, if present, has the following attributes:

- Rows

    A display may have one or more *rows* of characters.

- Columns

    A display may have one or more *columns* of characters. Typically displays are 10 to 80 columns in width.

4-10. Flink — A broken flutter is referred to as a flink (the combination of a flutter and a wink).

Figure 4-7. Display

Display

- Character Set

    A display's *character set* refers to the way that the data representing each character in the display, including spaces, is translated into a graphical symbol. The amount of data required to represent each character is determined by the number of different symbols supported by the display. A display's character set is typically ASCII or Unicode.[4-11]

- Contents

    The *contents* of the display is the data representing the information to be displayed. The contents may be thought of as a long sequence of characters representing the concatenation of each row in the display. The size of this sequence is fixed and is determined by the number of rows

4-11. ASCII and Unicode — ASCII (pronounced "as-kee") stands for American Standard Code for Information Interchange. The ASCII character set uses one byte per character and can represent 256 characters of which the first 128 are standardized. The Unicode character set is designed to allow representation of the characters used by all the languages of the world, including Chinese, Japanese, and Korean. It uses two bytes per character and standardizes tens of thousands of characters.

and columns in the display and the amount of data required to represent each character (size = rows x columns x character size).

A physical device element may have, at most, one display.

The telephone system display associated with a physical device element should not be confused with other types of graphical displays that may be present on various types of products. A video phone, for example, typically is a station device with physical device elements such as buttons, an auditory apparatus, etc. Although a video phone typically has a video display that shows decompressed video data from the media stream, this display presents information from the media stream and not telephone system feedback. It therefore is not a physical element display component, although a physical element display could be implemented by superimposing it on top of the video display. In this example, the video display is modeled as a type of media access resource associated with the physical device element.

4.2. Network Interface Devices

Up to this point we have been using the term *device* to refer to a general class of telephony resources capable of having a connection to a call. The first example of a device was the telephone set, which is known as a station device. Station devices actually terminate or originate media streams and can be thought of as being *local*, or directly attached, to a particular set of telephony resources.

In Chapter 3, section 3.2.5 we saw that establishing a call across a network typically involves a number of distinct telephone systems each of which use a set of their own local devices to communicate with other telephone systems. Figure 3-9 illustrates how these *network interface devices* are transparent to the two devices at either end of a call which spans the network.

A *network interface device* is an endpoint with respect to one set of telephony resources, but corresponds to a transmission facility that connects to another set of telephony resources. Figure 4-8 shows just one set of telephony resources and abstracts the rest of the network. In this case, device D3 is a network interface device.[4-12]

Figure 4-8. Network interface device

From the perspective of the telephony resources shown in Figure 4-8, the network interface device D3 is a proxy for the remote device D2 on the other side of the network. These telephony resources cannot manipulate D2, so they act on D3 in its place. For example, to drop connection D2C1 from the call, the switching resources drop connection D3C1, which has the same result. The relationship of D3 as a proxy to D2 may be represented in graphical notation as shown in Figure 4-9. The brackets indicate that D3 is a proxy for D2.

Figure 4-9. Network interface device representation

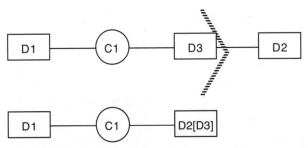

4-12. Network interface devices — Depending upon the implementation involved, network interface devices may be referred to as trunks, direct lines, or CO lines.

4.3. Call Routing Resources

Call routing refers to the movement of a call from device to device or, from the call's perspective, the sequence in which connections to new devices are created and cleared. In Figure 4-10 call C1 is routed first to D2, then to D3, and finally to D4.

Figure 4-10. Call routing

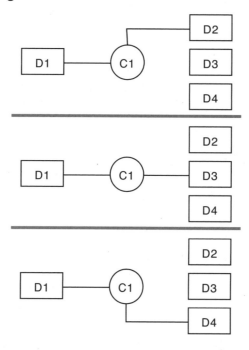

4.3.1. Call Processing

Call processing is ultimately responsible for all call routing activity as it directs the switching resources that establish and clear connections. Call processing establishes and clears calls as a result of:

- Commands from devices

- Commands from CTI interfaces

- Expiration of certain timers

155

- Feature settings (stored rules such as *call forwarding*[4-13] that override normal call progress)

- Default rules that determine "normal" call progress

When a device creates a call, it specifies a desired destination for the call. By default, call processing will attempt to connect the specified device to the call; the call may be redirected many times, however, before being connected. The various features and switching services that relate to call routing are explained in Chapter 5.

Among the telephony resources that may be present in a given telephone system are special devices that may be associated with a call strictly as an additional step in the routing of that call. As always, the call is associated with these special devices through connections, but the states of these connections generally are either *queued* or *alerting*.

4.3.2. Park Device

A *park device* is a special device with which calls may be associated in order to set them aside, or "park," them temporarily. Connections associating a call and a park device, if present, are in the *queued* state. A call associated with a particular park device later can be *picked* and connected to any other device in the telephone system.

4.3.3. Pick Group Device

A *pick group device* is a single special device associated with two or more devices (typically stations) that form a *pick group*. The pick group device is comparable to the park device, except that calls are not explicitly parked to it. In fact, it is a unique type of device that never interacts with calls directly. Instead, the pick group device simply keeps track of all connections associated with all of the devices in the

---

4-13. Call forwarding — Telephony features, and call forwarding in particular, are discussed in Chapter 5. Call forwarding is a feature that allows a rule to be established for the automatic redirection of a call in the event that certain criteria are satisfied.

pick group that are in the *ringing* mode of the *alerting* state, the *hold* state, or the *queued* state. Depending on the implementation of the telephone system, any device in the telephone system or just those from the pick group in question, can then request that a call be *picked* from those being tracked by a particular pick group device. Devices in the group pick their calls from the pick group by default.

4.3.4. ACD Device

An *Automatic Call Distribution* device, or *ACD* device, distributes calls that are presented to it. Depending on the implementation of the ACD, it may use any algorithm or logic for determining what device in the telephone system to which to redirect a call. An ACD device is sometimes referred to as a *split*.

When a call is presented to an ACD device, a connection in the *entering distribution* mode of the *alerting* state usually is created. The ACD then specifies the device to which it wants the call redirected, and the connection to the ACD is cleared. If the ACD cannot immediately identify an available device to which to direct the call, it may either leave the call in the *alerting* state or transition it to the *queued* state until a suitable destination is identified.

An ACD may employ other devices, including media access devices (for playing messages, etc.) and other ACDs or ACD groups. Implementations of ACD devices may be modeled one of two ways. The first model involves *visible ACD-related devices*. In this model, any devices employed by the ACD for interacting with the call are modeled independently from the ACD device itself (with separate connections to the call). In the second model, *non-visible ACD-related devices*, the additional resources used are all considered to be within the ACD device itself, so only a single connection to the call is used.

Telephony Devices

4.3.5. ACD Group Device

An ACD group device distributes calls to other devices like an ACD device but it only distributes calls to a specific group of devices that are specifically associated with its group. In every other way, an ACD group device operates in the same fashion as an ACD device. An ACD group device is sometimes referred to as a split or *pilot number*.

ACD Agents

The association between an ACD group device and a device to which calls can be redirected is called an *agent*. Just as connections are dynamic entities that represent the temporary relationship between a device and a particular call, an agent represents a similar temporary relationship between a device and one or more ACD group devices.

Agents are created when a device associates itself with a particular ACD group. This is referred to as *logging onto* an ACD group. An agent is cleared after the corresponding device *logs off* the ACD group.

Just as connections have a connection state, agents have an *agent status*. An agent's status may be one of the following:

- *Agent null*
  No relationship exists between the agent and the ACD group; that is, the agent is logged off.

- *Agent logged on*
  A relationship between the device and the ACD group has been established, but no distribution of calls to the agent has yet begun.

- *Agent not ready*
  The device associated with the ACD group is not prepared to receive calls distributed by the ACD.

- *Agent ready*
  The device associated with the ACD group is prepared to receive calls distributed by the ACD.

- *Agent busy*

  The device is dealing with a call that was directed to the device by the associated ACD group.

- *Agent working after call*

  The device has completed working with a call that was directed to the device by the associated ACD group, but it is not yet prepared to deal with another call.

Agent status is used by the ACD group device for two things.

1. Routing

   The ACD group will only distribute calls to devices associated with agents that have a status of *agent ready*. It should be noted that agent status relates only to the status of the agent and not to the device itself. So a device that is busy with a call delivered by some mechanism other than through the ACD group may have a status of *agent ready*. Likewise, only the ACD groups associated with the device through the association represented by the agent are affected by the agent's status. This means that a device still may receive calls from any other devices.

2. Statistics

   Most ACD group implementations provide detailed statistics on the activity of agents. The amount of time that an agent had a given agent status, along with the times each call spent queuing before being redirected to a particular device, are generally captured so that the efficiency of the system and its users can be optimized.

4.3.6. Hunt Group Device

A *hunt group device* distributes calls to a specific group of devices like an ACD group device. However unlike the ACD group device, the group of devices served by a hunt group's distribution function is fixed in some fashion so the concept of agents does not apply. In every other way it behaves in the same fashion as an ACD device or an ACD group device. Hunt group devices are also sometimes referred to as pilot numbers.

4.4. Logical Device Elements and Appearances

Every device that is able to participate directly in a call has a portion referred to as a *logical element*. The concept of logical and physical device elements was illustrated in Figure 4-1.

A logical device element is a resource that is used to manage call control and switching functionality. Such elements represent the media stream channels and signal processing facilities used by a device while participating in calls. All telephony features and services, except those that specifically act on components of a physical device element, act on the logical element of a device or a set of logical elements in a device configuration. (Device configurations, or the relationship between multiple logical and physical elements, are described in section 4.5.)

 A logical device element has a set of one or more *call appearances* that represent a device in a given connection. The rectangular symbol that was introduced earlier to represent devices in graphical notation reflects the participation of that device through a particular appearance within the corresponding logical device element.

Logical device elements have the following attributes:

- Appearance type
- Appearance addressability

All of the appearances within a particular logical device element have the same type and addressability.

4.4.1. Call Appearances

A call appearance, or just *appearance* for short, may be thought of as a "connection handler" within a logical device element. One appearance can be associated with, at most, only a single call at a time (through a connection).

All switching services that manipulate a device's association with a call act on the appropriate appearance within the logical element of that device. Appearances inherit all of the properties, attributes, and feature settings that apply to their logical device element.

The concept of appearances within logical device elements is illustrated in Figure 4-11. In this case logical device element D1 (the logical element portion of device D1) has two appearances labeled "A1" and "A2." Appearance A1 is representing device D1 in connection D1C1, and appearance A2 is representing device D1 in connection D1C2.

Figure 4-11. Logical device elements and appearances

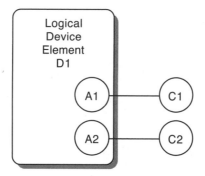

The state of an appearance is the state of the corresponding connection. If no connection is present at a given appearance, it is said to be *idle*.

4.4.2. Addressability

The *addressability* of an appearance refers to whether or not call processing is able to act specifically on a particular appearance within a logical device element.

Addressable Appearances

If an appearance is *addressable,* call processing can explicitly manipulate and observe connections associated with the appearance.

Addressable appearances are permanent and fixed in number once set and cannot be created or destroyed. At any given time, a particular addressable appearance may or may not have an associated call.

Non–addressable Appearances

If an appearance is *non–addressable,* call processing is unable to distinguish a particular appearance from the encompassing logical device. In this case, the logical device itself is responsible for associating each connection with one of its appearances.

Non–addressable appearances are created and destroyed as needed to handle new connections that are associated with a particular logical device element. This behavior is illustrated in Figure 4-12.

Figure 4-12. Non–addressable appearance behavior

4.4.3. Appearance Types

The type of appearances supported by a particular logical device element plays a very significant role in determining the functionality and behavior associated with that device. There are two appearance types:

- Standard

- Bridged

The appearance type supported by a logical device element determines if it may be associated with, at most, one single physical device element, or if it may be shared by multiple physical device elements.

There are six behavior-type combinations (listed in increasing order of complexity):

- Selected–Standard

- Basic–Standard

- Exclusive–Bridged

- Basic–Bridged

- Independent–Shared–Bridged

- Interdependent–Shared–Bridged

Standard Appearances

Logical device elements containing *standard appearances* may not be shared by multiple physical device elements, and thus also are referred to as *private appearances*. All of the media streams associated with calls corresponding to the standard appearances are available only to the physical device element (if any) associated with the logical element in question.

Logical device elements with standard appearances exhibit one of two behaviors referred to as *selected* and *basic*.

Selected–Standard Appearance Behavior

When calls are presented to a logical device element that has *selected–standard* appearances, the new call is presented only to a single "selected" idle appearance.

Selected–standard behavior is illustrated in the example shown in Figure 4-13. In this example, D2 is a logical device element with four addressable standard appearances. Connection D2C1 is being handled by appearance A2. When a new call, C2, is presented to the device, selected–standard behavior dictates that a new *alerting* connection be created for a selected appearance, in this case A3.

Figure 4-13. Selected–standard appearance behavior

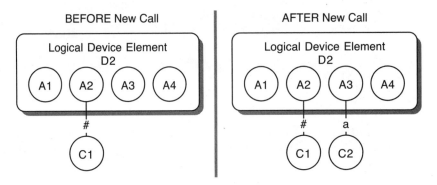

Basic–Standard Appearance Behavior

When calls are presented to a logical device element that has *basic–standard* appearances, the new call is presented simultaneously to all of the idle appearances.

When the call is answered by one of the appearances (i.e., the connection state of the corresponding connection transitions from *alerting* to *connected*), all of the other connections associated with the other standard appearances are cleared and the appearances return to being idle.

Basic–standard behavior is illustrated in the example shown in Figure 4-14. In this example, D3 is a logical device element with four addressable standard appearances. Connection D3C1 is non-*null* and is handled by appearance A3. When a new call, C2, is presented to the device, basic–standard behavior dictates that *alerting* connections are established for each of the (previously) idle appearances. After appearance A1 answers[4-14] the call, the connections that were created for appearances A2 and A4 are cleared.

Figure 4-14. Basic–standard appearance behavior

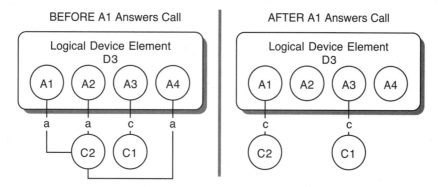

Bridged Appearances

Bridged appearances differ from standard appearances in that logical elements with bridged appearances may be associated with multiple physical device elements.

In the case of addressable bridged appearances, the logical device element can be associated with as many physical device elements as it has bridged appearances. Each bridged appearance is permanently associated with a particular physical device. The media stream associated with a call at a particular bridged appearance is available only to the corresponding physical device element.

4-14. Answer switching service — The answer switching service will be described fully in Chapter 5. It involves taking a connection in the alerting or queued state and making it active by transitioning it to the connected state.

In the case of non–addressable bridged appearances, the logical device element can be associated with any number of physical device elements, and it creates new appearances as needed. Each bridged appearance is associated with a particular physical device as long as a connection is present.

Typically each bridged appearance in a logical element is associated with a different physical device element, but this is not a requirement.

Bridged appearances are illustrated in Figure 4-15. In the example shown, logical device element D4 has three bridged appearances. It is associated with physical device elements D1, D2, and D3 through bridged appearances A1, A2, and A3, respectively. The association between a device and an appearance is shown graphically using double-headed arrows. The media stream associated with call C1 is accessible to physical device D1 by virtue of the fact that it is connected using bridged appearance A1.

Figure 4-15. Bridged appearances

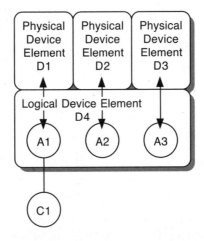

Logical device elements with bridged appearances exhibit one of four behaviors:

- Basic
- Exclusive

- Independent–shared

- Interdependent–shared

Basic–Bridged Appearance Behavior

When a new call is presented to a logical device element that has *basic–bridged* appearances, the call is simultaneously presented to all of the appearances.

When the call is answered by one of the appearances (the connection state of the corresponding connection transitions from *alerting* to *connected*) all of the other connections associated with the other bridged appearances are cleared.

Basic–bridged behavior is illustrated in the example shown in Figure 4-16. In this example, D4 is a logical device element with four bridged appearances. When a new call, C1, is presented to the device, basic–bridged behavior dictates that *alerting* connections are established for each of the (previously) idle appearances. After appearance A4 answers the call, the connections that were created for appearances A1, A2, and A3 are cleared.

Figure 4-16. Basic–bridged behavior

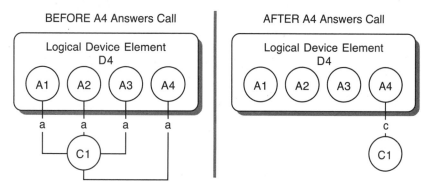

Exclusive–Bridged Appearance Behavior

When a new call is presented to a logical device element that has *exclusive–bridged* appearances, the call is presented simultaneously to all of the appearances.

When the call is answered by one of the appearances (i.e., the connection state of the corresponding connection transitions from *alerting* to *connected*), all of the other connections associated with the other exclusive–bridged appearances are *blocked* from further use until the connection to the appearance that answered is cleared.

Exclusive–bridged behavior is illustrated in the example shown in Figure 4-17. In this example, D4 is a logical device element with four bridged appearances. When a new call, C1, is presented to the device, exclusive–bridged behavior dictates that *alerting* connections are established for each of the bridged appearances. After appearance A4 answers the call, the connections that were created for appearances A1, A2, and A3 are blocked from further use by transitioning to the *fail* state.

Figure 4-17. Exclusive–bridged behavior

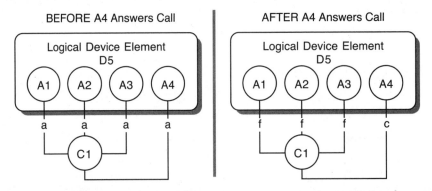

If the exclusive appearance places its connection to the call on *hold*, all of the connections at the bridged appearances transition to *hold*. Exclusivity is restored once again when any one of the connections transitions to the *connected* state and the others transition to the *fail* state.

Shared–Bridged Appearance Behavior

Shared–bridged appearances behavior (also known as *shared bridging*) has two cases, referred to as *independent* and *interdependent*. Shared bridging is the most important form of bridged appearances because it is the most common.

As with the other cases of bridging, when a new call is presented to a logical device element that has shared–bridged appearances, the call is presented simultaneously to all of the appearances.

In the case of shared bridging, however, when one appearance answers a call it becomes the appearance *participating*[4-15] in the call; the other shared–bridged appearances become *inactive*. The connections associated with the inactive appearances all transition to the *queued* state. This is illustrated in Figure 4-18. In this example, A4 answers the call and the connections handled by A1, A2, and A3 all transition to the *queued* state.

Figure 4-18. Shared–bridged behavior

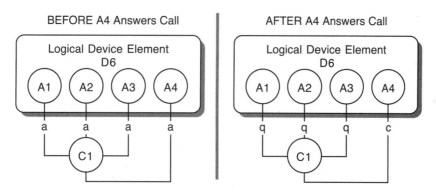

The key feature of shared bridging is that any of the appearances that are in the inactive mode (i.e., are handling a connection in the *queued* state) can answer at any time and be added to the call. Figure 4-19 continues the example from above. Shared–bridged appearance A2 answers call C1 and joins A4 as another participant in the call.

4-15. Participation — An appearance is considered to be participating in a call if its connection is in the connected or hold state.

Appearances A1 and A3 remain in the *queued* state. Additional appearances can be added to the call, up to a limit set by the telephone system implementation.

Figure 4-19. Shared–bridged behavior: adding a second appearance

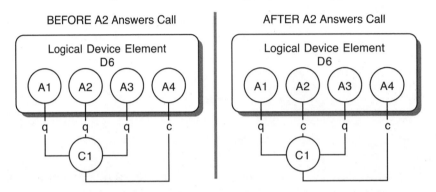

When an appearance drops[4-16] out of a call, it becomes *inactive* and the corresponding connection transitions to the *queued* state. This is illustrated in Figure 4-20. Here the example continues with A4 dropping from the call.

The call remains with the set of shared–bridged appearances represented by a given logical device element as long as at least one of the appearances is participating in the call.

The two forms of shared–bridged appearance behavior, *independent* and *interdependent*, differ in how appearances are affected when another appearance redirects a call or puts a call on *hold*.

4-16. Clear connection switching service — Dropping from a call is synonymous with clearing a connection to the call. The clear connection switching service will be described fully in Chapter 5. Normally the clear connection service transitions connections to the null state; in the case of shared–bridging, however, it transitions them to the queued state if other appearances in the same logical element are participating in the call.

Figure 4-20. Shared–bridged behavior: dropping an appearance

Independent–Shared–Bridged Appearance Behavior

In the case of *independent–shared–bridged* appearance behavior, when an appearance that is participating in a call moves the call away from the device, it has no effect on the other appearances unless it was the last participating appearance in the logical device element on the call. If it was the last, then the call is dropped from the whole logical device element. These two cases are illustrated in Figure 4-21 and Figure 4-22.

Figure 4-21. Independent–shared–bridged behavior: A2 and A4 active

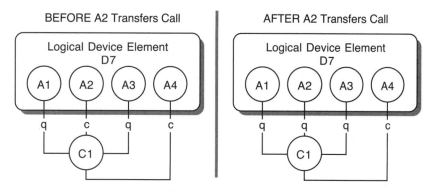

In Figure 4-21, both A2 and A4 are participating in the call when A2 transfers[4-17] it. Afterwards, A2 is *queued* and D7 remains associated with the call.

Figure 4-22. Independent–shared–bridged behavior: only A2 is active

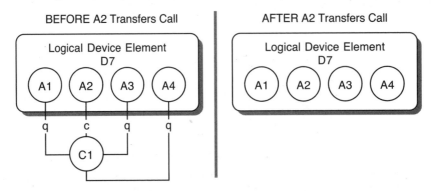

In Figure 4-22, only A2 is participating in the call when A2 transfers it. Afterwards, the call has left device D7 entirely.

Hold[4-18] works similarly in the case of independent–shared–bridged appearances. If an appearance places a call on *hold*, and there are other participating appearances in the logical device element on the call, then only the connection corresponding to the appearance in question is affected. All the connections are placed in the *hold* state only if the appearance was the last participating one.

Interdependent–Shared–Bridged Appearance Behavior

In the case of *interdependent–shared–bridged* appearance behavior, when any appearance that is participating in a call moves the call away from the device, the call is dropped from the whole logical device element. This is illustrated in Figure 4-23.

Hold works similarly. If any appearance places a connection in the *hold* state, all the connections are placed in the *hold* state.

4-17. **Transfer switching service** — The transfer switching service will be described fully in Chapter 5, section 5.12. It involves moving a call to a different device.

4-18. **Hold switching service** — The hold switching service will be described fully in Chapter 5, section 5.11.1. It involves suspending the media stream associated with a particular connection.

Figure 4-23. Interdependent–shared–bridged behavior

4.5. Device Configurations

A *device configuration* describes the arrangement of the various elements and appearances that can be associated with a given device. An endless variety of different device configurations may be formed from the possible combinations of physical elements, logical elements, and different appearance types.

Device configurations are described in terms of a specific *base* device. The following attributes determine the device configuration for a particular device:

- Physical element (Yes/No)
- Physical element's other associated logical devices
- Logical element (Yes/No)
- Logical element's other associated physical devices
- Logical element's addressability
- Logical element's behavior-type
- Maximum/total appearances

Collectively these attributes can be thought of as the device configuration attribute for a specific device.

This section looks at a number of typical, or foundational, examples that illustrate how device configurations may be formed. (In Chapter 10 we will look at how these are applied in telephone system implementations.)

4.5.1. Logical Element Only

A *logical element only* device configuration consists, as the name suggests, of only a logical device element containing non–addressable standard appearances. Generally all devices other than station devices have *logical element only* device configurations.

The *logical element only* device configuration illustrated in Figure 4-24 has the following attributes:

- Physical element: No
- Physical element's other associated logical devices: N/A
- Logical element: Yes
- Logical element's other associated physical devices: None
- Logical element's addressability: Non–addressable
- Logical element's behavior-type: Selected–Standard
- Maximum/Total appearances: unlimited[4-19]

Figure 4-24. Logical element only device configuration

4-19. Maximum appearances — In a product implementation, the maximum number of non–addressable standard appearances is typically limited to some predetermined maximum.

4.5.2. Basic

A *basic* device configuration consists of a single physical device element associated with a single logical device element that contains non–addressable standard appearances. This device configuration applies to simple station devices.

The *basic* device configuration illustrated in Figure 4-25 has the following attributes:

- Physical element: Yes
- Physical element's other associated logical devices: None
- Logical element: Yes
- Logical element's other associated physical devices: None
- Logical element's addressability: Non–addressable
- Logical element's behavior-type: Selected–Standard
- Maximum/total appearances: Unlimited[4-19]

Figure 4-25. Basic device configuration

In this illustration, the labels "L" and "P" are used to denote logical and physical device elements respectively. In Figure 4-25 the appearance "well" symbol represents a pool of non–addressable standard appearances.

Another variation of the *basic* device configuration involves two different devices that are associated with each other: one with only a logical device element and one with only a physical device element. This is illustrated in Figure 4-26. From the perspective of the physical device element P1, the device configuration in this example can be represented as follows:

- Physical element: Yes
- Physical element's other associated logical devices:
    - L2 (Non–addressable/Selected–Standard)
- Logical element: No
- Logical element's other associated physical devices: None
- Logical element's addressability: N/A
- Logical element's behavior-type: N/A
- Maximum/total appearances: N/A

Figure 4-26. Basic device configuration consisting of two devices

4.5.3. Multiple Logical Elements

A *multiple logical elements* device configuration consists of a single physical device element associated with multiple logical device elements that each contain standard appearances. This is a common device configuration for multiple line station devices.

One example of a *multiple logical elements* device configuration is illustrated in Figure 4-27. It has the following properties:

- Physical element: Yes
- Physical element's other associated logical devices:
    - L2 (Non–addressable/Selected–Standard)
- Logical element: Yes
- Logical element's other associated physical devices: None
- Logical element's addressability: Non–addressable
- Logical element's behavior-type: Selected–Standard
- Maximum/total appearances: 1

Figure 4-27. Multiple logical elements device configuration

As we saw with the *basic* device configuration, none of the logical device elements in this device configuration need be part of the same device as the physical device element.

Multiple logical element device configurations allow a single physical element (a telephone set) in a telephone system supporting only one appearance per logical device element to have access to multiple calls simultaneously. Other telephone system features usually are set so that if a new call arrives for a logical element with an appearance that is not idle, it will be redirected to another logical device element within the same device configuration.

4.5.4. Multiple Appearance

A *multiple appearance* device configuration consists of a single physical device element and a single logical device element containing two or more addressable standard appearances.

One example of a *multiple appearance* device configuration is illustrated in Figure 4-28. It has the following attributes:

- Physical element: Yes
- Physical element's other associated logical devices: None
- Logical element: Yes
- Logical element's other associated physical devices: None
- Logical element's addressability: Addressable
- Logical element's behavior-type: Selected–Standard
- Maximum / total appearances: 3

Figure 4-28. Multiple appearance device configuration

Multiple appearance device configurations are another way to allow a single telephone set to have access to multiple calls simultaneously.

4.5.5. Bridged

A *bridged* device configuration involves, as the name implies, bridged appearances. The form of a *bridged* device configuration therefore depends on whether the base of the device configuration is a physical or logical element.

In the example presented in Figure 4-29, the device configuration shown is for logical device element L3, which has bridged appearances. It has the following attributes:

- Physical element: No
- Physical element's other associated logical devices: N/A
- Logical element: Yes
- Logical element's other associated physical devices:
    - P1 (using appearance A1)
    - P2 (using appearance A2)
- Logical element's addressability: Addressable
- Logical element's behavior-type: Shared–Bridged
- Maximum/total appearances: 2

Figure 4-29. Bridged device configuration for a logic device element

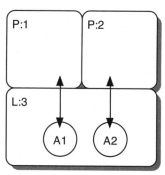

The device configuration for one of the physical device elements in this example is shown in Figure 4-30. It has the following attributes:

- Physical element: Yes
- Physical element's other associated logical devices:
    - L3 (using bridged appearance A1)
- Logical element: No
- Logical element's other associated physical devices: N/A
- Logical element's addressability: N/A
- Logical element's behavior-type: N/A
- Maximum/total appearances: N/A

Figure 4-30. Bridged device configuration: for a physical device element

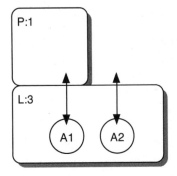

4.5.6. Hybrid

A physical device element associated with multiple logical device elements, each of which have different types of appearances, is said to have a *hybrid* device configuration.

In practice, most station devices are one of the typical device configurations described earlier. *Hybrid* device configurations are not at all uncommon, however, given the steady increase in telephone systems that support both standard and bridging appearance functionality simultaneously.

An arbitrary example of a *hybrid* device configuration is shown in Figure 4-31. It has the following attributes:

- Physical element: Yes
- Physical element's other associated logical devices:
    - L2 (with non–addressable standard appearances)
    - L3 (using bridged appearance A1)
    - L3 (using bridged appearance A2)
- Logical element: Yes
- Logical element's other associated physical devices: None
- Logical element's addressability: Addressable
- Logical element's behavior-type: Selected–Standard
- Maximum/total appearances: 3.

Figure 4-31. Hybrid device configuration

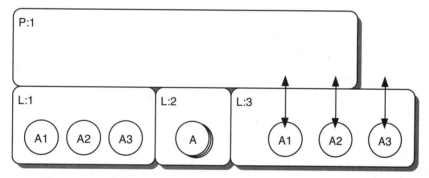

4.6. Addressing Devices

Now that we have seen how telephone systems, networks, and their resources allow the interconnection of device elements, and how complex device configurations can be formed, we turn to the subject of addressing. This refers to the means for referencing a particular device, element, or appearance of interest somewhere in a telephone network, or in a specific set of telephony resources.

4.6.1. Directory Numbers and Dial Plans

The addressing scheme for a telephone network is referred to as its *number plan* or *dial plan*. Telephone networks are hierarchical in nature, so each subset of a telephone network incorporates and expands on the dial plan of the larger network. Telephone networks that interconnect either encapsulate their neighbors' dial plans or provide a means of translating to them.

The address for a particular logical device in a telephone network is a *directory number* or, less precisely, a *phone number*.

If a telephone system allows a number of devices to share one or more network interface devices for access to an external network, then these devices are referred to as *extensions*. The extensions share the directory numbers assigned by the external telephone network to the shared network interface devices. Extensions have their own directory number inside the telephone system (typically one to five digits in length), but may or may not be directly addressable by devices outside the telephone system.[4-20]

A *prefix* is a digit or sequence of digits that indicates that a given *dial plan rule* is to be used for interpreting the rest of the sequence. Typically this is used to indicate that a call is to be established using a particular network interface device, carrier, or some other specific resource. For example, the digit "9" is often used to indicate that a sequence is to be interpreted as an external phone number.

4-20. DID and DIL — DID or Direct Inward Dialing is a service provided by the external network in which a whole block of directory numbers is associated with a particular network interface device (or group of network interface devices). Any call at that network interface device is then connected to a particular extension in the telephone system, based on the originally requested directory number. With DIL, or Direct-In-Line, all calls to a particular network interface device are always connected to the same internal device. For more information on DID and DIL, see Chapter 5, section 5.6.2 and Chapter 10, section 10.9.6.

4.6.2. Addressing in the Public Network

The International Telecommunications Union, or ITU, is responsible for setting standards for interoperability in the PSTN (the worldwide telephone network). The ITU has defined the basic elements that make up a telephone number. These are described below:

- Country code

  The *country code* portion of a telephone number is a one-, two-, or three-digit code that indicates the top-level network to which the number belongs. It refers either to a country like Bosnia (country code 387) or India (country code 91), for example, or a region like North America (country code 1) covered by an integrated numbering plan. The ITU is responsible for assigning[4-21] the unique country codes that make up the first tier of the worldwide numbering plan.

- Area code

  The *area code* portion (which may or may not apply to a given network, and is optional in certain networks) refers to a city or a particular region within the network specified by the country code. For example, the area code for British Columbia is 604 and the area code for Moscow is 095. The rules governing area codes are formulated uniquely within each network designated by a distinct country code. For example, in the North American numbering plan (NANP), area codes are always three digits long and, until recently, the middle digit had to be a "0" or a "1".[4-22] In some networks, area codes are all of uniform size; in others they may vary from area to area. In some networks every addressable device has an area code, but in others this not a requirement. In France, for example, until recently all numbers within Paris had area code 1 and all numbers outside of Paris had no area code.

4-21. Country code definitions — Refer to ITU-T-E.163 for the official standard country code definitions. Your local telephone book should include an up-to-date list of country codes.

- Subscriber number

    The *subscriber number* specifies a particular logical device within a particular local network. The restrictions on the formation of subscriber numbers are once again specific to the network represented by a given combination of country code and area code. In the North American numbering plan, subscriber numbers are always seven digits.[4-23] A North American subscriber number might be 555-1234, but in a small village in China a subscriber number might be just 12.

- Sub-address

    The *sub-address* portion of a telephone number, which may or may not apply to a given network, refers to a particular device that is sharing a subscriber number with other devices through bridging. It allows reference to a specific bridged appearance.

Both ITU and ECMA (formerly the European Computer Manufacturing Association) also have defined a collection of digit sequence notations that define how telephone systems may interpret the sequences of digits representing telephone numbers exchanged between components in a telephone network. The specific notations are described in the sidebar "Standard Telephone Number Exchange Notations."

4-22. North American Numbering Plan — Until January 1, 1995, the number of area codes in the North American Numbering Plan was limited to 152 because 2 through 9 are the only legal values for the first digit (0 and 1 are reserved) and the second digit had to be 0 or 1. The restriction on the middle digit was applied to ensure that there would be no overlap between the set of numbers used as area codes and the first three digits of subscriber numbers (referred to as the central office code). After January 1, 1995, the "middle digit restriction" was relaxed, which resulted in a set of 792 area codes and expanded the overall potential size of the North American network from approximately 1 billion subscriber numbers to over 6 billion.

4-23. North American Numbering Plan — The first three digits of a subscriber number in the North American Numbering Plan are referred to as the central office code or exchange code. The next four digits are referred to as the subscriber code. All of the devices with a common exchange code are generally part of the same telephone system and have a common switching resource.

Standard Telephone Number Exchange Notations

ITU and ECMA specifications provide the following list of notations that telephony resources in a public or private telephone network can use to convey how addressing information associated with a call is to be interpreted. Each of these notations is associated with a raw, unpunctuated sequence of digits intended for machine, not human, consumption.

The notation names use the acronym TON, which stands for Type of Number. The formats defined are:

Implicit TON
Example: "01133112345678"

This notation indicates that the address is a digit string that includes all of the digits originally dialed to specify the destination for the call, including any prefixes. In this example the first three digits are the prefix for international direct dialing ("011"). The next two digits are the country code ("33" for France). This is followed by the single digit area code ("1" for Paris), and finally the eight digit subscriber number ("12345678").

Public TON – unknown

This notation indicates that the meaning of any digits presented is unknown.

Public TON – international number
Example: "33112345678"

This notation indicates that the digits presented include the complete digit sequence for an international telephone number. This is the same destination used in the first example, but in this case the sequence does not include prefixes. It is simply the concatenation of country code, area code, and subscriber number.

Public TON – national number
Example: "112345678"

This notation applies only to calls that remain within a country network, because it specifies that the country code is omitted. In this example, the country code is assumed, so the sequence is a concatenation of only the area code and subscriber number.

Standard Telephone Number Exchange Notations (Continued)

Public TON – subscriber number
Example: "12345678"

This notation is for local calls. In this example we are left with only the digits of the subscriber number; everything else is assumed.

Public TON – abbreviated
Example: "611"

This format is used when the public network supports an abbreviation of telephone numbers. This includes "super speed-dial" numbers and special network numbers. In this example a call originating in North America is being placed to a local carrier's repair number. By specifying that abbreviated notation is being used, the network recognizes "611" as a short form, or abbreviation, of the carrier's repair number.

Private TON – unknown; – level 3 regional; – level 2 regional; – level 1 regional; – local; – abbreviated
Example: "245645678"

There are private network versions of each of the public network formats described above. The only difference between the private and public versions are that the knowledge of the dial plan needed to decode the private network formats are specific to a particular private network or portion thereof. The digit string in the example might be used with any of these notations. Suppose the notation was specified as level 3 regional. The telephone system would need to know that for the private network in question the level 3 regional codes are 1-digit, the level 2 codes are 2-digit, the level 1 codes are 1-digit, and the remainder of the digits are a subscriber number. With this knowledge, plus the indication that level 3 regional notation was being used, the system would know that the number in question was: level 3 "2", level 2 "45", level 1 "6", and subscriber number "45678".

For more information on these standards, consult the following standards documents: ITU-T-E.160, ITU-T-E.131, and ECMA-155.

4.6.3. Dial Strings

When a call is initiated, the most common way to specify the desired destination device is through a *dial string* or a sequence of *dialable digits*. A sequence of dialable digits may include:

- Digits making up all or part of a destination device's directory number;

- Embedded commands that direct the telephone system to use a particular external network, carrier, or routing logic;

- Embedded commands that direct the telephone system to wait for a period of time to pass, wait for silence, wait for a dial tone, wait for billing tones, or wait for additional digits; and

- Digits that represent billing or other information distinct from the destination device address.

The characters that may be used in a dial string in a given instance depend on the interface being used for issuing the dial string and the subset of possible characters supported by a particular implementation. See the sidebar "Dialable Digits Format" for the complete set of characters in a dial string, their meanings, and some examples. The examples illustrate both the power of referencing devices using dial strings, as well as the disadvantages of this approach.

Dialable Digits Format

Dialable digits format consists of a sequence of up to 64 characters from the set of characters listed below.

Each character in the sequence represents a digit to dial or an embedded command of some sort. Every telephone system implementation supports the twelve characters found on all telephone keypads ('0' through '9', '*', '#') but the rest of the characters defined are optional.

The defined characters are:

'0'–'9' '*' '#' Digits to be dialed or DTMF tones to generate

'A'–'F' DTMF tones to generate

'!' Flash the hookswitch

'P' All digits that follow are to be pulse dialed

'T' All digits that follow are to be tone dialed

',' Pause

'W' Pause until dial tone is detected

'@' Pause until ringback followed by a period of silence is detected

'$' Pause until billing tone (bong tone) is detected

';' Pause for another dial string to follow

These characters can be placed into the sequence in any order or combination. For example, a sequence of multiple comma characters would result in a long pause.

The following are some typical examples of dialable digits format. In each case a call is being placed to the same destination: extension 789 behind the directory number 555-1234 in New York City.

Example: "789"

In this example the caller is using the same telephone system, so the call is internal to the telephone system itself.[a] The digit sequence consists of only the digits of the extension number "789".

Dialable Digits Format (Continued)

Example: "321789"

In this example, a call is being initiated from a telephone system in Portland that is part of the same private network as the telephone system in New York. The "321" indicates that a tie line[b] between the two telephone systems is to be used. This is followed by "789", the extension number. No embedded commands for pausing are needed because the private network allocates the tie line very quickly and all signaling is managed transparently.

Example: "19,12125551234@789"

In this example, the call is being placed from a public telephone in France. The digits "19" instruct the telephone system to allocate a network interface for an international call. The comma indicates that the dialing should be paused for the last embedded command to complete. The digit "1" indicates that the call is directed to North America, the digits "212" indicate the destination is New York City (specifically Manhattan), and the seven digits "555-1234" indicate the desired subscriber number. The "@" character tells the telephone system to wait until it detects that ringing has stopped, the call has been answered, and there is silence again. The telephone system at 555-1234 answers the incoming call using an auto-attendant that waits for the desired extension to be dialed. When silence is detected, the final three digits in the sequence, "789", are dialed.

Example: "9,1028802125551234$88844466669999@789"

In this example the caller is using a telephone system in Los Angeles. The digit "9" indicates that the telephone system should allocate the first available network interface device associated with the local carrier. As before, the comma means to wait; we need to give the previous command some time. The digits "102880" indicate that the call is to be placed using an AT&T credit card. The digits "2125551234" are the area code and subscriber number. The "$" indicates that the telephone system should pause until it hears the bong tone from the AT&T credit card system. The dialing then continues with the digits "88844466669999", which comprise the caller's credit card number. Finally the telephone system waits for the destination to answer and go silent; it then continues with "789", the extension number.

Dialable Digits Format (Continued)

Example: "4W5551234@789"

In this example, the call is being placed from a telephone system in Toronto. The "4" indicates that the telephone system is to use a special foreign exchange line that connects directly to the local network in New York City. Once again we need to pause while this is going on, however because the delays in setting up this connection vary, and because this system delivers dial tone once the connection to New York is established, the "W" command is used. The call is being placed as a local call on the New York City network (even though it actually is being initiated from Toronto), so the number dialed includes only the subscriber number "5551234". The example then proceeds as before, with a wait for silence and the dialing of the extension.

a.**Internal calls** — Internal calls are calls local within a private telephone system and are sometimes referred to as intercom calls.

b.**Tie lines** — A tie line is a dedicated transmission facility linking two telephone systems in a private or virtual private network.

A disadvantage of referencing devices using dial strings is that determining what dial string to use to refer to a particular destination depends on location, knowledge of the dial plan at that location, and knowledge of the supported subset of the dialable digits format implemented in the telephone system about to be used. A second disadvantage is that the embedded commands in dialable digits format are applicable to the placing of calls but not to monitoring or manipulating a device in some other way. This means that separate references to the same device may be expressed differently when using dialable digits format.

Dial strings are the only form of device reference that can be entered using any telephone keypad, however, so they are the one format, and in many cases the only format, supported by every telephone system.

4.6.4. Canonical Phone Numbers

Canonical phone number format was developed in order to overcome the limitations of dial strings. It allows for the specification of location-independent, absolute references to a particular device in a telephone network.

Canonical numbers are meant for use by people (on stationery, business cards, personal calendars, etc.), by computers (in scheduling programs, personal information managers, databases, etc.), and by telephone systems. As a result, the format uses punctuation that is easy for a person to write and for a computer or telephone system to interpret.

Phone numbers expressed in canonical representation can be absolute because they allow the inclusion of all the information necessary to identify a particular device, regardless of the context. For example, a particular device could be called from within the same telephone system, from a different telephone system, or from anywhere in the public network using the same canonical number. In each case, the information necessary to establish a call from the location in question is intelligently extracted from the canonical phone number.

A device reference in canonical phone numbers may include:

- A country code, area code, and subscriber number;

- A sub-address and/or extension number; and

- A name

Various portions of the canonical format are optional. Incomplete canonical numbers are interpreted based on other context information. For example, if a given canonical number included only a name, a telephone system with access to a directory could simply look up the name and establish a call to the corresponding person. Other portions not provided are appropriately interpreted as either not being applicable to the device, or as being the same for the location of the caller. See the sidebar "Canonical Phone Number Format" for the complete set of symbols used in a canonical phone number and their meanings.

Canonical Phone Number Format

Canonical phone number format consists of a sequence of up to 64 characters using the following syntax:

O< [**DPR1**, **DPR2**, ... , **DPRn**] +**CC** (**AC**) **SN** *__SA__ x**EXT**>**NM**

The bold items are symbols that are replaced by the appropriate data. The other symbols are delimiters used to punctuate the string. The different portions of the canonical number are as follows:

'O' The leading 'O' indicates that this number is in canonical format. In practice, the 'O' is only used when communicating canonical numbers electronically. People do not require this character, so it will not appear in what they see and use.

'<' '>' The angle brackets indicate that a name (**NM**) is included in the canonical number. If no name is present, the angle brackets are not present. If a name is present, they enclose the rest of the canonical number.

'[' ']' The square brackets indicate that a list of one or more dial plan rule strings (**DPR**s) is included in the canonical number. A dial plan rule specifies how the other portions of the canonical number are to be interpreted. Most often the canonical number refers to a device in the public telephone network, so no dial plan rule is required. If the device is in a private network, however, the dial plan rule identifies the private network involved. If multiple dial plan rules are required, they are separated by commas.

'+' The '+' character indicates the presence of a country code (**CC**).

'(' ')' The parentheses enclose the area code (**AC**) portion of the canonical phone number. The presence of a dial plan rule for a private network might indicate that this portion is to be interpreted as a subnetwork.

SN This represents the actual subscriber number portion of the canonical phone number. It is required for use with the public network.

'*' The '*' character indicates the presence of a sub-address (**SA**).

'x' The 'x' character indicates the presence of an extension number (**EXT**).

NM This represents the name associated with the device being referenced.

Canonical Phone Number Format (Continued)

All portions of a canonical phone number are optional. However, a canonical number is required to include at least a subscriber number, extension number, or name. If the area code is not included and there are no dial plan rules indicating this is a private network reference, the area code will be interpreted either as not applicable, or as being the same as that of the device interpreting the number (depending on the country code). If the country code is not included and there are no dial plan rules indicating this is a private network reference, the country code will be interpreted as being the same as that of the device interpreting the number.

The following are two examples of complete device references in canonical phone number format for extension 789 behind the directory number 555-1234 in New York City.

Example: "O+1(212)555-1234x789"
Example: "O<+1(212)555-1234x789>John Smith"

The following are other examples of device references in canonical phone number format:

Example: "O[MYNET](4)x40220"

In this example, a dial plan rule ("MYNET") indicates that the device referenced is in the private network MYNET. The area code ("4") is then interpreted as a particular subnetwork in MYNET and "40220" is a particular extension on that subnet.

Example: "O+33(1)12.34.56.78*06"

In this example, the canonical phone number refers to a particular bridged device with a sub-address of "06" at the subscriber number "12.34.56.78" in Paris (country code "33" and area code "1").

Example: "O555-1234"

In this example the canonical phone number is only partially specified. It is a valid device identifier in canonical format, but it requires the interpreter to assume the missing information. In this case, if the interpreter were in New York City, the country code would be assumed as "1" and the area code would be assumed as "212".

Example: "Ox40220"

This is another example of a partially specified device in canonical number format. In this case the device in question is extension "40220" in the telephone system being used.

Example: "O<>John Smith"

Canonical numbers support call by name, that is, specifying the destination as a name that is then resolved by a computer or by the telephone system.

At the moment, canonical representation is used primarily by computers, which translate them into appropriate dial strings for telephone systems. This will change as telephone systems become more intelligent.

4.6.5. Switching Domain Representation

Another phone number format is the *switching domain*[4-24] *representation*. This representation is used to specify a particular device, element, or appearance within a particular telephone system.

When this representation is used, the directory number notation (see the sidebar "Standard Telephone Number Exchange Notations") must also be specified along with the phone number to allow interpretation of the directory number portion.

This format for referencing devices is the preferred format for references to devices within a given telephone system (as opposed to those in an external network) because it allows precise references down to the level of specific appearances and agents. See the sidebar "Switching Domain Representation Format" for the complete set of symbols used in this format and their meanings.

4-24. Switching domain — The term switching domain is defined in Chapter 6.

Switching Domain Representation Format

Switching domain representation format consists of a sequence of up to 64 characters using the following syntax:

N<DN *SA &CA %AID>NM

The bold items are symbols that are replaced by the appropriate data. The other symbols are delimiters used to punctuate the string. The different portions of the switching domain representation are as follows:

'N' The leading 'N' indicates that this reference is in switching domain format.

'<' '>' The angle brackets indicate that a name (**NM**) is included in the device reference. If no name is present, the angle brackets are not present. If a name is present, they enclose the rest of the device reference.

DN This represents the directory number portion of the phone number.

'*' The '*' character indicates the presence of a sub-address (**SA**).

'&' The '&' character indicates the presence of a call appearance reference (**CA**).

'%' The '%' character indicates the presence of an agent ID reference (**AID**).

NM This represents the name associated with the device being referenced.

All portions of this representation are optional. However, a device reference in this format is required to include at least a directory number or agent ID.

The following are examples of device references in switching domain representation for extension 789 behind the directory number 555-1234 in New York City.

Example: "N789"

In this example, the given directory number notation is "Private TON - local." The digits "789" reference the device as an extension with directory number 789 in the telephone system in question.

Switching Domain Representation Format (Continued)

Example: "N789&02"

In this example, the given directory number notation is "Private TON – local." "02" references a specific call appearance on the device in the telephone system in question with the directory number identified by "789".

Example: "N<12125551234>John Smith"

In this example, the given directory number notation is "Public TON – international." This is the device reference that a telephone system outside of North America would use if a call from this device had been received. Extension 789 shares the network interface device identified in the external network by "5551234" so it would be referenced using that number across an external network.

Example: "N5551234"

In this example, the given directory number notation is "Public TON – subscriber." This is the device reference that another telephone system in New York City would use if a call from this device had been received. As with the previous example, "5551234" refers to the subscriber number used because extension 789 shares the network interface device and only this number is visible in the public network.

The following are other examples illustrating the use of this format in reference to devices within a particular telephone system. The given directory number notation is therefore "Private TON – local" in each case.

Example: "N%604"

This example is a reference to the device associated with agent ID "604".

Example: "<N%604>John Smith"

This example is a reference to the device associated with agent ID "604" and the name "John Smith".

Example: "N40220*02"

This example is a reference to bridged appearance "02" on the extension with number "40220".

4.6.6. Device Numbers

The fourth type of device reference is the *device number*. In most telephone systems, every single device is internally assigned a unique device number. Some implementations allow devices to be referenced directly using these numbers.

See the sidebar "Device Number Format" for a description of the components of this format.

Device Number Format

Switching domain representation format consists of a sequence of up to 64 characters using the following syntax:

$$\backslash N$$

The bold items are symbols that are replaced by the appropriate data. The other symbols are delimiters used to punctuate the string. The different portions of the switching domain representation are as follows:

'\' The leading '\' indicates that this reference is in device number format.

N This represents the number associated with the device being referenced.

All portions of this representation are optional. However, a device reference in this format is required to include at least a directory number or agent ID.

4.7. Review

Devices are resources responsible for consuming and generating the media streams and control information associated with calls. A *network interface device* can be used to establish calls between one telephone system and an external telephone network. Other device *types* include *station, park, pick group, ACD, ACD group,* or *hunt group.* Devices may include a *logical element,* or a *physical element,* or both.

Physical device elements consist of *components* that make up the physical interface of a device. These components include *hookswitches, auditory apparatuses, buttons, lamps, message waiting indicators, displays,* and *ringers.*

Logical device elements represent sets of *call appearances* of a particular type. Each call appearance (*appearance* for short) that a logical element possesses enables it to participate in a call. Appearances have associated *types* and *behaviors.* The set of type-behavior combinations are *dynamic, basic–static, selected–static, basic–bridged, exclusive–bridged, independent–shared–bridged,* and *interdependent–shared–bridged.* Appearances may be *addressable* or *non–addressable.* An appearance that is not involved in a call is *idle.* Exclusive–bridged appearances may be *blocked* from a call and shared–bridged appearances may be either *participating* or *inactive* with respect to a call.

Device configurations describe the relationship between a particular device element and the device elements associated with it.

Device addressing refers to the ways that a particular device or appearance can be referenced. Addressing formats include *dial strings, canonical phone numbers, switching domain representation,* and *device numbers.*

Now that a complete abstraction of telephony resources and their attributes has been defined, we can explore the features and services that can be implemented by manipulating these resources in well-defined ways.

5.
Call Processing Features and Services

This chapter presents the call processing features and services commonly available in telephone systems independent of any CTI interface. These are referred to as *call control, call associated,* and *logical device* features and services. (Telephony features and services that are specific to a CTI interface are covered in Chapter 6.)

The preceding chapter explored a universal abstraction of telephone systems as telephony resource sets. Now we will breathe some life into these resources by seeing how they work together to deliver standard telephony features and services. As was true of the telephony resources themselves, these features and services are the result of international and industry standardization efforts. The resulting abstraction presented here represents a superset of the most frequently used telephony functionality. In other words, every vendor's product has a different combination of features that differentiates their product, but all of the significant features are represented here. Every telephony vendor has at least a few unique names for standard features, so popular alternative or proprietary terms are also referenced wherever appropriate.

Telephony features and services available to a given device or telephone system depend upon the telephony features and services provided by the telephone system or by an external network respectively. Making telephony features available (and billing for them if appropriate) is referred to as *subscribing* to a feature or service. The set of features and services available to a given device from its telephone system, or to a telephone system from an external network, is generally referred to as its *class of service*.

5.1. Basic, Supplementary, and Extended Services

Telephony services are operations that a telephone system may perform in response to commands from the system's devices or from a CTI interface. (Additional services that are specific to the CTI interface are presented in Chapter 6.)

Services are frequently grouped into three classes:

- Basic

  Basic telephony services refer to the lowest common denominator for all telephone service, specifically the ability to place, answer, and hang up calls.

- Supplementary

  Supplementary services refer to all the other well-defined services beyond the basic services.[5-1]

- Vendor specific extensions or Extended

  Vendor specific extensions, also called *extended* services, are the services beyond the supplementary services unique to a particular vendor or product.

5-1. Service naming — The term supplementary services specifically applies to the functionality of the services. The names used by different telephony and operating system vendors to refer to specific supplementary services vary. This book uses the terminology established by ECMA, Versit, and other standards-setting groups. Wherever appropriate, alternate names are provided.

This chapter presents the basic services, followed by the most popular supplementary services. Features unique to the products of a particular vendor are not represented here (though CTI interface support for vendor specific extensions is described in Chapter 6.)

5.2. Features

In addition to a telephone system's ability to respond to commands for switching services, the telephone system may track certain pieces of information about a call or device, manage timers associated with certain activities, or establish settings that govern call progress in some way. These are all referred to as *features*.

The term *feature interaction* refers to the ways that normal call progress may be affected by the certain features that may be in effect for a particular device.

In this chapter we will be describing all of the most popular telephony features in terms of the concepts presented in Chapter 3.

5.3. Basic Services

Basic services often are described as being the telephony functions available with a POTS[5-2] phone. It is the lowest common denominator for all telephone service:

- Placing calls

- Answering calls

- Hanging up calls

This section will explore each of these basic services. Later we will take a closer look at each one with respect to its use in the context of richer supplementary services.

5-2. POTS — POTS stands for Plain Old Telephone Service. It refers to traditional tip-and-ring analog telephone lines with only dial-answer-hangup functionality. POTS service is described further in Chapter 8.

5.3.1. Make Call

The formal name for the service that places calls is *make call*. This service, in its simplest form, does the following:

- Creates a new call

- Creates a new connection between the *calling device* and the call

- Routes the call to the destination (the *called device*) specified

In the basic case of the *make call* service, the operation is complete when the calling device has a connection to the new call in the *connected* state[5-3] and the called device has a corresponding connection in the *alerting, connected, fail, null,* or *queued* state. This is referred to as the service's *completion criteria*. Every switching service has a set of completion criteria that can be used to determine if the service succeeded or not.

A *make call* service may not succeed (or complete) for a variety of reasons. The most likely is that the calling device is already in use with another call and is therefore unable to place a new call.

It is very important not to confuse the concept of a telephony service that does not succeed with the concept of the *fail* connection state. The connection state of *fail* indicates that call progress associated with a particular call has stalled and that a certain appearance may be blocked from the call. This is quite distinct from a telephone system's switching resources being unable to perform a specific operation.

After the *make call* has completed, the telephone system will attempt to connect the new call to the specified destination. Due to feature interactions and the effects of other routing mechanisms, however, the call could be redirected many times before actually arriving at a device

---

5-3. Make Call basic case — The completion criteria for the basic case of the make call service is that it transitions to the connected state. In the general case, however, it may transition to (or retransition to) the initiated state in order to support multi-stage dialing. This is explained in section 5.4.2.

where it can be answered. The call also could get stalled, for example, because the telephone network is congested or because the destination device is busy.

Make call is described in graphical notation in Figure 5-1.

Figure 5-1. Make Call service (single-step dialing)

In the simplest case of the *make call* service, there is initially no connection (i.e., the connection is implicitly in the *null* state) between the calling device (D1) and the new call C1. (In fact, the new call doesn't yet exist.) There is a second, more common case, however, where the call already exists and the connection is in the *initiated* state. The difference between the two cases is that the first represents *on-hook dialing* and the second represents *off-hook dialing*. An example of on-hook dialing is what you are doing when you place a call from a cellular phone: You dial the number and press the Send button. Off-hook dialing happens when you pick up the receiver (or go off-hook in some way), hear dial tone, and then begin dialing. With off-hook dialing, the new call is created when you hear dial tone. The dial tone indicates that the connection is in the *initiated* state.

The *make call* service is complete in either case when the connection to the calling device transitions to the *connected* state as shown. At this point the device is said to have *originated* the call. The connection to the called device (D2) may be in any of the following states:

- *Alerting* – the call has already reached D2 and the telephone system is attempting to connect the call.

- *Connected* – the call not only has reached the called device, but has already been answered, possibly because D2 has the auto-answer feature[5-4] active. (An alternative case is when the call is an external call and it is connected to a network interface device as a proxy for D2. Making external calls is explained in section 5.4.3.)

- *Fail* – call processing stalled trying to connect to D2, most likely because it was busy or no network interface device was available.

- *Queued* – the call is being queued at the called device, probably because it is some type of distribution device.

- *Null* – the call was redirected so there is no connection to D2. There probably is a connection to the new call at some other device, however.

5.3.2. Answer Call

The service that answers calls is appropriately named *answer call*. This service is quite simple: It causes a connection in the *alerting* (or *queued*) state to transition to the *connected* state. The service also activates any appropriate auditory apparatus.

Typically the *answer call* service is triggered by picking up a telephone handset or pressing an appropriate hookswitch button to activate a headset or speaker phone auditory apparatus. This service also can be supported through the CTI interface if the physical element involved has a hookswitch that can be set to off-hook automatically by the telephone system.

The *answer call* service may not succeed for a variety of reasons. Two likely cases could be that the connection is in the wrong initial state (not *alerting* or *queued*), or that the answering device does not support going off-hook automatically.

Answer call is presented graphically in Figure 5-2.

5-4. Auto-answer — The auto-answer feature is described in section 5.9.1.

Figure 5-2. Answer Call service

BEFORE   AFTER

Before the service connection D2C1, the answering connection was in the *alerting* or *queued* state. After the call is answered it is in the *connected* state.

5.3.3. Clear Connection

Dropping a call, or "hanging up" on a call, is accomplished by clearing the associated connection. The service for doing this is appropriately named *clear connection*. This service removes just one connection from a multi-way call and is also used to inactivate a shared–bridged appearance.

The *clear connection* service transitions a connection from any non-*null* state to the *null*, *queued* state and sets the hookswitch on the physical device element to on-hook if appropriate. When the service completes, ordinarily the connection will be cleared and will transition to the *null* state. The exception is when the connection is one of a set being used by shared–bridged appearances. In this case, the connection will be cleared only if it corresponds to the last participating appearance in the set; otherwise it will transition to the *queued* state. (This behavior was described in Chapter 4, section 4.4.3.)

After the service completes, the other connections in the call are unaffected, assuming there are two or more connections remaining. If only one connection remains in the call after the service completes, the call itself may be cleared, in which case the remaining connection also would transition to the *null* state.

The *clear connection* service is illustrated in Figure 5-3. Before the service the *clearing* connection is D3C1, which is a connection in any non-*null* state. The other connections in the call may be in any state.

After the service completes, the D3C1 is in the *null, queued,* or *fail* state and the other connections are unaffected. (The *clear connection* service is discussed in further detail in section 5.15.1.)

Figure 5-3. Clear Connection service

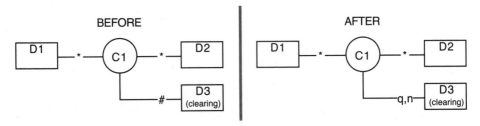

5.4. Placing Calls

Now that we've looked at all the basic services, we'll explore further the additional features associated with creating new calls. This section deals with the general case of the *make call* service (including prompting and multi-stage dialing), the concepts surrounding *external calls,* the *make predictive call* service, and the *last number dialed* service.

5.4.1. Make Call and the Initiated State

The *initiated* connection state plays a number of important roles in the implementation of the *make call* service.

If the calling connection (D1) is in the *initiated* state, the call has already been created, so there is dial tone and no instructions have been provided yet for the call. This is referred to as *off-hook dialing*.

Under certain circumstances the *make call* service may invoke *prompting* for the calling connection, which is indicated by having the connection state transition, or retransition, to the *initiated* state. (Prompting is explained in section 5.4.6.)

Finally, if the dial string supplied to the *make call* service was incomplete, the service will initiate multi-stage dialing. In this case the service completes when the connection state of D1C1 transitions (or, retransitions) to the *initiated* state.

5.4.2. Dial Digits for Multi-Stage Dialing

Multi-stage dialing, or *delayed dialing*, refers to the situation where a dial string is provided in increments rather than as a single string of digits.

In section 5.3.1 we saw the case of the *make call* service for single-stage dialing. The other case of the *make call* service is shown in Figure 5-4. The difference between the version in Figure 5-1 and the one in Figure 5-4 is that in the multi-stage case of the service, connection D1C1 is in the *initiated* state at completion.

Figure 5-4. Make Call service

There are two ways to initiate multi-stage dialing.

1. Provide an incomplete dialing sequence to a *make call* service.

2. Manually take a telephone off-hook, which creates a new call with a connection in the *initiated* state.

In either case, additional digits in the dial string sequence are supplied using the *dial digits* service. The *dial digits* service is illustrated in Figure 5-5.

Figure 5-5. Dial Digits service

Connection D1C1 remains in the *initiated* state until a complete dial string has been provided. It then transitions to the *connected* state and the call is said to have been *originated*. Figure 5-6 shows a multi-stage dialing sequence.

Figure 5-6. Multi-stage dialing sequence

5.4.3. External Outgoing Calls

External calling is a feature that allows devices and calls inside a given telephone system to be connected to devices in an external network. (For a telephone system consisting of a single telephone, all calls are considered external.)

Support for *external outgoing* calls is a feature that allows network interface devices to be connected to new calls being generated inside the telephone system. Network interface devices are the proxies within the telephone system that represent a distant called device's participation in a call. This is shown in Figure 5-7. In this illustration, the "?" symbol indicates that because the connection between the network interface device (D3) and the called device (D2) is entirely outside of the telephone system (and potentially is made up of many different connections), it cannot be controlled directly and the external network may or may not provide state information.

Figure 5-7. External outgoing call

When the telephone system determines that a destination is external,[5-5] it routes the call through a network interface device. The point when it actually connects a network interface device to the call,[5-6] referred to as the *network reached* point, is shown in Figure 5-8.

Figure 5-8. Network reached

When dialing takes place in a single step using just the *make call* service, the call has already been originated, so connection D1C1 is in the *connected* state before the network interface device is connected. During multi-stage dialing, however, before receiving the entire dial string the telephone system may have enough of it to know that the call is an external outgoing call and place D1C1 into the *connected* state. In this case, D1C1 is still in the *initiated* state even though D2C1 is in the *connected* state.

5-5. Identifying external calls — A call may be identified as being for an external destination in a variety of ways. In some cases all calls are external, so the call is identified as being external automatically as soon as it is determined that a new call is being placed. Otherwise, the presence of a special prefix in a dial string or the information in a canonical phone number determines whether a particular destination is external or internal. See section 5.4.5 for more on dial plan management.

5-6. Seizing — When a network interface device is allocated for a call using the external network, it is known as seizing the network interface or seizing the trunk.

211

5.4.4. Network Interface Groups

If the telephone system has more than one station device, it typically has fewer network interface devices than stations. The stations share the network interfaces for more efficient utilization of these resources.

When an external outgoing call is made by a device that shares a pool of network interface devices, that call must be routed through the next available network interface. As we have already seen, a special device exists for just this purpose: The hunt group device redirects calls to the next available device in its group. A *network interface group*[5-7] is therefore a hunt group device that is set to distribute outgoing calls to the next available network interface device. If all network interfaces are busy, the call may either fail or wait for one to become available, depending on the implementation.

Figure 5-9 illustrates this sequence. In this example, D5 is placing an external outgoing call, C2. The new call is directed to the network interface group device D1. There are three network interfaces in the group: D2, D3, and D4. D2 is busy with call C1, so D1 finds the next available network interface device, D3, and redirects the call there.

5.4.5. Dial Plan Management and Least Cost Routing

The appropriate network interface device or group to use for placing a call, and the selection of an appropriate long-distance carrier (if appropriate), is determined from the address provided to identify the called device. The rules set up to manage the process of transforming a calling device's address into routing decisions for a call, and the process of performing these transformations, is referred to as *dial plan management*. One aspect of dial plan management is *least cost routing* functionality. Least cost routing involves determining the cheapest available way to place a given call, given the destination of the call, the

---

5-7. Trunk groups — Network interface devices (see Chapter 4, section 4.2) are commonly referred to as trunks, so network interface group devices are commonly referred to as trunk groups.

Figure 5-9. Network interface device group behavior

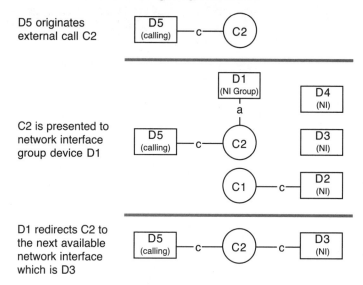

D5 originates
external call C2

C2 is presented to
network interface
group device D1

D1 redirects C2 to
the next available
network interface
which is D3

available private network and public carrier options for routing the call, the location from which the call is originating, the date and time of day at that location, and current rate plan information.

One or more of the following techniques typically are used to implement dial plan management:

- Prefixes

- Leading-digit translation

- Canonical number translation

Depending on the techniques implemented for a given telephone system, the users of the system may have to make more of the routing decisions themselves.

Prefixes

Prefixes are the most commonly used mechanism for indicating that a call is to a device on an external network. Prefixes vary from single to multiple digits, and are used at the beginning of a dial string to indicate that a particular network interface, network interface group, or routing mechanism is to be used to route a particular call. In North America the most common prefix in private telephone systems is "9". This digit usually indicates that a particular call is an external outgoing call using the local carrier. Other prefixes may be set up to correspond to network interfaces associated with connections to private networks, specific carriers, or frequently called destinations. (Options for network interfaces are covered in more detail in Chapter 8.)

If a telephone system relies entirely on prefixes to perform dial plan management, most of the routing effort is left to the people using the telephone system. To place a call they must enter (in varying orders):

- The most appropriate prefix for a call. Often they just dial the prefix for the local carrier even though this may not be the most cost-effective routing.

- The appropriate carrier selection code, if necessary. If they don't do so, and the call is a toll call, it will be routed by the default carrier.

- Any applicable billing codes.

- The desired destination number in an appropriate form for the external network associated with the selected network interface device. Depending on the network interface used, the caller might not have to dial a long-distance prefix or an area code, or even the full subscriber number.

The scenario described above is the worst case, but is commonly encountered by users of hotel, small business, and home telephone systems.

Leading Digit Translation

Leading digit translation involves analyzing the first few digits of the called device's number.

Simple systems rely on just the area code to determine how to route a call. In North America this is referred to as *three-digit translation* because the first three digits of a ten-digit telephone number are its area code. These systems route all local calls to the local carrier and use least cost routing tables to determine the least cost route for each long-distance area code.

Many area codes in the North American dial plan have unique dialing rules for long-distance dialing within an area code and to adjacent area codes. Some require that the digit "1" be dialed and others do not. Some allow the number to be dialed with or without the leading "1" but will charge more for the call depending on how it is dialed. Some allow calls to be placed to different areas codes using seven-digit numbers. All these exceptions make it necessary for robust systems to be based on *six-digit translation* that relies on a database of dial plan rules.

Dial plan rules vary dramatically throughout the world, and leading-digit translation typically must be designed for each country individually, so a system built for use in one country will not work in another.

Canonical Number Dialing

Canonical number dialing, or full-number dialing, means working with numbers in canonical form (as described in Chapter 4, section 4.6.4) rather than just scanning leading digits. Dial plan managers built on canonical numbers may support the parsing of arbitrary numbers into canonical numbers based on context rules, but then do their routing and translation based on the resulting canonical number.

By working from canonical numbers, network interface devices can be added and removed with ease, and a single international dial plan database can be constructed and maintained more easily. Most important, users of the system can always dial a canonical number rather than worry about the right prefixes to use.

Using Dial Plan Management

Dial plan management and least cost routing are a significant area where CTI solutions can play a significant role, both in making telephone systems of all sizes easier to use and in reducing telephone expenses. Examples include:

- An individual traveling with a notebook computer can use CTI functionality to dial calls from hotel rooms, airports, and meeting locations. Using the dial plan management capabilities of the CTI software in the personal computer—rather than having the traveler figure out access codes, carrier codes, billing codes, etc.— saves a great deal of time and money.

- A home business owner can use CTI software in a personal computer to automate all outbound dialing. Although the home might have only two phone lines, the CTI software ensures that every voice, fax, and data call that is made is dialed correctly, uses the cheapest carrier given the time of day and the destination, and is placed on the most appropriate line (business versus personal).

- Larger telephone systems typically have dial plan management built in. Unfortunately, as the number of new area codes, country codes, carriers, and routing options increases, these built-in dial plan management features must be reprogrammed frequently. Using a CTI-based dial plan management system greatly simplifies the task of updating the dial plan information and provides much greater control to the telephone system's manager.

5.4.6. Prompting

Prompting refers to a feature that allows the *make call* service and certain other services to be used for on-hook dialing on station devices that don't have the ability to go off-hook without manual intervention. Prompting will delay the point where the call is originated until the device goes off-hook. Figure 5-10 shows the sequence of transitions for prompting in the case of the *make call* service.

Figure 5-10. Prompting feature

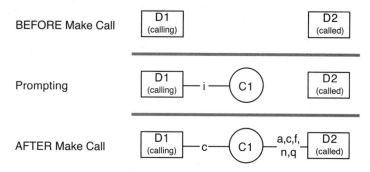

In this sequence, initially there is no connection (and no new call) because the *make call* is being dialed on-hook. When the dialing is complete, the call is created and connection D1C1 is placed in the *initiated* state while prompting takes place. When the station device is then taken off-hook manually, connection D1C1 transitions to the *connected* state, signifying that the call has been originated and satisfying the completion criteria of the *make call* service.

Prompting involves a telephone set making an indication, typically audible, to indicate that someone should take it off-hook in order to progress out of the *initiated* state. It is distinct from ringing, which is an indication made by a device that a connection is in the *alerting* state. Prompting is used whenever some service is initiated for a device but it is unable to go off-hook automatically.

5.4.7. Make Predictive Call

The *make predictive call* service is a very powerful capability that allows the telephone system to make calls on behalf of the calling device without actually involving it in the call until the call is delivered or connected to the desired destination and other criteria are satisfied.[5-8]

5-8. Predictive dialing — The name of the make predictive call service derives from the fact that the device or computer requesting the service must predict the availability of someone to handle the call if and when it completes, based on call completion and call duration statistics.

This service typically is used in situations where one or more people are making large numbers of calls and want to spend every possible moment talking to "live" people, and not be tied up dialing telephones, waiting for the calls to be placed, waiting for the calls to be answered, getting answering machine recordings, etc.

With predictive dialing, the telephone system typically is instructed to make the calls automatically and to wait until the call is either delivered to, or answered by, the called device. If the option to wait for the call to be *connected* is chosen, the telephone system may be further instructed to differentiate between a human, a recording, and a fax machine. Depending on the criteria set, the telephone system then either drops the call or delivers it to the original calling device. This sequence is illustrated in Figure 5-11.

Figure 5-11. Make Predictive Call service example

This service is even more powerful when the calling device is some type of routing device (such as an ACD, ACD group, or hunt group device). In the case of an ACD group, when the telephone system successfully places a call to the desired destination and then delivers the new call to the ACD group, it in turn redirects the call to the first available agent. This combination of the *make predictive call* service and an ACD group device allows a set of agents to maximize the time spent talking rather than waiting.

5.4.8. Last Number Dialed and Redial

The *get last number dialed* feature involves the logical device remembering the number of the called device for the last call it initiated. Combining this feature with the *make call* service is referred to as *redial*.

5.5. Call Associated Information

Whenever a new call is created, the telephone system begins associating various important pieces of control information with it. In the services we have explored so far, we have already seen that two key pieces of information associated with a new call are the called device and the calling device. Another key piece of information associated with external calls is the network interface device that is involved.

5.5.1. CallerID and Automatic Number Identification (ANI)

Aside from knowledge of a call's intended destination, the most important piece of information associated with a call is the identity of the calling device. This information is used both by the telephone network for billing purposes and by those being called.

Depending on the features of a particular telephone system and the features subscribed to from an applicable external telephone network, the feature that provides this information to the called device is referred to as either *callerID* or *automatic number identification (ANI)*.[5-9] (These specific features of the external telephone network, or *service offerings*, are described in Chapter 10.)

The called party in a call may use this information to determine how a call should be handled in an endless variety of money- and time-saving ways. Uses of this information include:

- Determining the name of the person calling by automatically looking up the number in a directory before answering;

- Deciding whether or not to accept or answer the call;

- Deciding what information should be automatically presented to the called party in order to best handle the call;

- Creating new database entries for tracking information provided by this caller;

- Logging information about the call (e.g., caller and length of call) for future reference or billing;

- Deciding if the call should be redirected and, if so, the destination; and

- Deciding what person is best able to answer the call.

Developing solutions that take advantage of callerID and ANI information is one of the most readily achievable ways to benefit from CTI technology.

Get and Set CallerID Status Services

The *Get CallerID Status* and *Set CallerID Status* services relate to a logical device element's ability to suppress the delivery of callerID information with calls that it places. If this feature is used to activate

---

5-9. ANI — ANI is slightly different from callerID in that it delivers the billing number of the calling device, that is, the calling number that is used for billing purposes. Typically the two numbers are the same. When they are not, it is usually because a shared network interface device is being used to place a call on the network but the call is being billed to a single main number. The actual device originating the call from the network's point of view is the network interface device, but in this case calls from any of the shared network interface devices are billed to a single main number; ANI will reflect the main number and thus will be different from (and more useful than) callerID.

CallerID blocking, then calling device information for calls made by the logical device element will be sent to the called device only in the case of toll-free calls and emergency calls (such as 1-800, 1-888, 1-900, and 911 in North America).

CallerID blocking may be *complete blocking* or *selective blocking*. With complete blocking, every new call made is blocked from sending callerID until the service is deactivated. With selective blocking, the callerID blocking only applies to the next call that is placed and must be reactivated for each subsequent call.

5.5.2. Dialed Number Identification Service (DNIS)

Depending on the features of a particular telephone system and the features subscribed to from an applicable external telephone network, *dialed number identification service* (*DNIS*) may be available. (DNIS service offerings are described in further detail in Chapter 10.)

With calls that are delivered to a particular destination device, dialed number identification service provides the actual number that was dialed by the calling device. This feature is used in scenarios where multiple numbers in one or more networks are all redirected such that calls to any of the numbers are routed to a single designated device.

One way of taking advantage of the DNIS feature is by centralizing all calls through a single routing device (typically an ACD, ACD group, or hunt group). In this way the utilization of people and telephony resources for handling calls can be maximized. The DNIS feature allows the telephone system to know the actual number dialed, even though all calls are being delivered to the same device. This information then can be used in much the same way that CallerID and ANI information is used. For example, you could subscribe to a set of numbers in the external network (such as 800-BUY-FOOD, 800-BUY-CARS, 800-BUY-HATS) and have each one directed to the same group of operators using your telephone system. They could then handle all of the calls to any of these numbers. Using DNIS information, the telephone system could:

- Direct the call to the best available operator for the appropriate product category;

- Inform the operator whether to say "Hello, Acme Grocery," "Hello, Cars-Are-Us," or "Hello, Mad Hatter's" when answering each call;

- Present the appropriate order form for the right sales activity; and

- Log the call activity in the appropriate database.

Developing solutions that take advantage of DNIS information is another excellent way to benefit from CTI technology.

5.5.3. Last Redirected Device

The *last redirected device* information associated with a call identifies the last device that rerouted this call. Like CallerID and DNIS information, this is very useful for interpreting why a particular call is being delivered to a particular device. In fact, the last redirected device actually is used, or interpreted, as a combination of called device and calling device. It is a device to which the call was previously directed, and it is the device that is responsible for redirecting the call to its current destination. Typical uses of this information include:

- Answering a call differently, depending on whether or not it was already answered by an operator or assistant.

- Identifying which of several forwarded calls came from a particular forwarder, independent of what number was originally dialed. For example, an assistant answering calls on behalf of two different managers could identify which manager forwarded the call.

- Allowing a voice mail system to play the message for the correct person when their call is redirected to voice mail because it was not answered.

- Determining whether or not to answer a call based on whether it has been screened by an appropriate person.

Support for the last redirected device information is another essential feature in any system.

5.5.4. Account and Authorization Codes

One feature associated with many systems is support for *account codes* and/or *authorization codes*. These are codes that are associated with the call in some way when it is created; these codes (in addition to information about the calling device) may be used for billing the call to the correct person, project, or organization. These pieces of control information also may be used to determine whether or not a call is external, what network interface device should be used, and which other call-related features might be applied.

5.5.5. Correlator Data

Correlator data is an arbitrary block of use-specific information that can be attached to a call. Correlator data is attached to calls in order to allow the many different devices that may interact with a call as it is routed through a telephone system to have access to some piece of common information, so that the caller need not provide it again and again.

Typical examples of the use of correlator data include:

- A customer's credit card number is attached to a call so that, as her call is transferred between different agents responsible for selling different products, she does not need to provide this information again and again.

- A man calls a local store, asking if a particular product is in stock. After describing the desired product to a clerk, the product's part number is determined and attached to the call. Unfortunately the store doesn't have it on hand, but the clerk transfers[5-10] the call to another store that is likely to have the product in stock. The clerk at the second store does not need to repeat the interrogation process to determine the product number because it is attached to the call.

- A woman calls a travel agent about some vacation plans. The receptionist answers the call and discovers that the woman does not yet have an account. He then creates a new database entry for the woman and gathers all the pertinent information from her. The woman's newly created customer number is attached to the call as correlator data. All the travel planners in the office are busy with other callers, so the receptionist puts the call on hold[5-10] for the next available agent. The next person available retrieves[5-10] the call and, thanks to the correlator data attached, can immediately pull up her customer record. In fact, when it is discovered that the woman's travel plans involve South America, the travel planner transfers the call to the agency's expert on the subject and once again the woman's information travels with her call.

Correlator data allows information to travel with a telephone call. While it is useful by itself (as information displayed on a telephone display), it is very powerful, if not essential, when used in a CTI solution.

5.5.6. User Data

User data are blocks of information that are broadcast to all the devices in a call at any point in the life of a call. Its association with a call may or may not be in conjunction with a call control service. In any case, however, it is distinct from correlator data because it is a "one-shot" mechanism and the data is not persistent.

5-10. Transfer, Hold, and Retrieve switching services — The transfer, hold, and retrieve switching services is described fully later in this chapter. The transfer switching service involves moving a call to a different device. Hold and retrieve involve suspending and restoring the media stream associated with a particular connection.

Typical examples of user data usage include:

- Delivering VersitCard electronic business card information to other participants in a call;

- Providing vendor, customer, or account number information;

- Authenticating callers through password or cryptographic exchange; and

- Exchanging an encryption key to be used in a subsequent data transfer over the Internet.

Like correlator data, user data can be a very powerful feature in the implementation of a CTI solution.

5.6. External Incoming Calls

Support for *external incoming* calls is a feature that allows calls from an external network to be connected to devices inside a telephone system, using network interface devices. Network interface devices are the proxies within the telephone system that represent the calling device's participation in a call. This is shown in Figure 5-12. In this illustration, the "?" symbol indicates that because the connection between the calling device (D1) and the network interface device (D2) is entirely outside of the telephone system, the external network may or may not provide information about its state. External incoming calls are different from external outgoing calls in that the calling device is outside the telephone system for the incoming case, and is inside the system for the outgoing case.

Figure 5-12. External Incoming calls

Once a network interface device has been associated with an external calling device, the new incoming call must be presented to a device inside the telephone system. The destination for the call is chosen in one of three basic ways, depending on the specified called device and the subscribed external network features:

- Fixed network interface device association
  All incoming calls on a specific network interface are connected to a predetermined device.

- Selectable network interface device association
  Each incoming call on a network interface indicates its own desired destination device.

- Attendant
  Incoming calls are presented to an attendant of some sort that provides assistance in connecting to the desired destination.

5.6.1. Fixed Network Interface Device Association

A telephone system that supports *fixed network interface device association* allows specific network interface devices to be configured so that, when a new external call is connected to them, they originate a call to a single specific device inside the telephone system. This feature is sometimes referred to as *DIL* or *direct-in-line*.

An example of fixed network interface association is shown in Figure 5-13. In this case, D1 places a call to the telephone system in question by dialing the telephone number in the external network (555-1234, for example) associated with network interface D2. In turn, D2 has a fixed association with device D3, so it delivers the call from D1 by originating a call to D3 inside the telephone system.

The relationship between the network interface device and its associated device also can be based on predetermined rules. For example, the association could be fixed with one device during the day and with a different device at night.

Figure 5-13. Fixed network interface device association

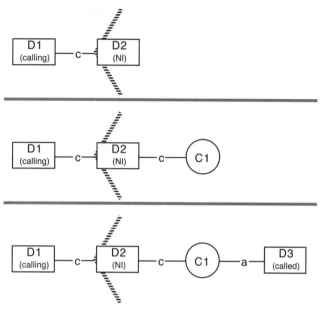

In any case, the device accepting the calls could be an individual station device, or it could be a routing device such as a hunt group, ACD group, or ACD that will, in turn, deliver a call to its destination.

5.6.2. Selectable Device Association

A telephone system that supports *selectable device association* allows each new external call to select its own destination within the telephone system by communicating this information when it connects to the network interface device.

There are three common forms of this functionality:

- Sub-addressing

- Direct Inward Dialing (DID)

- Direct Inward System Access (DISA)

In each case, information flows from the external network to the telephone system after a call has been presented to the network interface device involved.

Sub-addressing

Sub-addressing is a feature of certain telephone networks[5-11] that allows a sub-address to be dialed along with the telephone number of the network interface device. The sub-address effectively provides a "hint" as to which device in the telephone system is preferred.

The sub-address may be handled in different ways by the receiving telephone system, but the typical implementation is a simple variation on the fixed network interface device association. In this case, the network interface device that accepts the call has a fixed association with a bridged logical device element. The new call triggers all the bridged appearances to begin alerting; audible ringing is suppressed, however, for all physical devices other than the one indicated by the sub-address. Each device also can use the sub-address information to determine if it wants to answer or not. This approach is illustrated in Figure 5-14. Here L3 is the bridged logical device element, and P4 and P5 are the associated physical devices. If the sub-address dialed corresponds to P5, only it will ring.

Figure 5-14. Typical sub-addressing implementation

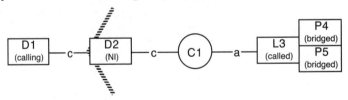

5-11. Sub-addressing availability — Sub-addressing is an ISDN feature that is available in most European countries and in many private networks based on ISDN.

DID

Direct inward dialing (*DID*) is a very popular feature that involves assigning telephone numbers in the external network directly to selected devices inside the telephone system. These devices are then referred to as *DID extensions.*

When an external caller dials the telephone number corresponding to one of the DID extensions, the external network delivers the call to one of a set of "root" DID network interface devices. The external network then dials the last two, three, or four digits of the DID extension so that the network interface device can determine the appropriate destination device.

The sequence for DID may be implemented as shown for DIL in Figure 5-13 or as shown for DISA in Figure 5-15. In both cases, D1 places a call to the telephone system in question by dialing the telephone number in the external network for the particular DID extension (555-4220, for example). The network connects the call to an appropriate network interface device (555-4000 for purposes of this example) and then sends the digits "220". The network interface, D2 in this example, originates a call to D3 (which is identified as "N40220" inside the telephone system and has been assigned the number 555-4220 externally).

DISA

Direct inward system access (or *DISA*) is similar to DID but, rather than having the network specify the desired device, the caller dials the desired number directly. In fact, DISA can be much more than merely the ability to dial an extension. As the name implies, it represents complete access to the telephone system. Once a remote device connects to a DISA network interface device, the remote extension is treated as if it were a basic station device directly attached to the telephone system. Any commands that such a device can send to the telephone system can be issued by the external device, subject to class-of-service restrictions.

An example of a DISA sequence is shown in Figure 5-15. In this example, D1 places a call to the telephone system in question by dialing the telephone number ("555-4444", for example) in the external network for the special DISA network interface device. This DISA device, D2 in this example, then initiates a new call in the system. D1 will then hear dial tone just as if it were a station device that had just gone off-hook. In this example, D1 then dials the appropriate number for D3 ("40220", for example) and just as if it were a locally dialed call, the new call is originated and delivered to device D3.

Figure 5-15. Direct inward system access (DISA)

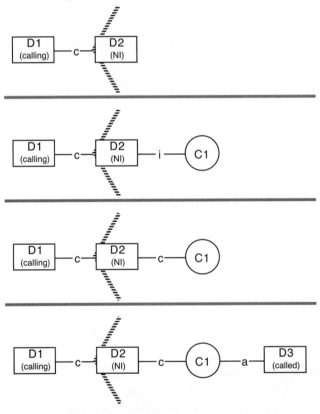

DISA potentially provides external callers with access to every feature and capability of your telephone system. As a result, the DISA feature is a serious source of toll fraud problems. (See the sidebar "Toll Fraud.")

Toll Fraud

Toll fraud refers to the misuse, or theft, of telephone services.

Telephone services are either billed implicitly, based on the calling device's network address, or explicitly, based on billing information (such as credit card number) provided during a call.

Toll fraud can therefore take place in two basic ways:

- Theft of billing codes

- Unauthorized access to facilities for placing long-distance calls

Billing codes (like calling card numbers) are primarily stolen by so-called "shoulder-surfers" who look over your shoulder as you punch in your billing information. They are also stolen by hackers breaking into phone company computer records, and by employees who steal and sell this information.

The principal way to steal access to a telephone system is through the DISA feature. Unless restrictions are placed on the features to which a DISA user has access, someone can call a DISA number and then, with full access to all of a telephone system's functionality, can place long-distance calls that will be billed to the owner of the telephone system. This means that if you are implementing support for DISA in your telephone system, you should make sure that you appropriately restrict the class of service that applies to it. Another way to secure DISA is to use callerID to restrict use to only specifically identified external devices. Many companies are switching to auto-attendants as an alternative to DISA for more control and security.

The latest form of toll fraud involves thieves who monitor cellular telephone traffic and intercept the code that uniquely identifies your cellular phone. They then program their cellular phone to be a clone of yours, and they can bill their calls to you because their phone has become almost indistinguishable from yours.

5.6.3. Attendant

The traditional approach for routing telephone calls between one system and another has involved using a human being who, with access to appropriate controls, can route calls manually. These people are traditionally called *telephone operators* or just *operators*. The term *attendant* refers to a designated operator (or one of a group of operators) for a private telephone system.

The attendant case is basically the same as the fixed network device association case except that the caller cannot directly dial the desired device. If the attendant feature applies to a particular network interface device (the default case for most systems), all calls arriving at that network interface are delivered to the designated attendant device or attendant group device.

The use of an attendant to complete delivery of a call is shown in Figure 5-16.

Attendant Consoles

The station device designated as the attendant for handling incoming calls (and other special attendant-related features) is referred to as the *attendant console*. (They used to be called *switchboards* when attendants were called operators.) The rate of call delivery to an attendant console typically is quite high, so they generally are designed and configured as multi-appearance devices. (See Chapter 10, section 10.4.7 for one example of an attendant console.)

Auto-attendant

Auto-attendants, or *automated attendants*, are a popular alternative to the traditional human attendant. An auto-attendant is a device that automatically answers external incoming calls, plays an appropriate greeting, and requests that the caller choose from a specific list of options where the call is to be directed. Depending on the implementation, the auto-attendant may allow the caller to specify the

Figure 5-16. Attendant operation

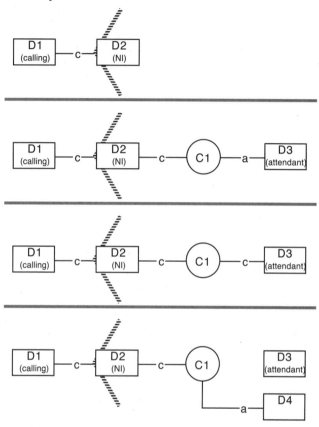

actual extension desired, allow the caller to dial the desired extension using the person's name, or look up the desired person in an interactive directory.

Auto-attendants provide a more secure and user-friendly "self-service" interface than DISA, so many people are switching from DISA to auto-attendants. Another benefit of auto-attendants is that they can be programmed to automatically handle fax and modem calls correctly, so that dedicated network interface devices are not required.

For most small telephone system owners, who generally do not have human attendants due to the obvious expense, auto-attendants are an extremely cost-effective alternative. It might even be desirable for a telephone system with only a single network interface device to have an automated attendant to screen calls.

Building auto-attendants is one of the fastest growing areas of CTI solution development. This is another area where the power and benefits of CTI technology are evident. CTI technology allows for the customization and day-to-day optimization of a system's automated attendant. The addition of technology such as text-to-speech and speech recognition allow for very simple and compelling interactions with callers.

5.7. Call Routing

Routing refers to all the ways that a call progresses through a telephone system. Each step in a routing sequence involves presenting a call to a device and having it either answer the call or direct it elsewhere. As long as a call exists, every new destination it reaches becomes part of its routing history.

5.7.1. Do Not Disturb

A feature called *do not disturb* can prevent the routing process from even starting at a particular device. If the do not disturb feature is activated for a particular logical device element, then all calls made to the device that satisfy specified criteria are rejected before they are ever presented to the destination. The rejected call may either fail or be redirected elsewhere. Do not disturb is illustrated in Figure 5-17. In this example D2 has activated the do not disturb feature; when D1 places a call to D2, the connection D2C1 immediately transitions to the *fail* state.

Figure 5-17. Do Not Disturb

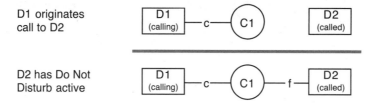

D1 originates
call to D2

D2 has Do Not
Disturb active

The do not disturb feature is used frequently in conjunction with the call forwarding feature (described later in this chapter), which allows the call to be redirected instead of transitioning to the *fail* state. This feature is very popular in hotel solutions, where guests do not want to be awakened by the telephone.

The do not disturb feature is managed with the *get do not disturb* and *set do not disturb* services. The feature can be activated to apply to all calls, or just those satisfying certain criteria relative to the origin of the call.

Internal / External

One option of the do not disturb feature is the ability to specify that the feature is to apply only to calls originating inside or outside the telephone system, rather than to all calls.

Selective Blocking

Selective blocking is a particular option of the do not disturb feature that allows one or more specific devices to be blacklisted. Calls made from these devices, and only these devices, will be rejected or redirected before they are presented to the destination.

Selective blocking is an example of a feature that is made largely obsolete by CTI. With CTI technology, call blocking can be implemented on your own computer system by setting it up to drop calls from blacklisted callers immediately. Unlike the telephony service, however a CTI-based implementation would have virtually no limit to the number of blocked callers and could be endlessly customized. For example, it could log attempts by blocked callers before rejecting their calls, it could block different callers at different times, it could play a prerecorded message to blocked callers before dropping their calls, etc.

5.7.2. Alerting

When call processing attempts to connect a device to a call, the connection between the call and the new device is in the *alerting* state. There are three modes of the *alerting* state through which a connection may transition during the process of attempting to establish a connection. A telephone system implementation may support one or more modes for a given device. The modes of *alerting*, in the order through which they transition, are the following:

1. *Entering distribution* mode

2. *Offered* mode

3. *Ringing* mode

The caller hears ringback as long as the call is in the *ringing* mode of the *alerting* state. Whether or not the caller hears ringback in the other modes depends upon the telephone system and/or the external network.

The first mode of the *alerting* state, *entering distribution*, refers to a state in which the call is being associated with a device with the specific intent that it be routed elsewhere. This is the mode normally used when a call is presented to a device for routing purposes. The use of the offered and ringing modes are described later in this chapter.

5.7.3. Queuing

Another connection state often observed in routing sequences is the *queued* state. This state reflects the fact that call progress has been suspended but not stalled. Typically connections are transitioned to the *queued* state while waiting for some resource to become available or some routing decision to be made.

Most implementations do have timers associated with *queued* calls to ensure that a call is not *queued* indefinitely; these timers are much longer than those for states such as *alerting*, however.

5.7.4. ACD Features

In many telephone systems, the first device that an incoming call is presented to is an ACD device. For example, auto-attendants that are built into the telephone system are ACDs.

An ACD device has built-in rules for distributing calls and may make decisions based on any of the call control information discussed earlier, as well as the time of day, the last device the call was redirected from, and information that it captures directly from the caller (through DTMF digit detection, for example).

ACDs that have *visible ACD-related devices* use independent devices (with separate connections to the call). A sequence using ACDs with this model is shown in Figure 5-18. In this example, the ACD device D2 is programmed to queue its connection if it can't immediately find an appropriate destination. It then enlists the help of a second ACD device, D4, which is also queued to the call. A media access device, D3, is connected to the call to play music to the caller while waiting for an available destination. When an appropriate destination (D5) is found, the "helper" devices are dropped from the call and it is diverted to D5.

With ACDs that have *non-visible ACD-related devices*, on the other hand, all of the media resources (and other resources) that they rely upon are internal to the ACD device itself. Using ACDs with this model, the same ACD activity appears as the sequence shown in Figure 5-19.

5.7.5. ACD Group and Hunt Group Features and Services

ACD groups and hunt groups behave in a fashion identical to that described for ACDs. The only difference is that these devices can only distribute calls to a finite set of devices. In the case of hunt groups, the group of devices is fixed and the only criterion for presentation of new

Figure 5-18. Visible ACD-related devices model

calls is availability. In the case of ACD groups, the group consists of logged-on agents and the status of an agent is the basis for presenting a call.

Get and Set Agent Status

The *set agent status* service is used by a device to log on to a particular ACD group, log off again, or indicate some other agent status during the cycle of handling calls presented by the ACD group. The ACD group will present a new call to a device only if its corresponding

Figure 5-19. Non-visible ACD model

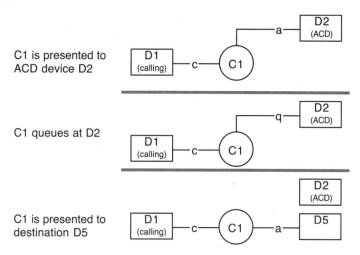

C1 is presented to
ACD device D2

C1 queues at D2

C1 is presented to
destination D5

agent status is *agent ready*. If there are no agents with this status, the
ACD group will queue the call. The *get agent status* service can be used
by a device to observe the status of the agent association.

5.7.6. Parking and Picking

Another service that involves queuing calls as part of the routing
process is the *park call* service. This service allows a call to be *parked*, or
queued, at a particular device. The *park-to* device may be a station
device or a park device from which the call can later be *picked* using the
directed pickup call service. The *park call* service is shown in Figure 5-20.

Figure 5-20. Park Call service

Park to a Station

The most common use of parking involves an attendant queuing a call on behalf of someone. Figure 5-21 shows an example that might be found in a hotel. The hotel's attendant, D2, answers a call from someone, D1, who wants the occupant of room 3002. The device in that room, D3, is busy; the caller indicates that she wants to wait, so the attendant parks the call.

Figure 5-21. Park Call to a station

Park to a Park Device

A call parked at a park device can be accessed using the *directed pickup call* service (also referred to by some as the *unpark* or *retrieve park* service in this context). It is shown in Figure 5-22.

Figure 5-23 shows an example that might be found in an environment where people are not in fixed locations and do not have assigned telephones. In this example, the caller D1 asks for Fred Jones, who is roving somewhere in a large warehouse. The attendant, D2, parks the

Figure 5-22. Directed Pickup Call service

call to a special park device, D3. The attendant then makes an announcement on the public address system for Fred Jones to pick up the call on D3. Fred finds the nearest station, D4, and uses the directed pickup feature to connect to the call.

Figure 5-23. Parking and picking a call

5.8. Forwarding and Coverage

Forwarding is a very popular telephony feature. It allows rules to be established for the automatic redirection of calls associated with specific logical device elements.

The forwarding feature is managed using the *set forwarding* and *get forwarding* services. Activating a particular forwarding rule involves specifying that the combination of a particular forwarding rule and a particular corresponding *forward-to* destination device should be activated. If the rule is then satisfied, the call is redirected to the corresponding forward destination.

5.8.1. Forwarding Types

Forwarding types are the different cases, or rules, for which forwarding can be activated. There are four basic forwarding types and three different *origination types* that allow for a total of twelve different forwarding types.

The basic forwarding types are:

- Immediate
- Busy
- No Answer
- Do Not Disturb

Origination type refers to where a call was originated. It may be one of the following:

- Internal
- External
- All

Immediate Forwarding

If *immediate forwarding* is active for a particular logical device element, then calls to that device (given the appropriate origination type) are immediately redirected to the corresponding forward-to destination. This is illustrated in Figure 5-24.

Figure 5-24. Immediate forwarding

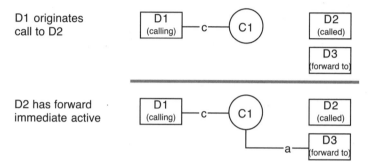

Immediate forwarding typically is activated when users will be away from their phones for a period of time and would like to have calls redirected, either to a temporary location (such as someone else's office or a cellular phone) or to someone who will be covering for them.

Busy Forwarding

If *busy forwarding* is active for a particular logical device element, then calls intended for that device while it is busy and therefore incapable of accepting calls (given the appropriate origination type) are forwarded to the corresponding forward-to destination. *Busy* forwarding is illustrated in Figure 5-25.

This type of forwarding is occasionally also referred to as *roll-over*. A call is said to "roll over to coverage" when this rule is triggered. The term *coverage* refers to the device responsible for covering for the device being called in the event that a call is not answered.

Figure 5-25. Busy forwarding

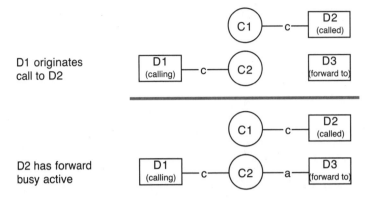

D1 originates
call to D2

D2 has forward
busy active

Busy forwarding is typically activated all the time for people that have voice mail, an answering service, or a full-time assistant that can handle calls that overflow.

No Answer Forwarding

In the case of *no answer forwarding*, the rule includes the number of rings to wait before giving up. If a call is presented (in the *ringing* mode of the *alerting* state) to a logical device with this type of forwarding active, it remains associated with the device for the number of rings specified. If it has not been answered, it is redirected to the forward-to device specified. This feature is illustrated in Figure 5-26.

No answer forwarding, like *busy* forwarding, is typically activated at all times for those with voice mail or an answering service.

Do Not Disturb Forwarding

The trigger for *do not disturb forwarding* is that the logical device element in question has activated the do not disturb feature and a call is about to be rejected as a result. In this case, the call is redirected to the specified forward-to device rather than being transitioned to the *fail* state. Contrast this behavior, shown in Figure 5-27, with the basic do not disturb behavior in Figure 5-17.

Figure 5-26. No answer forwarding

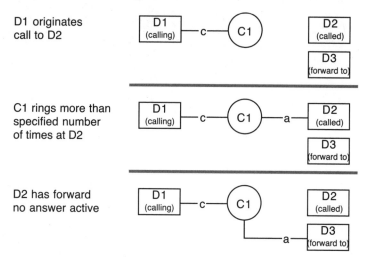

D1 originates
call to D2

C1 rings more than
specified number
of times at D2

D2 has forward
no answer active

Figure 5-27. Do not disturb forwarding

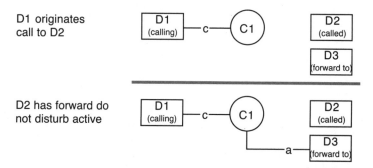

D1 originates
call to D2

D2 has forward do
not disturb active

Do not disturb forwarding never triggers unless the do not disturb feature is active, so the latter acts as a switch to turn the actual forwarding rules on and off. As a result, *do not disturb* forwarding typically is activated at all times for those with voice mail or an answering service. When users wish not to be disturbed, they activate the do not disturb feature, which in turn activates forwarding. This saves having to activate the forwarding each time.

5.8.2. System Default and User Specified Forwarding

There are often a number of default settings associated with implementations of the forwarding feature. A telephone system may have a set of *system default* forwarding rules for each logical device element in the system. This is often the case because the settings for *busy, no answer,* and *do not disturb* are typically all the same and all typically set to forward to a voice mail system if one exists. To eliminate the need to set all of these rules individually for each device, they are usually set up as system defaults.

Each device then can be managed individually to override these defaults, or turn them off altogether, on an as-needed basis with *user specified* forwarding. User specified forwarding also has associated defaults. If the *set forwarding* service is used, and a forward-to device is specified but a forwarding type is not, the telephone system will assume that its default forwarding type (normally *immediate–all*) should be used. If the *set forwarding* service is used and a forwarding type is specified without a forward-to device, the default forward destination (normally a voice mail system or the attendant) is used.

5.9. Offering

Once a call has been routed to its intended destination, the second mode of the *alerting* state, the *offered* mode, may come into play.

With the *offering* feature, the telephone system presents a new call in offered mode to a device in order to give the device the opportunity to accept, reject, or deflect the call before it transitions to the *ringing* mode. This gives the device the opportunity to screen calls before a human being is given the opportunity to interact with them.

If no action is taken, the *offered* mode times out and transitions automatically to the *ringing* mode. Depending on the implementation of the offering feature in a particular telephone system, the device may be able to answer the call directly while it is still in the *offered* mode.

5.9.1. Accepting

The *accept call* service is used to accept calls in the *offered* mode of the *alerting* state. When the service completes, the specified connection has transitioned from the *offered* mode to the *ringing* mode. *Accept call* is shown graphically in Figure 5-28.

Figure 5-28. Accept Call service

5.9.2. Deflecting

Another option for dealing with a call that is presented in the *offered* mode is to *deflect* it using the *deflect call* service.

The *deflect call* service is not limited to the *offering* case. It can operate on any call that is *alerting* or *queuing* at a device and divert it to a new destination device. If the service completes successfully, there is no connection at the deflecting device and the connection at the new destination device is one of the following:

- *Alerting* – the call has already reached the new destination device and the telephone system is attempting to connect the call.

- *Connected* – if the new destination is on an external network, this indicates that the call is connected to a network interface device; otherwise, the call not only reached the called device but has already been answered.

- *Fail* – call processing stalled while trying to connect to the new destination device, most likely because it was busy.

- *Queued* – the call is being queued at the called device, probably because it is some type of distribution device.

- *Null* – the call was redirected once again, so there is no connection to the new call at the destination device; there probably is a connection to the new call at some other device, however.

This service is described graphically in Figure 5-29.

Figure 5-29. Deflect Call service

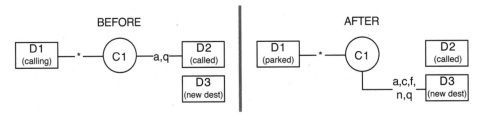

An example of an offering sequence in which the *deflect call* service is used appears in Figure 5-30. In this example, D1 places call C1 to device D2. The call is presented in the *offered* mode and D2 decides that it would be better handled by D3, so the *deflect call* service is used.

Figure 5-30. Deflect Call service in offered scenario

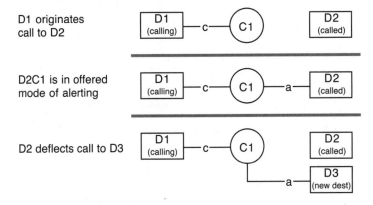

5.9.3. Rejecting

A device may reject an offered call. This involves simply dropping the call using the *clear connection* service.

Rejecting a call in the *offered* mode allows a device to implement selective blocking locally. Rather than rely on the telephone system or the telephone network to provide selective blocking, the device itself checks the information associated with a particular call against its list and drops the call before it transitions to the ringing state.

5.10. Answering

In section 5.3.2 we looked at the basic implementation of the *answer call* service. Answering a call involves making the transition from the *alerting* (or *queued*) state to the *connected* state. In this section we'll look at other ways that calls can be answered.

5.10.1. Auto Answer

The *set auto answer* service instructs a device to answer calls automatically after a certain number of rings. When a connection to the device has been in the *ringing* mode of the *alerting* state for the appropriate number of rings, the system automatically answers it as if an *answer call* service had been used. If the number of rings is zero, the call is auto answered the instant it is presented to the device.

This service typically is used by voice mail systems, auto-attendants, and fax machines that operate in an autonomous fashion.

5.10.2. Pickup

Another way that calls may be answered is through the *pickup* services. These services allow a device to answer a call that is at a different device.

Directed Pickup Call

In section 5.7.6 we saw how the *directed pickup call* service is used in conjunction with the *park call* service to park and pick calls. This service also can be used to answer a call that is *alerting* at a different device. This use of the *directed pickup call* service is also referred to as *dial pickup* and *reverse transfer*.

Directed pickup call typically is used when a person hears someone else's phone ringing and wants to answer it. The most common situation where this occurs is in an office building after hours. There is no attendant to answer incoming calls, so people working late can use *directed pickup call* to answer the calls ringing at the attendant console.[5-12]

An example of the use of *directed pickup call* is shown in Figure 5-31. Note that this sequence is very similar to the sequence for *deflect call* in Figure 5-30. The key difference is that the device picking the call is actually answering it, so when the service completes, D3 is in the *connected* state.

Figure 5-31. Directed Pickup Call for an alerting device

D1 originates call to D2

D2C1 is in ringing mode of alerting

D3 picks call from D2

5-12. Night bell — Some installations have an especially loud ringer, referred to as the night bell, that is mounted in a central place in the office. The night bell can be activated in the evening so that the presence of a pickable call is heard throughout the office.

Group Pickup Call

The *group pickup call* service is very similar to the *directed pickup call* service. Rather than specifying a particular device from which to pick the call to answer, however, the call is picked from another device by specifying the pick group device with which it is associated. By default, the pick group device used is the one with which the new destination device is associated. The *group pickup call* service finds an appropriate call associated with one of the devices in the group and redirects it to the new destination device. This is illustrated in Figure 5-32.

Figure 5-32. Group Pickup Call service

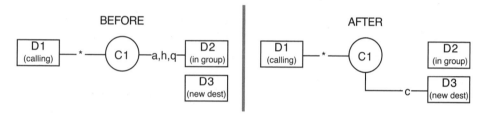

5.11. Suspending Calls

Once a connection has been answered and is in the *connected* state, the media stream(s) associated with the call flow through the established media stream channels. This continues until the call is placed into the *hold* state or is cleared.

When a connection is in the *hold* state, the media stream(s) associated with it are suspended and the media stream channels for it usually are deallocated. Generally there are six reasons to place a connection in the *hold* state.

1. As an alternative to muting the speaker and microphone.

2. To allow the call to be picked up somewhere else (using pickup services or by taking advantage of a bridged device configuration).

251

3. To reallocate the media stream channels being used in order to issue a command to the telephone system (if applicable for a given system).

4. To reallocate the media stream channels being used in order to participate in an additional call without dropping the first.

5. To reallocate the media stream channels being used in order to contact a second device to which the first call is to be redirected (*transferred*).

6. To reallocate the media stream channels being used in order to create a second call with which the first is to be joined (*conferenced*).

The term *hard hold* refers to an implementation of the *hold* state in which the media stream is suspended but the corresponding media stream channels cannot be reallocated for some other purpose. A hard hold supports the first two of the uses above.

The term *soft hold* refers to an implementation of the *hold* state in which the corresponding media stream channels can be reallocated for some other purpose. A soft hold is able to support all of the uses above, but some telephone systems require that the purpose for the *hold* state be specified in advance and then used only for the specified purpose.

Some telephone systems support optional *channel reservation*. If this feature is invoked when a connection is placed in the *hold* state, the associated media stream channels are not deallocated. If additional media stream channels are available, they can be used to allow for a soft hold; otherwise the result is a hard hold.

5.11.1. Hold

The *hold call* service places an active (e.g., *connected*) connection into the *hold* state. This single transition is the only stipulation of the completion criteria for this service. The *hold call* service is shown graphically in Figure 5-33.

Figure 5-33. Hold Call service

By default, the *hold call* service results in a hard hold, but many telephone system implementations attempt to provide a soft hold if possible. The sequence for a typical soft hold is illustrated in Figure 5-34. When call C1 is placed on hold by D1, the telephone system automatically creates a new call, C2, and reuses the media stream channels previously used by connection D1C1 for connection D1C2.

Figure 5-34. Soft hold implementation

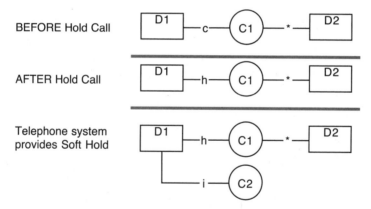

5.11.2. Consult

The *consultation call* service places an active (e.g., *connected*) connection into the *hold* state and originates a new call to specified destination. The name derives from the fact that this service typically is used to allow one person to leave a call and consult with someone else. This is shown graphically in Figure 5-35.

Figure 5-35. Consultation Call service

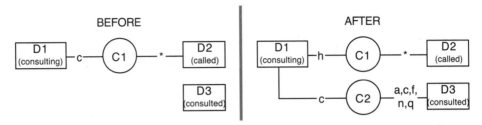

With respect to this service's ability to originate a new call, it operates in a fashion identical to the *make call* service. Therefore, in addition to placing the connection D1C1 into the *hold* state, the completion criteria are identical to those for *make call*. Multi-stage dialing and external outgoing calls are supported in the same way. Invoking the *consultation call* service without providing any digits to dial (thus initiating multi-stage dialing) is effectively a soft hold.

The *consultation call* service sometimes is referred to as a *compound service* because in many implementations it is equivalent to a a *hold call* service followed by a *make call* service. This is true for implementations that support soft hold and do not require any indication of the purpose for a transition to the *hold* state.

Consult Purpose

Consultation call is distinct from a *hold call* / *make call* combination in implementations where the *hold call* service only applies to the first two or three uses of hold listed earlier. With these telephone systems, *consultation call* is intended specifically for reallocating the media stream channels used for the first call in order to perform one of the last three uses listed. In addition, these implementations typically require that the *consult purpose* be specified. Consult purpose is one of:

- consult only;

- transfer;

- conference; or

- conference and transfer.

Each of these consult purposes can be thought of as a different variation of the service itself. Telephone systems of this variety typically have buttons on their telephone sets that correspond to these different variations of the *consultation call* service, and the individual services have names such as *setup transfer* and *setup conference*.

5.11.3. Retrieve

The *retrieve call* service, also referred to by some as the *unhold* service, transitions a connection from the *hold* state into the *connected* state. This single transition is the only stipulation of the completion criteria for this service. The *retrieve call* service is shown graphically in Figure 5-36.

Figure 5-36. Retrieve Call service

It is important to note that there are two common reasons for the *retrieve call* service to be unsuccessful:

1. An active connection for a voice call already is present at the physical device element represented by D1. Physical device elements can interact with only one active voice call at a time, so if an attempt is made to retrieve a call while another voice call is active, the service will not be successful.

2. No media stream channels are available to allocate to the connection. Even if there is no active voice connection tying up the physical device element concerned, the media stream channels needed to retrieve the previously held connection may have been used for digital data connections or for other physical devices in an independent–shared–bridged configuration. This situation can be avoided by invoking the channel reservation feature to ensure that the connection will be retrievable.

The *retrieve call* service is frequently used in bridged configurations where a call is placed on hold at one physical device element and is retrieved from a different one.

5.11.4. Alternate

The *alternate call* service operates on two connections at the same time. It takes one connection in the *connected* state and transitions it to the *hold* state. It transitions the other connection from the *hold* state to the *connected* state. The *alternate call* service is shown graphically in Figure 5-37.

Figure 5-37. Alternate Call service

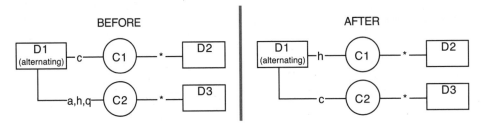

The *alternate call* service sometimes is referred to as a compound service because it is functionally equivalent to a *hold call* service followed by a *retrieve call* service.

The *alternate call* service typically is used in conjunction with the *consultation call* service. After consulting on the second call, *alternate call* can be used to flip back and forth between the two calls.

5.11.5. Reconnect

The *reconnect call* service is similar to the *retrieve call* service except that it typically ensures that the necessary media stream channels will be available to retrieve a connection previously placed on *hold* by clearing another connection. The *reconnect call* service is shown graphically in Figure 5-38.

Figure 5-38. Reconnect Call service

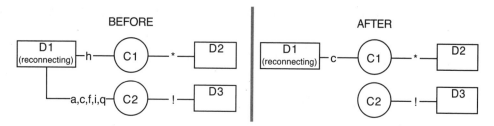

The *reconnect call* service typically is used in conjunction with the *consultation call* service. After consulting on the second call, *reconnect call* can be used to return to the first call in one step.

5.12. Transfer

Transferring (or *extending* as it is sometimes known) a call to a new destination is different from deflecting or forwarding a call because it involves a connection that has been answered. Deflecting and forwarding manipulate the routing of a call before any media stream channels have been allocated to the call at a given destination.

5.12.1. Transfer with Consult

A *two-step transfer* operation involves putting the call to be transferred on hold and consulting with the intended recipient of a call (typically using the *consultation call* service) and then executing the *transfer call* service. This service drops the transferring device from both calls and connects the other device(s) from the first call and the device(s) from the second call into a single new call. The *transfer call* service is shown graphically in Figure 5-39.

Figure 5-40 illustrates a typical sequence in which the *transfer call* service is used.

The *transfer call* service may be unsuccessful if the telephone system requires that the intent to transfer be indicated during the first step in the process and a consult purpose of *transfer* or *conference and transfer* was not provided. (Refer to consult purpose in section 5.11.2.)

257

Figure 5-39. Transfer Call service

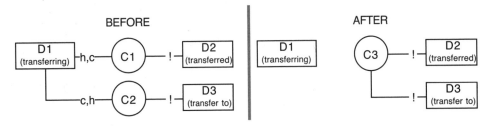

Figure 5-40. Two-step transfer call sequence

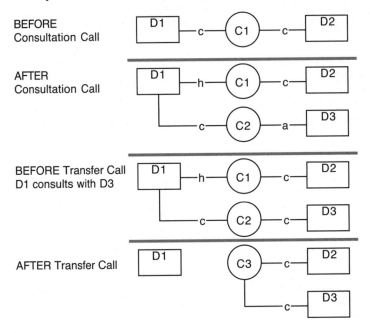

5.12.2. Single Step Transfer

Single step transfer call, also referred to as *blind transfer* by some, is similar to *transfer call* but it skips the consultation step. The *single step transfer call* service is shown graphically in Figure 5-41.

In one step this service drops D1 from the call it is transferring, C1, places a new call, C3, to the transferred-to device D3, and merges the remaining device(s) from call C1 into C3. At the completion of this

Figure 5-41. Single Step Transfer Call service

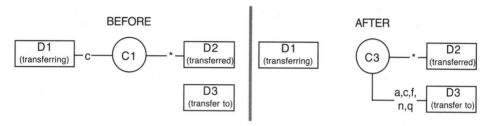

service, the result is very much as if D2 had used the *make call* service to call D3 directly. The state of connection D3C3 is the same as described for the called connection after successful completion of a *make call* service.

5.13. Multi-party Calls

All the services we have looked at so far have involved the creation and manipulation of two-device[5-13] calls. In this section we look at the services that allow the addition of multiple devices into calls to form multi-party calls.

5.13.1. Conference

The *conference call* service is very similar to the *transfer call* service in that it involves a two-step process that merges an original call with a second call to form a new call. The difference is that the conferencing device remains in the resulting call along with all of the devices involved in both the calls, so the result is a three-way (or more) call. The *conference call* service is shown graphically in Figure 5-42.

5-13. Two-device calls — Calls involving only two devices represent the majority of all calls, and they are frequently referred to by a number of different names. These include: point-to-point calls, two-party calls, basic calls, and simple calls.

Figure 5-42. Conference Call service

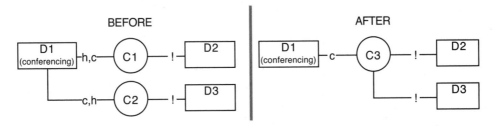

Typically the *consultation call* service is used to set up the second call; the *alternate call* service also may be used to flip between the calls before actually conferencing them together. Figure 5-43 shows an example of a conference call sequence.

Figure 5-43. Two step conference call sequence

Like *transfer call*, the *conference call* service may be unsuccessful if the telephone system requires that the intent to conference be indicated during the first step in the process and a consult purpose for conferencing was not provided. (Refer to consult purpose in section 5.11.2.)

5.13.2. Single Step Conference

The *single step conference* collapses the two steps of the conference call process into one, as shown in Figure 5-44.

Figure 5-44. Single Step Conference Call service

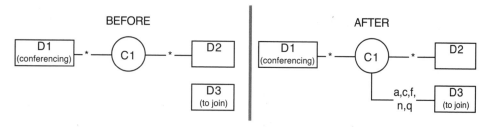

By specifying D3 as the destination for a single step conference involving call C1, the connection D3C1 is created in exactly the way that it would if any of the devices already in C1 had just placed a new call to D3 using the *make call* service. The difference is that all of the devices already in C1 remain in the call.

Traditionally single step conferencing has not been very popular because if the new connection to the call is not answered, all the participants in the call may hear endless ringback or busy tone. Unless the device to join is known to be expecting the call, or the telephone system provides support for the ability to drop an individual connection from a call (see *clear connection* in section 5.15.1), *single step conference call* is not a recommended service. On the other hand, if equipped with the ability to drop individual connections, *single step conference call* is a very powerful service because it allows a call to be established quickly with a group of selected devices.

5.13.3. Join

The *join call* service is very closely related to *single step conference call* but differs in one very important way. The *join call* service involves a device attempting to establish a new connection to an existing call, in

261

contrast to the *single step conference call* service, which attempts to connect an existing call to a device. The *join call* service is shown graphically in Figure 5-45.

Figure 5-45. Join Call service

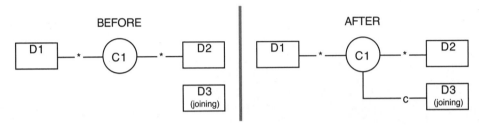

D3 uses the *join call* service to add itself to call C1. If the service is successful, the connection D3C1 is created in the *connected* state.

5.13.4. Silent Participation

Telephone systems may support the *silent participation* feature in conjunction with *single step conference call* and *join call*. Silent participation refers to creating the new connections as unidirectional, so that the added devices can listen into the call but cannot themselves be heard.

One situation where the silent participation feature is used with *single step conference call* is when one party in the call wants to record the conversation.

In another example, shown in Figure 5-46, D3 uses *join call* to add itself into call C1 with silent participation. D3 could be a supervisor who is quality-monitoring the calls being handled by a team of customer support agents. Using the *join call* service with silent participation allows the supervisor to listen discreetly into these calls in order to verify that customers are being treated in an appropriate fashion.

Figure 5-46. Join Call with silent participation

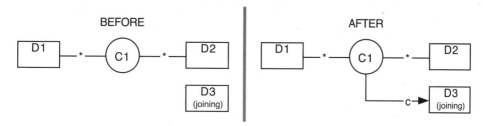

5.14. Call Failure

When call processing is unable to route a call and ultimately connect it to a destination for some reason, then call progress is said to be *stalled*. This section describes the ways that telephone systems handle calls in these situations, and the services that may be available when they occur.

5.14.1. Call Failure Handling

The *fail* state indicates that call progress was stalled for some reason and an attempt to associate a device and a call (or keep them associated) failed. The most common example of this state is attempting to connect to a device that is busy.

In most cases, all of the other active (e.g., *connected*) connections to the call will hear busy tone, or another appropriate failure tone, while a connection in the *fail* state is associated with the call.[5-14]

When a call ultimately stalls, the device for which it was intended may or may not (depending on the telephone system implementation) be aware of the connection attempt. Normally the connection is created and it transitions to the *fail* state. If the attempt to create the connection failed, however, there is no connection and the intended device will be

---

5-14. Blocked — One case where the fail state is not associated with an audible tone is in the case where a bridged connection is blocked from a call. This is described in Chapter 4, section 4.5.3.

unaware of the attempt. This is illustrated graphically in Figures 5-47 and 5-48. In both cases D1 has placed a call, C1, to device D2. In the first implementation (Figure 5-47), D2 is aware of the failed connection. In the second implementation (Figure 5-48), D2 is unaware of the connection attempt and its failure, in this case the call alone effectively failed.

Figure 5-47. Failed connection

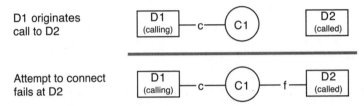

Figure 5-48. Failed connection attempt (failed call)

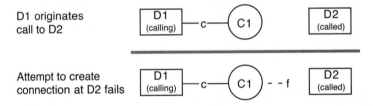

In both the scenarios illustrated, certain services still may be applied to the call (specifically to the connection D2C1 in the *fail* state in the first case, and to the *pseudo-failed* connection in the second case). One option is simply to drop the call or specific connections on the call (described in section 5.15). The other applicable services are called *call completion services* and are described in this section.

5.14.2. Camp On Call

One service that can be applied to a connection in the *fail* state is the *camp on call* service. This service queues a call that has failed for a given called device. The calling device remains connected to the call in the *connected* state and a connection in the *queued* state is established for the called device. Figure 5-49 shows the *camp on call* service in graphical form.

Figure 5-49. Camp On Call service

When the called device becomes available (and has dealt with any previously queued calls), the called device's connection transitions to the *alerting* state and it then may be answered by the called device.

An example showing how the *camp on call* feature works is illustrated in Figure 5-50. In this example, D1 places a call to D2 but the call fails because D2 is unavailable. D1 uses the *camp on call* service to queue the call at D2 (which results in the previously incomplete connection D2C1 being created in this case). When D2 becomes available, connection D2C1 transitions to the *alerting* state and D2 may answer it.

Figure 5-50. Camp On Call example

Typically this service, sometimes called *off-hook queuing*, is used when a call is placed and the call fails because the called device is busy. This feature is very useful not only for waiting on station device availability, however, but also for queuing to use network interface devices. When a large number of station devices share a small number of network interface devices, there will be times when no network interface devices are available. At these times, devices attempting to make external outgoing calls will have calls fail at the corresponding network interface group (described in section 5.4.4). By invoking the *camp on call* service, these calls are effectively queued for the next available network interface device in the group. When a network interface device becomes available, the call proceeds.

5.14.3. Call Back

Another service that can be applied to a connection in the *fail* state is the *call back call-related* service, referred to as the *ring again* service by some. This service activates a feature that will cause the unavailable called device to be called again automatically when it becomes available.

The *call back call-related* service can be applied to the calling device's connection if the call itself failed, or if the connection to the called device is in the *fail, alerting, queued,* or *null* state. (It can be in the *null* state if the call was forwarded or deflected prior to failing.) When the service completes successfully, it clears the call and all its connections and leaves the call back feature set on the called device. This service is shown graphically in Figure 5-51.

Figure 5-51. Call Back Call-Related service

Figure 5-52 provides an example illustrating how the call back feature works. In this example, D1 places a call to D2 but the call fails because D2 is busy. D1 uses the *call back call-related* service, which clears the call and sets up the call back feature. When D2 becomes available, the call back feature is triggered and the telephone system attempts to place the appropriate call between D1 and D2. First it creates a new call, C2, with a corresponding connection to D1 and prompts D1 to go off-hook to accept it. (This is indicated by placing D1C2 in the *initiated* state.) When D1 goes off-hook, the call is originated to D2, so D1C2 transitions to *connected* and D2C2 transitions to *alerting*. (We know it will be *alerting* because it is available.)

Figure 5-52. Call Back example

This feature is sometimes referred to as *on-hook queuing* because it allows a calling device to wait in queue for the called device to become available, without having to maintain a connection. In other words, it can wait for the call back while on-hook and not consume switching resources. Like *camp on call*, this service is typically used in conjunction with busy station devices, but it is also very good for queuing to use a network interface device. When a network interface device becomes available, the device is called back and the previously originated call can be attempted again. The trade-off between the two is the extra complexity of the call back feature versus the resource consumption of the camp on call feature.

Another variation of the call back feature is the *call back non-call-related* service. With the call-related version of the service, the device from which a call back is desired is indicated by the intended called device. If a particular device is known to be busy, however, the *call back non-call-related* service allows a call back to be registered against it without having to first place a call.

The *cancel call back* service allows one or all registered call backs to be cleared from the queue for a given device.

5.14.4. Call Back Message

The *call back message* feature, also referred to as the *call me message* feature on some systems, is a variation on the call back feature wherein a special call back indicator is set for the called device so that it can initiate the call back, rather than having the telephone system initiate the call back.

The actual call back "message" can take any form. Examples include:

- A special call back lamp that is lit on the physical element.

- A special message that appears on the display of the physical element.

- The name of the person who wants to be called (and the date and time they called) appears on the display of the physical element.

- A special tone is played just prior to dial tone when the handset goes off-hook.

In many implementations, a special "call back" button is provided on the physical device element, which triggers a *make call* service to place a call to the first device that registered an outstanding call back message.

The *call back message call-related* service allows the call back message feature to be set for the called device associated with a particular call. This service is otherwise identical to the *call back call-related* service (see Figure 5-51). The *call back message non-call-related* service allows the call back message feature to be set for any given device.

The *cancel call back message* service allows one or all registered call back messages to be cleared from the queue for a given device.

5.14.5. Intrude

Yet another way that failed calls may be handled is with the *intrude call* service. The *intrude call* service allows a device to "break into" another call in order to reach the called device if a call fails. There are two variations of the *intrude call* service, depending on the telephone system implementation. These are shown graphically in Figures 5-53 and 5-54. In both cases D1 placed a call to D2 that failed because D2 was busy in call C2 with D3.

Figure 5-53 shows the first case of the *intrude call* service, where the result is that D2's connection to its original call, D2C2, is placed on *hold* and connection D2C1 is transitioned directly into the *connected* state.

Figure 5-53. Intrude Call service (case 1)

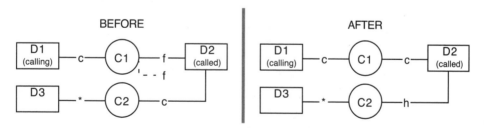

Figure 5-54 shows the second case of the *intrude call* service, where the result is that a new call C3 is created in which D1, D2, and D3 are all participants.

Figure 5-54. Intrude Call service (case 2)

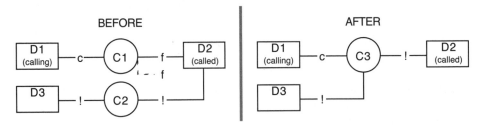

This second case also is referred to as *barge in* and the intruding device may have the option to join as a silent participant. This is illustrated in Figure 5-55. In this case, the intruding device can hear the conversation but cannot participate.

Figure 5-55. Intrude Call service (case 2) with silent participation

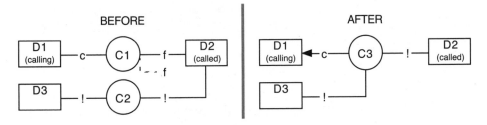

5.14.6. Recall

Telephone systems that support the *recall* feature have an alternative to failing calls in certain circumstances. It is a feature wherein a timer is associated with calls after certain services have been applied to them. The services that recall applies to include the following:

- Hold Call
- Transfer Call
- Single Step Transfer Call
- Deflect Call
- Park Call

If a device moves a call away, or places it on hold, using one of these services and that call fails to connect to a device before the associated timer expires, the recall feature is invoked and the call is returned to the device with a connection in the *alerting* state. Figure 5-56 illustrates an example of the recall feature in operation.

Figure 5-56. Recall feature

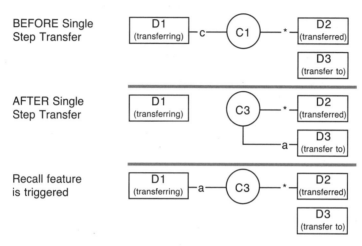

Recall is a very important feature because it ensures that callers are not "lost" or forgotten in the telephone system.

5.15. Dropping Calls and Participants

Now that we have looked at all of the ways to create, manipulate, and combine calls and their connections, we turn to disposing of calls and connections.

5.15.1. Clear Connection

The *clear connection* service either drops a specific connection from a call or inactivates a connection in a shared–bridged device configuration. The connection in question may be in any non-*null* state initially, and the state of all other connections in the call are not considered or directly affected by the *clear connection* service.

271

If the *clear connection* service completes successfully, the connection in question is transitioned to either the *null* state or the *queued* state. The *queued* state applies to a call at a device with shared–bridged appearances where at least one other appearance is participating in the same call. (This was illustrated in Figure 4-20 in Chapter 4, section 4.4.3.) In all other cases the connection will be completely disposed of with a transition to the *null* state.

The *clear connection* service is shown graphically in Figure 5-57.

Figure 5-57. Clear Connection service

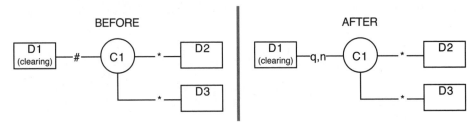

The connection being cleared may transition through the *fail* state on its way to *null* state under certain circumstances. Specifically, if a connection is cleared but the associated hookswitch on the physical element is off-hook, the connection will enter the *fail* state. It will remain in *fail* state until the physical element is set on-hook. When this occurs, the connection will transition to the *null* state and the service will have completed.

If there were only two devices with connections in the call prior to one of the connections being cleared, the call itself generally will be cleared. If the second device in the call was a network interface device, there may be multiple devices in the external network that remain in the call, even though the call no longer exists inside the telephone system and the network interface device is no longer being used.

The *clear connection* service is very useful in multi-party calls because it allows individual devices to be dropped from the call as needed. An example of an excellent CTI solution involves conferencing software that allows participants in a conference to see the names of all of the people involved in a conference call. The chairperson for the conference can use this software to add people to the conference (through the *single step conference call* service) and drop people (through the *clear connection* service) without affecting the others in the conference.

5.15.2. Clear Call

The *clear call* service tears down an entire call and all of its connections regardless of the state of any of its connections. The *clear call* service is shown graphically in Figure 5-58.

Figure 5-58. Clear Call service

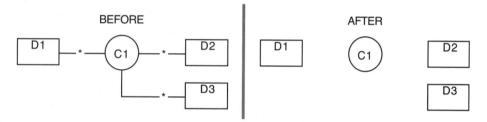

The *clear call* service is useful whenever it is desirable to shut down an entire multi-party call quickly and efficiently. Referring to the example of the conferencing software again, if the conference chairperson wanted to conclude the entire conference all at once, it would be quite painstaking if each individual participant had to be cleared one at a time using *clear connection*. The *clear call* service makes this a single step process.

5.16. Review

In this chapter we have seen the tremendous diversity of telephony functionality in the form of switching *services* and *features* that may be available in a telephone system.

This functionality ranges from the *basic* set of services of making, answering, and clearing calls (using the *make call*, *answer call*, and *clear connection* services), through to an extensive range of well-defined *supplementary* services. In addition, many telephone system vendors have implementations that include an *extended* set of functionality known as *vendor specific extensions*.

Telephony features and services can be described graphically by showing how the simple abstractions of device, appearance, call, connection, and connection state that we saw in the previous chapter combine to make even the most complicated functions easy to understand at a glance.

The *make call* service is used to originate calls in a single step. Multi-stage dialing is accomplished by combining it with the *dial digits* service. When a new call is created and its intended destination (the called device) has been identified, the call is connected to the calling device and is said to have been *originated*. If the calling device cannot go off-hook automatically, the telephone system may need to perform *prompting* to indicate that the device should be taken off-hook by a person. *Predictive dialing*, using the *predictive make call* service, allows the telephone system to place calls automatically and only uses a person's time if a call is successfully placed. A logical device that supports the *last number dialed* feature remembers the destination of the last originated call.

Calls may be *internal* to a telephone system, or they may be *external incoming* or *external outgoing*. External calls take advantage of network interface devices to connect to the telephone network outside a given telephone system. Special hunt group devices referred to as *network interface groups* (or *trunk groups*) route external outgoing calls to the next available network interface device. *Dial plan management* and *least cost routing* capabilities make telephone systems easier to use and cheaper to operate.

External incoming calls are routed to their intended destinations through one of three basic methods: fixed network interface device association (e.g., *DIL* service), selectable device association (e.g., *sub-addressing*, *DID* service, and the *DISA* feature), and through an attendant (e.g., a switchboard operator or an auto-attendant).

Associated with every call are critical pieces of information that can be put to good use by people and CTI systems. They include *CallerID* and *ANI, DNIS,* the *last redirected device, account* and *authorization codes, correlator data,* and *user data*. A range of services, such as *Get CallerID* and *Set CallerID* interact with the features that manage call information.

Calls may be delivered directly to a special routing device (such as an ACD, ACD group, or hunt group) that redirects the call to an appropriate destination based on call associated information and interaction with a caller.

Features such as *do not disturb, forwarding, call offering,* and *auto-answer* provide mechanisms for the automated handling of attempts to connect calls to specific devices.

Calls may be *parked* at specific devices and retrieved using services such as *directed pick up call* and *group pickup call*. Calls may be suspended, or placed in the *hold* state, using the *hold call* service or the *consultation call* service. The *consultation call* service creates a new call after suspending the first call. The *alternate call* service can be used to flip between the calls, and the *reconnect call* service can be used to drop one call and return to the one that is on hold. Depending on the *consult purpose* specified (if necessary for a given telephone system), the first call can be transferred to the second called device using the *transfer call* service, or all three devices can be merged onto the same call using the *conference call* service. The *single step transfer call* and *single step conference call* services provide single-step versions of these functions. If *silent participation* is supported, devices can be added to calls in such a way that they can listen but not participate in a call.

If call processing is unable to connect a call and its routing options are exhausted, the call may either be *recalled* to the last device that routed it, or it may be stalled. If call processing for a call is stalled, the call is said to have *failed*. The *camp on call, call back call-related, call back message call-related*, and *intrude call* services are all *call completion services* for handling a failed call.

Finally, when nothing further is required of a connection or an entire call, it can be cleared using the *clear connection* or *clear call* service, respectively.

6.
CTI Concepts

One telephony resource that has not yet been discussed in detail is the CTI interface. This telephony resource is the one that allows a particular telephony resource set, or telephone system, to offer CTI functionality and become part of a CTI system. In this chapter we will explore what CTI interfaces are, how they work, and what functionalities they provide.

As we have seen already, most telephone systems offer, or are capable of offering, a very rich suite of functionality. The vast majority of people traditionally use only a fraction of the functionality that their telephone system offers, however, because that functionality is locked away behind a difficult-to-use telephone keypad interface. The CTI interface represents an alternative means of reaching this functionality with all the power of a modern computer-based user interface design.

6.1. CTI Abstraction

An abstraction is a myth that we create in order to make something manageable. In the case of CTI, the abstraction of telephony functionality makes it possible to design software that will work with more than one telephone system and support more than one user's paradigm.

The huge diversity in telephone system implementations means that there is an equally huge diversity of very different implementations. In the past (before the development of a robust abstraction of telephony functionality), software developers, installers, and customers who wanted to build CTI systems had no choice but to be aware of the internal designs and proprietary terminology, concepts, rules, and behaviors of every telephone system.

The development of a universal abstraction allows any implementation, regardless of its size, to be described in the same terms. One way to think of this is that the telephone system presents a *façade* that appears to all observers to behave precisely as the universal abstraction dictates. Behind the façade, however, the telephone system is doing whatever is appropriate to translate between its own internal representations and those in the façade. To the observer of the façade, there is no difference between this telephone system and one with the same functionality that might be built with an internal representation based directly on the abstraction. This is illustrated in Figure 6-1. This translation between the universal abstraction (the façade) and the actual implementation is the role of the CTI interface.

6.1.1. Observation and Control

As we have already learned, computer telephony integration allows computer systems to both observe and control resources and entities in telephone systems.

This observation and control is not directly a function of the actual telephone system implementation, but of the abstraction or façade that the telephone system presents through a CTI interface. The internal

Figure 6-1. Telephony abstraction is a façade

implementations of a PBX handling an incoming call from a central office switch on a T-1 trunk and presenting it to a DID extension is very different from the internal implementation of a simple POTS telephone receiving a call on an analog line—but two computer systems using CTI interfaces to monitor each of these station devices would see the same thing through the abstraction: a new call being presented to the device in the *alerting* state.

Generally speaking, observation and control are closely linked. While there are some exceptions, virtually every useful application of CTI technology needs the ability to observe activity within the telephone system so that it can control it in some desired fashion or, at a minimum, provide logging and monitoring functions. In other words, the computer system has to be able to see what it is doing before it can manipulate resources and objects within the telephone system. Furthermore, because the telephone system is a dynamic, constantly changing facility in which every component has a state or status that affects what can and cannot be done, the observation of the telephone system must be continuous and ongoing.

6.1.2. Manual versus CTI interfaces

A good way to think of a CTI interface is as an alternative to the standard observation and control interface used with telephone systems, i.e., the telephone set.

In the example illustrated in Figure 6-2 a CTI interface is being used by a computer to observe and control all of the activity associated with a particular telephone. Like the human sitting next to the telephone, the computer can "see" all of the lamps that are lit, the buttons that may be pressed, the text on the display, etc. Just like the person using the telephone, the computer can place calls, answer calls, press buttons, etc. Bounded only by the capabilities of the CTI interface in question, the computer can, in fact, do anything the human can do with the telephone—and possibly more. It is as if the computer can reach out an invisible electronic arm and do the same things that the human can do to the telephone set. In this example, both the computer and the human have free access to the telephone set and can manipulate it independently.

Figure 6-2. Multiple interfaces to telephone functionality

This example illustrates an important point about the way CTI interfaces generally work. To the extent that the CTI interface allows observation of a particular device, a computer may be just one of many observers (either humans or other computers) of that device. To the extent that the CTI interface allows controlling the device, that control is not exclusive. The computer cannot prevent a human from pressing a particular button, lifting the handset at a given instant, or performing some other action. Both are effectively peers in the telephone system. While at first glance this might seem quite simple, it does pose a few technical challenges of which to be aware.

In application software development the concept of multiple simultaneous control interfaces is not a new one. A single application running on a computer might be controlled through its graphical user interface with inputs coming from a mouse and a keyboard, through an independent speech recognition interface, and possibly through a scripting interface such as DDE or Apple Events. The application must be prepared to combine all of theses requests into a single stream and deal with each in its turn. This is called application *factoring*. It separates from the core application code (which simply takes the next command from the outside world, processes it, and updates all the interfaces appropriately), all the code that creates and manages each of the different interfaces (different façades, if you will) and corresponding input paths. One result of application factoring is that requests that are obsolete occasionally may arrive from the interfaces. For example, if a voice command instructs a drawing application to close a file, while simultaneously a mouse gesture indicates an item is to be duplicated, the application would behave differently depending on the order in which it received the commands. If the command to close the file was received first, the command from the mouse would fail. This failure is not the result of a bug, a design flaw, or an error on anyone's part. It is just a natural consequence of an implementation in which function is factored away from the interface, and multiple interfaces are active simultaneously.

Multiple interdependent or conflicting inputs may be presented to a telephone system in a near-simultaneous fashion. To return to the first example, the computer might observe a call being presented in the *alerting* state and might react by requesting that it be deflected elsewhere. Meanwhile, however, the human sitting next to the phone (or some other computer) might already have answered it in the time it took for the computer to make its decision. This does not occur frequently in practice, but it is yet another important aspect of the CTI abstraction.

Observation and control are independent. While any request to manipulate a resource should be based on the last observed state of that resource, the request may or may not be successful because other activity may have taken place as the request was being issued. Therefore the results of a request should never be assumed; there is no substitute for observing what actually takes place.

6.1.3. Scope of Observation and Control

The portion of the telephony feature set to which a CTI interface has access varies between telephone system implementations. In telephone systems designed specifically to support CTI interfaces, the portion of functionality that can be accessed typically is much more than what can be accessed through the system's telephone sets. In systems where CTI functionality was not part of the initial design, the portion of the implemented telephony features that can be accessed through a CTI interface varies dramatically, from all to just a small subset. In some cases a system will offer both a proprietary CTI interface and a standards-based CTI interface, where the latter has less functionality than the former for historical and time-to-market reasons.

The resources making up a particular telephone system may in fact be distributed among different components, where each component may have a different view (or no view at all) of the telephony resources in the other components. Depending upon which component a particular CTI interface is associated with, its scope will be appropriately limited. Two different computers interacting with the same telephone system therefore might see different abstractions of the same system if they have access through different CTI interfaces.

6.1.4. Security

Security features can play an important role in determining what can and cannot be observed and controlled through a CTI interface.

If a particular telephone system authenticates the users of its CTI interface, the abstractions presented to two different computer systems may differ depending on their identities. A system administrator may be able to see all of the devices in a system, but another individual might only be able to see the device corresponding to his or her telephone. A secretary might be able to see the boss's phone but not have any control over it.

This type of security can be very important for avoiding the use of a CTI interface for toll fraud (see sidebar "Toll Fraud" on page 231). In practice, however, it can add a great deal of overhead to the administration and operation of the system. As a result, security features tend to be a key point of differentiation among the implementations of CTI system components from different vendors. Some customers need this type of security, but others might not.

6.1.5. Vendor Specific Extensions

No matter how extensive the abstraction, the diversity of the telephony industry dictates that individual telephone systems are likely to have features and capabilities that represent unique *vendor specific extensions*.

Most of the functionality provided and routinely used will be within the abstraction (the façade), and the majority of CTI solutions will limit their scope to just this portion of the telephone system's functionality. It is very important, however, that the CTI interface allow a "back door" or *escape mechanism* that provides direct access to these vendor specific extensions. Of course, this escape mechanism is useful only to computer systems that are aware of the precise identity of the telephone system, and are aware of precisely how its vendor specific extensions work.[6-1]

For example, the XYZ telephone system might offer the "persistent call back" feature.[6-2] This feature involves the ability to register a callback against a device so that, when the device becomes available, the telephone system calls it every 60 seconds and plays a prerecorded

6-1. Vendor specific extensions — One way around this limitation is using buttons. Pressing buttons on a physical device element is part of the general abstraction, but the action of a specific button is not defined in any way. Within the telephone system's abstraction, individual buttons are assigned labels to identify them and pressing the button triggers the corresponding vendor specific extension.

6-2. Vendor specific extension example — The "persistent call back feature" is purely fictitious. The author requests that no telephone system vendor try implementing this feature.

announcement requesting that the originating device be called back. The XYZ company could support this feature through the CTI interface as a vendor specific extension by using the appropriate escape mechanism. If a computer wanted to activate this feature, it would have to know it was using the right model of the XYZ system and the appropriate escape mechanism. Once activated, the persistent calling would be observed appropriately through the normal abstraction, and both the computer that set the feature and all other observers through the CTI interface would continue to accurately observe call activity.

6.2. The CTI Interface

 A *CTI interface* is a telephony resource that creates a portal through which other telephony resources can be observed by CTI components,[6-3] and through which these CTI components can request that features be set and services be carried out. CTI interfaces also may issue requests to the computer-side CTI component to perform certain tasks.

6.2.1. CTI Messages

 A CTI interface operates by generating, sending, receiving, and interpreting messages containing status information and requests for services to be performed. This is illustrated in Figure 6-3.

Messages are used in either direction both to provide information and to issue requests. The structure, content, and rules governing the flow of messages back and forth through a CTI interface is defined by a CTI protocol.

6-3. CT Components — The word component is used in two different contexts in this book. A CT component is a hardware or software module in a CT system that uses or provides a CTI interface or a media services interface. A physical element component is a lamp, button, display or some other part of a physical element.

Figure 6-3. CTI messages

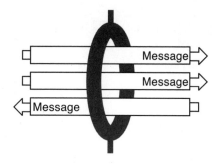

6.2.2. Parametrization

We have already seen that abstraction plays a key role in making CTI possible. In fact, with just a few simple constructs (devices, elements, components, appearances, calls, and connections) and a reasonably small vocabulary of types, states, and attribute values, we have been able to describe and model the vast majority of telephony functionality with little effort.

By translating into specific parameter values all references to parts of the abstraction and their state, status, or setting, any activity within a telephone system, desired or actual, can be described precisely.

Our abstraction of telephony resources, features, and services now can be expressed in concrete terms through parameters that are placed into messages. This is illustrated in Figure 6-4.

Figure 6-4. Parameters in a CTI message

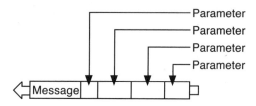

6.3. Modular CTI Systems

A general goal for implementing CTI interfaces is allowing any combination of hardware and software components to be assembled into a CTI system of any size.

Even the smallest CTI system is made up of many components, and this means that the system itself contains multiple CTI interfaces. A CTI interface is needed between each CTI component that must be integrated with another. Figure 6-5 builds on the example we viewed earlier. While this example is among the simplest of all CTI systems, it still involves three distinct components:

- The telephone

- The computer

- The CTI software running on the computer

Figure 6-5. CTI interfaces in a CTI system

Because there are three distinct CTI components, there are two different CTI interfaces at work in this simple CTI system:

- Between the telephone and the computer:
  This CTI interface uses a protocol.

- Between the CTI software and the computer:
  This CTI interface uses a programmatic interface.

Once a CTI system is assembled, it becomes difficult to say where the telephone system begins and the computer system ends. All the components that make up the system are working together in a cohesive fashion to form what can be viewed on one hand as a more sophisticated telephone system, or on the other as a more sophisticated computer system.

Graphical Notation for CTI System Abstractions

While Figure 6-5 illustrates a tangible CTI system configuration, the rest of this chapter deals with the abstraction of CTI systems and interfaces. CTI systems and the relationships between their components are described using a standardized graphical notation. Rounded blocks represent abstract CTI components and arrows represent the flow of messages. Other symbols are introduced as they are used. These symbols are also summarized on the inside of the front cover.

The graphical notation used for describing tangible CTI system configurations is defined in Chapter 11.

6.3.1. Inter-Component Boundaries

As noted earlier, CTI systems are assembled from many individual CTI components. Regardless of a particular component's form, its role in the overall system involves exchanging CTI messages with neighboring CTI components. Each CTI component communicates CTI messages to another component through an *inter-component boundary*.

Figure 6-6. Inter-component boundary

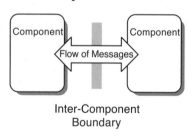

Figure 6-6 depicts an inter-component boundary through which two CTI components are exchanging messages. All the messages describing and directing CTI activity travel between two interoperable CTI system components across this boundary.

6.3.2. Logical Clients and Servers

In order to distinguish between two components exchanging CTI messages across a particular inter-component boundary one is referred to as the *logical client* and the other as the *logical server*.

 The *logical server* is the CTI component, relative to a particular inter-component boundary, that is making its CTI interface available for exchanging CTI messages.

 The *logical client* is the CTI component, relative to a particular inter-component boundary, that is making use of the CTI interface offered by a logical server across the inter-component boundary.

 The terms *logical client* and *logical server* are always used relative to a specific inter-component boundary. They should not be confused with terms such as *client implementation, server implementation, client computer,* and *CTI server.* These terms are used as absolute references to particular types of products and are defined later in this book.

6.3.3. Organizing Components into Systems

The simplest way to organize multiple components into a system is to chain them together as shown in Figure 6-7. In this arrangement, components play the roles of both logical server and logical client (except those at either end of the chain). In this example, the second component from the left is the logical server for the component on its right and the logical client for the component on its left. The result is a "pipeline" or "bucket-brigade" in which CTI messages are passed from one component to the next.

Figure 6-7. Multi-component chain

Each component in the system has some combination of the following roles:

- Generating CTI messages

- Conveying CTI messages from logical clients to logical servers

- Interpreting and responding to CTI messages

In this way components may embellish, simplify, merge, or manipulate in some fashion the CTI messages they deal with.

Figure 6-8 illustrates another arrangement of CTI components that allows for a CTI system to scale. It is referred to as a *fan-out* arrangement. A *fan-out component*, the second component from the left in the diagram, is able to simultaneously act as the logical server to multiple logical clients through a number of distinct inter-component boundaries. Each component is only aware of its counterpart across a specific inter-component boundary so the fan-out component's logical server is not aware of its logical clients and its logical clients are unaware of each other.

Figure 6-8. Fan-out component

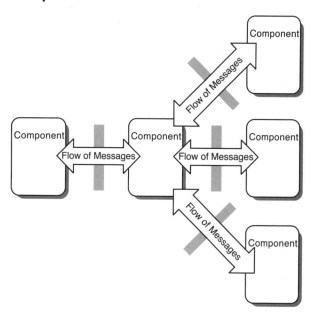

The role of the fan-out component is to maintain these relationships independently by interpreting CTI messages from each of its logical clients and either acting on them or conveying them to its own logical server. When a fan-out component conveys a CTI message in this fashion, it is effectively acting as a proxy for the logical client that generated the message in the first place. By keeping track of which CTI messages were conveyed on behalf of which clients, and in some cases by merging and separating CTI messages as they flow back and forth, the fan-out component is able to maintain all of these logical client-server relationships in a transparent fashion.

 The concept of inter-component boundaries makes possible the interoperability of actual CTI system components by standardizing only what travels between disparate components and not by standardizing their implementation.

Other benefits of this view of CTI system construction include the simplification of scaling a CTI system and the ease with which new functionality can be added to an existing CTI system. Interoperable CTI components can be added and removed from a system without affecting other components or the function of the system overall. In fact, new capabilities can be incorporated into a system by inserting CTI components with special features or capabilities into the appropriate place in a chain of other components.

6.4. Service Boundaries and Domains

In order to support a single definition for CTI interfaces that can be applied to any boundary in a CTI system of any size, configuration, or complexity, a single key insight is required:

 Regardless of the number of components in a CTI system, it can be broken down and analyzed in terms of each inter-component boundary that allows two adjacent CTI components to exchange CTI messages.

In creating or integrating different CTI components, the focus at any instant is on just one inter-component boundary through which a given component must interoperate. CTI messages and the CTI interfaces that work with these messages can therefore be described in terms of a simplified CTI system which consists of just three parts:

- Switching domain

    The *switching domain* is everything on the logical server's side of a particular inter-component boundary.

- Computing domain

    The *computing domain* is everything on the logical client's side of a particular inter-component boundary.

- Service boundary

    The *service boundary* is the inter-component boundary that lies between the computing domain and the switching domain in a given context.

This simplification of a CTI system is illustrated in Figure 6-9. All interaction between a given computing domain and a switching domain takes place through a CTI service boundary.

Figure 6-9. The CTI service boundary

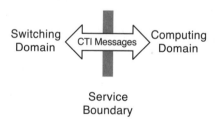

Switching Domain — CTI Messages — Computing Domain

Service Boundary

This abstraction applies to every inter-component boundary in a CTI system so care must be taken to specify what inter-component boundary is being referred to as a service boundary at any given instant. Specifying the service boundary not only identifies which pair of components is being discussed, but also the CTI messages that travel between them, and the CTI interface that interprets these messages.

6.4.1. CTI Service Boundary

 A *CTI service boundary* is the inter-component boundary through which CTI messages from a CTI component acting on behalf of its computing domain are communicated to the CTI interface of a CTI component acting on behalf of the switching domain it represents.

In concrete terms, a service boundary can take the form of either a protocol or a programmatic interface. A service boundary that lies between two software components running on the same hardware component (e.g., a computer) can take the form of a *programmatic interface* through which CTI messages are passed using function calls. Otherwise, the service boundary takes the form of a *CTI protocol* used for conveying CTI messages as a stream of data.

6.4.2. Switching Domain

 The term *switching domain* refers to all of the telephony resources that can be observed or controlled through a designated service boundary and all of the CTI components that provide this access.

The mechanism a switching domain uses to interact with a computing domain through the service boundary is its CTI interface[6-4]. The switching domain encompasses any telephony resources associated with components in the switching domain that can be accessed using this CTI interface. This is illustrated in Figure 6-10.

6-4. Switching domain versus switching function — CSTA phase III differs from ECTF C.001 and the Versit CTI Encyclopedia by differentiating between the switching domain in which resources are manipulated and an associated switching function which the CTI interface provides access to.

Figure 6-10. The switching domain

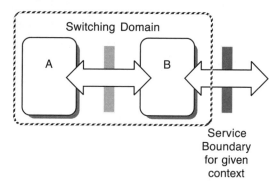

In concrete terms, the number and type of hardware and software components that might be found in a given switching domain is unbounded. However, at least one component must contain call processing functionality. Typically it will include one of the following:

- Switches
    - Front-end switches
    - KSUs or Hybrids
    - PBXs
    - Application specific switches

- Telephone station equipment
    - Telephone station
    - Telephone station peripherals

Any type of component that can be found in a CTI system may be found in a switching domain given the service boundary context. Other types of CTI hardware components are presented in Chapter 11 and software components are presented in Chapter 8.

6.4.3. Computing Domain

The term *computing domain* (as illustrated in Figure 6-11) refers to all of the CTI components that are involved in observing or controlling telephony resources in the switching domain on the other side of a designated service boundary.

Figure 6-11. The computing domain

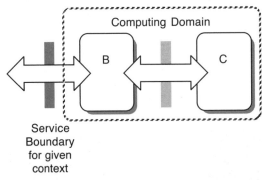

Service
Boundary
for given
context

Any type of CTI component may be found in a computing domain, given a particular service boundary context. However, at least one component is a software component that is attempting to observe or control telephony resources in the switching domain. (See Chapter 11 for details on CTI hardware components and Chapter 12 for details on CTI software components.)

6.4.4. Service Boundary Context

The abstraction of switching domain, computing domain, and service boundary can be applied to any inter-component boundary in a CTI system of any size. For a given context, the applicable service boundary always determines the roles of the components on either side of the boundary by whether they fall within the computing domain or the switching domain.

With respect to achieving interoperability between the two domains at a given point, the central concept is the service boundary: an inter-component boundary between the two domains over which well-defined CTI control and status messages pass.

 Regardless of the number of components in a given CTI system, each can be viewed as being interfaced to its neighbor through a particular service boundary.[6-5] This is illustrated in Figure 6-12.

6-5. CTI components and service boundaries — In a CTI system consisting of n interoperable components, there are n-1 service boundaries.

Figure 6-12. Service boundary contexts

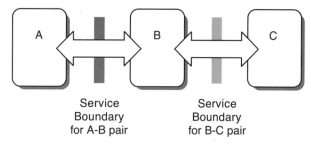

Service
Boundary
for A-B pair

Service
Boundary
for B-C pair

As shown in Figure 6-13, the particular service boundary for a given context can be established by referencing the pair of components which share it. When any given pair of CTI components are referenced (such as components "B" and "C" in the example), the implied boundary is considered the service boundary, with the computing domain on one side and the switching domain on the other, forming a domain pair.

Figure 6-13. Service boundary defines switching domain and computing domain

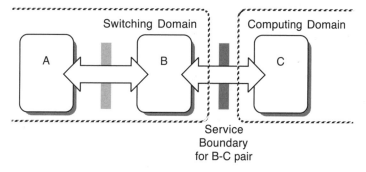

Switching Domain

Computing Domain

Service
Boundary
for B-C pair

In fact, given a particular division of components between the switching domain and the computing domain, the service boundary is unique. There is exactly one service boundary over which messages pass between any computing domain-switching domain pair.

Applying these concepts, an unlimited variety of CTI configurations can be supported; from the simple interconnection of two CTI products, to a whole system of interconnected CTI components. CTI control and status messages are communicated over these service boundaries, and each component may add value to the services provided by its neighbor or simply act as a conduit. The CTI system as a whole is able to function because of the individual service boundaries that hold its components together. (See Chapter 11 for examples of CTI system configurations made possible by this modularity.)

6.4.5. Protocols

CTI protocols are specifications of the structure, contents, use, and flow of CTI control and status messages that travel between CTI system components over well-defined communication paths. CTI protocols are high-level protocols, like the protocols used to send electronic mail, print to a printer, retrieve files from a file server or, browse the World Wide Web. Like these other protocols, they are designed to be transmitted over any type of reliable communication path.[6-6] CTI protocols are applicable to all types of communication paths and to all types of CTI configurations.

As shown in Figure 6-14, implementations of components that interoperate with other components using CTI protocols include a subcomponent referred to as a *CTI protocol encoder/decoder*, which is responsible both for establishing communication paths that carry the CTI protocol, and for interpreting the CTI protocol that flows across the communication path.

6.4.6. Programmatic Interfaces

Programmatic interfaces are mechanisms, typically made up of function calls, that allow two software components present on the same hardware component (generally a computer of some sort) to link to one another and exchange messages.

6-6. CTI Protocols, OSI layers — In the terminology of Open Systems Interconnection (OSI) layering, CTI protocols are layer 7 protocols with defined layer 6 encodings. They do not assume any particular underlying protocol stack. Communication paths provide OSI layers 1–5 in the form of functionally acceptable session/transport protocol stacks.

Figure 6-14. CTI protocols

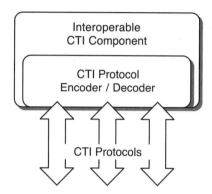

There are three distinct categories of programmatic interfaces of concern to CTI systems.

- Read/Write interfaces

  Read/write interfaces, or just *R/W interfaces* for short, are simple interfaces that allow a software component to obtain access to a communication path carrying a CTI protocol stream.

- Procedural interfaces

  Procedural interfaces allow a software component to obtain CTI functionality through a set of procedural function calls.

- Object interfaces

  Object interfaces allow a software component to obtain CTI functionality by manipulating software objects.

The use of programmatic interfaces for CTI is described extensively in Chapter 12.

6.4.7. Fundamental CTI System Configurations

To illustrate the concepts of switching domains, computing domains, service boundaries, protocols, and programmatic interfaces, we'll look briefly at two fundamental examples of actual CTI system configurations. The following sections show examples of basic *direct-connect* and *client-server* CTI system configurations.[6-7] (A detailed discussion of these and other CTI configurations is presented in Chapter 11.)

CTI interfaces, accessed through CTI service boundaries, allow the integration of CTI system components with varying functionality and from different vendors (such as the client computers, telephone stations, CTI servers, and switches shown in the examples below).

Direct-Connect Configuration

This first example, depicted in Figure 6-15, involves a simple direct-connect[6-8] CTI system configuration. Two CTI hardware components are connected with a serial cable. One is a personal computer running a CTI application and the other is a telephone station.

There are two service boundaries in this configuration. The first is a protocol boundary that lies between the telephone station and the personal computer. The second boundary is a programmatic interface within the personal computer that supports the CTI application.

Figure 6-16 shows the same configuration in terms of abstract CTI components. The telephony resources in this example are all ultimately accessed through a CTI interface associated with the telephone station. The telephone station is a CTI hardware component that uses telephone resources in the switch. However, the switch is not

6-7. Configurations versus call control models — The terms direct-connect and client-server (which describe configurations) should never be confused with the concepts of first-party and third-party call control. (The terms first-party and third-party are explained in sections 6.5.2 and 6.5.3.)

6-8. Direct-connect configuration — A direct-connect configuration involves direct interconnection of telephone station equipment with computer equipment. Communication paths generally take the form of serial cables/buses or add-in cards. See Chapter 11 for details on CTI system configurations.

Figure 6-15. Service boundaries in a direct-connect configuration

Switch   Telephone   Client
              Station   Computer

Figure 6-16. Direct-connect example CTI components

considered a distinct component of the CTI system because no CTI interface exists between the telephone station and the phone and thus no boundary is present.

The telephone station exchanges CTI messages with the personal computer using a protocol. Within the personal computer, it is actually a software component known as a *CTI client implementation*,[6-9] that is interacting with the telephone station. It interprets the protocol from the telephone station and, in turn, directs CTI messages to the application (and vice versa) using the programmatic interface.

From the telephone station's perspective, the CTI client implementation is in the computing domain; from the application's perspective, however, the CTI client implementation is part of the switching domain.

Client-Server Configuration

This second example (Figure 6-17), involves a client-server[6-10] CTI system configuration. This configuration involves one more component than in the direct-connect example so there is one additional boundary

---

6-9. CTI client implementation — CTI client implementations are CTI software components in the CTI software framework that is presented in Chapter 12.

in the system. A personal computer running a CTI application is connected to a CTI server hardware component over a local area network. The CTI server is, in turn, connected to a switch.

There are three service boundaries in this configuration. The first is a protocol boundary between the switch and the CTI server. The second is another protocol boundary between the CTI server and the personal computer. The third is a programmatic boundary used by the application software.

Figure 6-17. Service boundaries in a client-server configuration

Switch   CTI   Client
          Server   Computer

Figure 6-18. Client-server example CTI components

Figure 6-18 shows this example in the form of abstract CTI components. Each service boundary represents a different pair of switching domain-computing domain combinations. The one being

6-10. Client-server configuration — A client-server configuration involves indirect control of telephony resources using a fan-out component in the form of a CTI server. The CTI server sits between client computers and the telephony resources being accessed. Despite the fact that there is no direct connection between the client computer and the telephone station being manipulated in this configuration, logical integration takes place through the indirect path of CTI messages. See Chapter 11 for details on CTI system configurations.

discussed at any given instant depends upon the context. If the service boundary in question is the programmatic interface between the application and the client implementation components, the switching domain implementation includes the switch, the CTI server, and the client implementation. On the other hand, if the service boundary in question is the cable between the CTI server and the switch, the switching domain consists only of the switch.

6.5. Switching Domain Abstraction

The emphasis in this chapter is primarily on the CTI functionality that the switching domain offers to the computing domain. (Chapter 12 will look more closely at the implementation and functionality of the computing domain.)

From the perspective of the computing domain, the switching domain is the CTI system that is being observed or controlled. The computing domain is completely unaware of the actual implementation of the telephony resources in question, or of the physical topology and components making up the CTI system. The reality of the implementation and the abstraction of the switching domain may be closely related or very different—but this is completely irrelevant as far as the computing domain is concerned. The only telephony resources, features, and services of which the computing domain is aware are the ones that can be observed in the switching domain.

Throughout the remainder of this book, the term *switching domain* will be used to refer to the abstraction, or version, of the CTI system that can be observed through CTI technology, and the term *CTI system* will be used in reference to the configuration and tangible components of the CTI system.

6.5.1. Switching Domain Scope

Switching domain *scope* refers to the set of telephony resources in a particular switching domain—devices and calls in particular—that can be observed or controlled. External networks represent other sets of telephony resources outside a given switching domain. The switching domain in question may or may not have direct access to a particular external network and its resources. This is illustrated in Figure 6-19.

Figure 6-19. Switching domain scope

Most resources inside a switching domain are visible, that is, they are directly observable and are within the scope of the switching domain. Certain telephony resources, however, may be invisible. This means that their presence may be known and indirectly used, but they cannot be directly referenced or observed. Observation of connections is assured only within the scope of the switching domain (i.e., for visible devices inside the switching domain). The status of other connections cannot be relied upon as they may or may not be reported, depending on the implementation of the switching domain and the nature of the external network.

Switching domain scope is a very important concept because it indicates what can and cannot be accomplished using a given CTI interface. For example, if a call is placed to a device outside the switching domain, the computing domain cannot obtain any information about the status (or even existence) of the called device, and it cannot be assured of connection state information for the called device.

6.5.2. First-Party Call Control

First-party call control is a call control model in which only a single device or device configuration can be observed and controlled. If a particular CTI interface supports first-party call control, the scope of the associated switching domain contains only a single device or device configuration. This is illustrated in Figure 6-20.

Figure 6-20. First-party call control

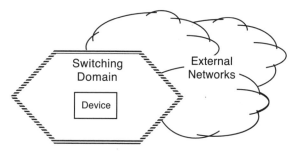

Because the switching domain has only one visible device configuration, all calls are either external incoming or external outgoing.

Though first-party call control may appear quite limiting with respect to the many functions actually being performed within a telephone system, this level of functionality is all that is required for the majority of CTI applications.

The CTI system configuration has no bearing on what telephony resources are in a given switching domain or vice-versa. The CTI interface presenting the single-device (i.e., first-party) switching domain may be an individual telephone, a CTI server, a switch, or any other CTI component in a given system configuration. Figure 6-21 shows a CTI system configuration involving a CTI server that is providing the computing domain with visibility of a single device only. (Refer to Chapter 11 for more information on CTI system configurations.)

Figure 6-21. First-party call control in a CTI system

6.5.3. Third-Party Call Control

Third-party call control is a call control model in which multiple devices, or device configurations, can be observed and controlled simultaneously. If a particular CTI interface supports third-party call control, the switching domain it presents is one comprising two or more visible devices or device configurations. This is illustrated in Figure 6-22.

Figure 6-22. Third-party call control

As noted earlier, the CTI system configuration has no bearing on what telephony resources are in a given switching domain. Figure 6-23 shows a CTI system configuration involving a CTI interface exposed by an individual telephone station. Despite this, the switching domain presented contains multiple devices and is thus an example of third-party call control. (Refer to Chapter 11 for more information on CTI system configurations.)

Figure 6-23. Third-party call control in a CTI system

Comparing Figures 6-20 and 6-21 to Figures 6-22 and 6-23, it is clear that the only difference between first-party and third-party call control is the scope of the switching domains presented by the CTI interface implementations. Third-party call control is effectively the general case, and first-party call control is a special case in which only one device is visible.

6.5.4. External Network

Any device or other telephony resource not within the switching domain is, by definition, part of an external network. Network interface devices are special devices that exist in both the switching domain and one or more accessible external networks. Network interface devices may or may not be visible within a switching domain. Because network interface devices exist simultaneously inside and outside the switching domain, they are used as proxies. (Refer to Chapter 4, section 4.2 for more information on network interface devices as proxies.)

If a given network interface device is invisible, so that it cannot be observed directly, information about the state of connections to the network interface device may or may not be provided.

6.6. CTI Service Requests and Events

There are four kinds of messages[6-11] that pass through a CTI interface:

- Events

- Service requests

- Positive acknowledgments

- Negative acknowledgments

Every CTI message is defined in terms of its kind, which message among those of its kind it is, and its parameters. For example, the CTI message corresponding to a request for the *set lamp mode* service can be shown graphically as in Figure 6-24.

Figure 6-24. Set Lamp Mode service request message

Each parameter in the list of parameters appropriate for a given message may be optional, mandatory, or conditionally mandatory.[6-12] Most parameters are either identifiers that reference a particular resource (or resource attribute) in the switching domain, or they are variables representing a state, status, or setting value.

6-11. CTI messages — CTI messages are implemented within CTI protocols as self-contained protocol data units (PDUs). On the other hand, procedural and object-based programmatic interfaces translate messages to sequences of functions, parameters, function return codes, call back routines, and data structures. In the remainder of this chapter, CTI messages are presented in the context of the standard CTI Plug & Play protocols. (See Chapter 11 for more information about standard and proprietary CTI protocols, and Chapter 12 for a discussion of standard CTI programmatic interfaces.)

6-12. Parameter optionality — The following annotations are typically used to indicate the requirements for a particular parameter: "M" means mandatory, "O" means optional, and "M/O" means conditionally mandatory.

6.6.1. CTI events

CTI event messages, or just *events* for short, are messages sent from the switching domain to the computing domain to indicate transitions of states and changes in the status or setting of an attribute in the switching domain. Events are the primary mechanism used by the computing domain to observe activity within the switching domain.

For example, if the connection state of a particular connection transitions from *alerting* to *connected*, the CTI event message *established* would be sent to the computing domain to indicate that the connection in question had transitioned to the *connected* state. The established event (with a partial parameter list) is shown graphically in Figure 6-25.

Figure 6-25. Established event message

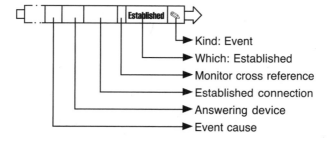

Kind: Event
Which: Established
Monitor cross reference
Established connection
Answering device
Event cause

In order to receive event messages that are relevant to a particular call or device in the switching domain, the computing domain must request them by *starting a monitor* on the item in question. (Monitoring is described in section 6.9.3 of this chapter.) The monitor cross-reference identifier parameter in the event identifies what previously established monitor caused this event to be sent.

The definition of event messages include:

- The meaning of the event and, in the case of events that reflect state changes, its context in terms of implications for other related connections;

- Mandatory, optional, and conditionally mandatory parameters; and

- The possible causes to which the event can be attributed. (The *cause code* of a given event is an essential parameter in the event message, and in many instances it represents a very important clarification of the meaning of an event.)

Event messages are defined for every type of item that has a state, status, or setting that can change. Events also are defined to indicate that new information has been received, such as an update to the correlator information associated with a call or the detection of a DTMF tone. (Refer to Table 6-1 for more examples of events.)

Table 6-1. Event message examples

Event messages	Resource	Event Indicates
Call Cleared	Call	Call no longer exists
Call Information	Call	Updated call associated information
Bridged	Connection	Transition to queued state (during shared bridging)
Connection Cleared	Connection	Transition to null state
Delivered	Connection	Transition to alerting state
Digits Dialed	Connection	Transition to initiated state (digits were dialed)
Established	Connection	Transition to connected state
Failed	Connection	Transition to fail state
Held	Connection	Transition to hold state
Offered	Connection	Transition to alerting state (offered mode)
Originated	Connection	Transition to connected state (after originating a call)
Queued	Connection	Transition to queued state
Retrieved	Connection	Transition to connected state (after retrieve)
Service Initiated	Connection	Transition to initiated state
DTMF Digits Detected	Connection	DTMF digits detected
Telephony Tones Detected	Connection	Telephony tones detected
Button Press	Button Component	Button was pressed

Table 6-1. Event message examples (Continued)

Event messages	Resource	Event Indicates
Display Updated	Display Component	Updated display contents
Hookswitch	Hookswitch Component	Change in hookswitch status
Lamp Mode	Lamp Component	Change in lamp mode
Microphone Mute	Microphone Component	Microphone mute attribute updated
Ringer Status	Ringer Component	Change in ringer attribute
Speaker Volume	Speaker Component	Speaker volume attribute updated
Agent Logged On	Agent	Transition to agent logged on status
Agent Not Ready	Agent	Transition to agent not ready status
Do Not Disturb	Logical Element	Do not disturb setting changed
Forwarding	Logical Element	Forwarding settings changed
Out of Service	Device Configuration	Device is out of service

The example in Figure 6-26 illustrates how event sequences communicate what is taking place in the switching domain. This example revisits the scenario involving parking a call, which was presented in Figure 5-23 of Chapter 5, section 5.7.6. The scenario involved an attendant, D2, parking a call to park device D3 and then having it picked up by device D4. Figure 6-26 shows what's happening to the devices and connections in the switching domain on the left, and the corresponding event sequence generated by a monitor started for device D1 on the right. It is important to note in this example that the event sequence shown is just the sequence for D1. If the other devices and/or the call itself were being monitored, each corresponding monitor would generate a similar sequence of events. In this case, *diverted* event messages indicate that the connection indicated in the event has been cleared, and that the call is about to be associated with the new destination device shown. The *queued* event indicates that the connection D3C1 is in the *queued* state and has transitioned to this state because it was parked. Finally, the *established* event indicates that the connection D4C1 is in the *connected* state because it was picked.

Figure 6-26. Park and pick scenario example event flow

6.6.2. Service Requests

Service requests are messages sent by either domain to request some service of the other. The vast majority of services are *switching domain service requests*. These correspond to services that the switching domain can carry out. *Computing domain service requests* are messages sent to the computing domain to request that it perform some function. *Bidirectional service requests* can be sent in either direction.

There are three categories of service requests:

1. Service requests associated with the telephony features and services explored in Chapter 5. These include:

   - Call control services

   - Call associated services

   - Logical device services

2. Service requests that manipulate or check the status of physical element components. These services include:

   - Pushing buttons

- Getting and setting button information

- Getting lamp information

- Getting and setting lamp mode

- Getting and setting display contents

- Getting and setting message waiting indicator

- Getting auditory apparatus information

- Getting and setting hookswitch status

- Getting and setting microphone gain and mute

- Getting and setting speaker volume and mute

- Getting and setting ringer status

3. Service requests that are specific to the CTI interface. These services will be discussed at greater length through the rest of this chapter. These include:

- Capabilities exchange

- System status services

- Monitoring

- Snapshot services

- Routing

- Media access

- Vendor specific extensions

Service Request Messages

The definition of service request messages include:

- The service that is invoked or the feature that is set;

- Required initial states and possible final states for any connections on which the service acts;

- Mandatory, optional, and conditionally mandatory parameters of the service request;

- Possible outcomes of the service;

- The sequence of events that should be expected if the service completes successfully;

- The completion criteria used to determine if a service completed successfully; and

- The possible reasons for an unsuccessful service request.

When a service request message is issued by one domain, the other domain responds[6-13] with a positive or negative acknowledgment. Independent of these acknowledgments, if the service results in any action that affects the state of one or more connections, or the status or setting of some resource or attribute, the switching domain will generate all the appropriate event messages.

The event sequence defined as part of the service completion criteria for each service is referred to as a *normalized event flow,* or just *flow* for short. It is very important because it allows the computing domain to verify that a service has taken place. As noted earlier, a given computing domain is not the only source of commands manipulating items within the switching domain. If another interface, such as a telephone set, is used to issue a command, the computing domain sees the results of this command as an event sequence. The event sequence observed allows it to determine what just took place. For example, the event sequence for the completion of a *consultation call* service (with single-step dialing) is shown in Figure 6-27. (Refer to Chapter 5, section 5.11.2 for a description of the *consultation call* service.) In this case the first step, placing the connection to the initial call in the *hold* state, is indicated with the *held* event. The fact that a new call is created for the consultation call is indicated through the *service initiated* event.[6-14] Finally, the *originated* event indicates that the second call has

---

6-13. Service request responses — There are a small set of service requests for which no response messages (positive or negative acknowledgments) are required (or defined).

been originated and connection D1C2 is in the *connected* state. Note that while the *consultation call* service specifies that the final state of connection D1C2 must be one of those shown, the completion criteria for the service are satisfied when the *originated* event is provided.

Figure 6-27. Consultation Call event sequence example

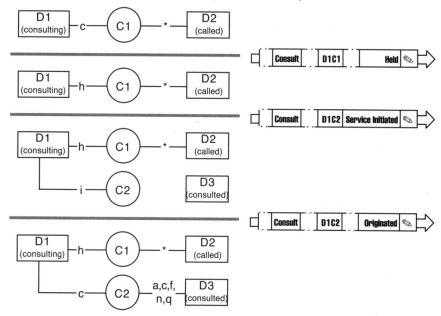

6.6.3. Negative Acknowledgments

If a service request is unsuccessful for some reason, the domain attempting to carry out the service indicates the lack of success by sending a negative acknowledgment message. This negative acknowledgment message contains an error value that provides an explanation as to what the problem was. An example of a negatively acknowledged service request is illustrated in Figure 6-28.

6-14. Service initiated — The service initiated event in this flow is optional. If no prompting of the device is required, the newly created connection could go directly to the connected state because single-step dialing was being used (i.e., the complete called device address was provided as a parameter to the consultation call service).

Figure 6-28. Negative acknowledgment sequence example

6.6.4. Positive Acknowledgments

A positive acknowledgment indicates that a service request is being or has been acted upon. There are two basic types of positive acknowledgments, depending on the nature of the service request:

- One type of service request is a direct request for the CTI interface to return information about something in the switching domain. In this case, the positive acknowledgment not only indicates that the request was completed successfully, it also includes the requested information.

- If the service request involved requesting that some manipulation of resources take place, the positive acknowledgment indicates that the CTI interface has passed the request to call processing to be carried out. If the service in question is carried out in an atomic fashion, the positive acknowledgment also indicates notification that the service was completed successfully.

An example of a positively acknowledged service request is illustrated in Figure 6-29. The service request in this example is a request for information.

The definition of a positive acknowledgment message for a given service may include mandatory, optional, and conditionally mandatory parameters.

Figure 6-29. Positive acknowledgment sequence example

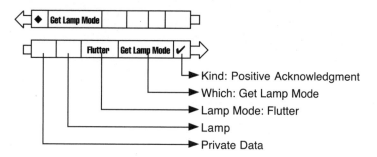

6.6.5. Atomic and Multi-Step services

The implementation of services in the switching domain may be atomic or multi-step. The nature of the implementation determines the correct interpretation of positive and negative acknowledgment messages.

If a particular service is implemented in an atomic fashion, the switching domain treats requests for it as follows:

1. Validation

    Are all of the parameters valid? Are all applicable connections in the correct states? Are all necessary resources available? If the service request is invalid for any reason, the switching domain sends back a negative acknowledgment message to indicate that the request has failed.

2. Execution

    The switching domain then attempts to carry out the service requested.

3. Acknowledgment of success or failure

    If the service request succeeded, the switching domain sends a positive acknowledgment message to indicate it has succeeded. Otherwise it sends a negative acknowledgment, indicating that the service did not succeed and why it did not.

If an atomic service does not succeed, no resources in the switching domain are affected in any way and therefore no events will be generated. If the service is successful and it affects states, statuses, settings, etc., then appropriate events are generated. For example, if the *consultation call* service is implemented as an atomic service, the complete flow might be as shown in Figure 6-30.

Figure 6-30. Atomic implementation of Consultation Call service

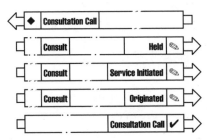

If a service is implemented in a multi-step fashion, the request is processed somewhat differently:

1. Validation

As in the atomic implementation, all of the parameters are first validated, initial states are checked, and the availability of all required resources is verified. In the multi-step implementation, the decision to attempt the service or to reject the service request is made at this point. If the validation does not succeed, the switching domain returns a negative acknowledgment. In a multi-step implementation, this is the only point at which a negative acknowledgment will be generated. If everything appears to be in order, the switching domain returns a positive acknowledgment. However, this positive acknowledgment indicates only that the request has been passed to call processing to be executed and that there is every expectation of success; it does not indicate that any execution has begun or that it has satisfied the completion criteria.

2. Execution

> The switching domain begins to carry out the multi-step service after the positive acknowledgment message is sent. As the execution of the service proceeds and connection states are affected, statuses are changed, etc., the switching domain generates appropriate events. The computing domain uses these events to determine when the service has completed by comparing the events received to those stipulated in the definition of the service request. Assuming that the service does succeed, there will be no other indication of success.

3. Reporting incomplete service execution

> If the service fails to succeed for any reason, the switching domain sends a special event message known as a *service completion failure* event to indicate that the execution of the service could not be completed.

This event is not related to the *fail* state in any way. It merely indicates that one or more of the completion criteria associated with a particular service could not be satisfied.

The multi-step implementation of *consultation call* is illustrated for both the case of success and the case of no success in Figures 6-31 and 6-32 respectively.

Figure 6-31. Multi-step implementation of Consultation Call service (succeeds)

The switching domain uses capability exchange services (discussed in section 6.8.2) in order to determine which services a given switching domain implements as atomic and which it implements as multi-step.

Figure 6-32. Multi-step implementation of Consultation Call service (does not succeed)

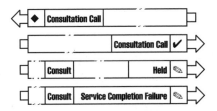

6.7. Identifiers: Referencing Switching Domain Objects

Identifiers are message parameters used to identify a particular resource or entity in the switching domain. They allow the abstraction represented by the switching domain to be conveyed across the service boundary via messages.

6.7.1. Device Identifiers

Device identifiers are parameters that refer to devices and device configurations. Devices are referred to by address, so device identifier parameters contain device addresses in a *device identifier format* corresponding to one of the address formats described in Chapter 4, section 4.6.

Device Roles

A particular service request message or event message may contain many references to different devices. Each of these devices has a different role with respect to the service desired or the event being reported.

These *device roles* include:

- Called device identifier

  The *called device identifier* is the destination address specified for a particular call. For an external incoming call, this parameter will contain DNIS information or the full DID number dialed.

- Calling device identifier

  The *calling device identifier* is the device that originally placed a particular call. For an external incoming call, this parameter will contain callerID or ANI information.

- Associated called device identifier

  For an external outgoing call, the *associated called device identifier* parameter refers to the network interface device being used. For an external incoming call, the associated called device identifier parameter refers to the device in the switching domain associated with the number originally called (e.g., the attendant, DID extension, DISA extension, etc.). If the call is not an external call, this parameter is not included in any messages.

- Associated calling device identifier

  For an external incoming call, the *associated calling device identifier* parameter refers to the network interface device being used. If the call is not an external call, this parameter is not included in any messages.

- Redirection device identifier

  The *redirection device identifier* parameter refers to the last device from which this call was previously routed. In addition to being Not Known (which means that the last redirection device is not visible), the redirection device identifier may be specified as Not Required to indicate that the call has not been redirected, or Not Specified to indicate that the switching domain doesn't know if the call has been redirected.

- Subject device identifier

    The *subject device identifier* parameter indicates the subject, or focus, of a particular event message.

Depending on the context, a particular device that must be referenced in an event message might not be visible within the switching domain. Parameters referencing these devices have the value *Not Known*.

Addressable and Non–addressable Appearances

Specific appearances within a logical device element may be referenced explicitly if the appearances are addressable. (Addressability of appearances is explained in Chapter 4, section 4.4.2.) A device identifier containing explicit appearance references is formed using appearance suffixes. (See the sidebar "Switching Domain Representation Format" on page 195 in Chapter 4, section 4.6.5.)

Physical, Logical, Appearance, and Device Configuration References

The interpretation of a device identifier parameter depends upon the value provided and the service request or event in which it appears.

In the case that a device identifier contains an appearance suffix (see Chapter 4, section 4.6.5), the context in which it is being used determines its interpretation. If the context is one where an appearance reference is appropriate, it is used for this purpose; otherwise the parameter is treated as a reference to the appearance's logical device element.

If the device identifier is used in a context in which only a physical element reference is applicable, it is interpreted as being a reference to the physical element portion of the device specified.

Likewise, if the device identifier is used in a context in which only a logical element is applicable, the device identifier is interpreted as referring to the logical element portion of the device specified.

Otherwise, the device reference is interpreted as being the device configuration which has the device referenced as its base.

6.7.2. Physical Element Component Identifiers

References to the components of physical elements are made by specifying a particular physical device element using an appropriate device identifier and providing the identifier for the desired component. The component's identifier may be one of the following:

- Hookswitch identifier

- Auditory apparatus identifier

- Button identifier

- Lamp identifier

- Ringer identifier

A reference to an auditory apparatus identifier is used whenever referring to a particular auditory apparatus, or when referring to either the speaker or microphone associated with that auditory apparatus.

6.7.3. Call and Connection Identifiers

Calls are referenced using parameters called *call identifiers*. They are numbers that uniquely identify individual calls in the switching domain.

Connection identifiers uniquely reference a particular connection in the switching domain. Because a connection represents the association of a call and a device, a connection identifier is formed simply by combining a device identifier and a call identifier. If the device is one that is not visible to the switching domain, or the connection involves a non-addressable appearance, the switching domain will form the connection identifier using a private or *dynamic device identifier*.[6-15]

6-15. Connection Identifiers — The device identifiers contained within connection identifiers are not to be extracted by the computing domain because they may or may not be valid outside the context of the connection identifier.

321

Call and Connection Roles

Services that involve multiple calls involve both service request and event messages that include multiple connection identifiers. In these cases (typically transfer- and conference-related services), the different calls referenced have different roles with respect to the service. These roles include:

- Primary call
  The *primary call* is the first of two calls being operated upon. It is the call placed on hold by a *consultation call* service request when setting up for a transfer or conference service.

- Secondary call
  The *secondary call* is the second of two calls being operated on. It is the new call created by a *consultation call* service request when setting up for a transfer or conference service.

- Primary old call
  The *primary old call* is a reference to the primary call that was merged into a new call as a result of a switching service involving the merging of two calls.

- Secondary old call
  The *secondary old call* is a reference to the secondary call that was merged into a new call as a result of a switching service involving the merging of two calls.

- Resulting call
  The *resulting call* is a reference to the new call that is the result of a switching service involving the merging of two calls.

6.8. CTI Interoperability

Building a CTI system involves assembling many independent CTI components by attaching one to another through an appropriate service boundary. As CTI technology becomes increasingly ubiquitous, CTI components (hardware and software) will become increasingly interoperable through the use of standard interfaces: CTI protocols (for hardware and software components), and programmatic interfaces (for software components).

A measure of increasing CTI component interoperability is the level of human intervention involved in getting them operational. Market forces will see to it that the use of standard CTI protocols assures system integrators, customers, and individuals of getting full benefit from the potential that CTI Plug & Play has to offer. Human involvement in connecting two CTI components should be limited to ensuring that the appropriate physical communication path is in place and instructing one component where to find the other. The rest of the information needed for the two components to interoperate should be determined dynamically by the two components through negotiation and exchange of capability information.

6.8.1. Protocol and Version Negotiation

Before two components in a CTI system can begin exchanging messages across their service boundary, they first must determine what protocol and/or CTI interface version to use. In addition, if the switching domain implementation supports vendor specific extensions, these have their own *private data version* that is independent of the version of the CTI interface used.

When two CTI Plug & Play hardware components first establish a communication path, they exchange *protocol negotiation packets*. The computing domain first sends the switching domain a protocol negotiation packet that indicates the range of protocols and versions supported, and an indication as to whether or not private data negotiation is desired. The switching domain then chooses the protocol and version it wishes to use (or indicates that there is no version that is supported in common, so the two components can gracefully disconnect). If private data negotiation was requested, the switching domain provides the information necessary for this as part of its response, and the computing domain responds with a service request indicating the private data version it chose. The negotiation process completes when the switching domain indicates its system status (see section 6.9.1) and the computing domain begins learning about the switching domain using the capabilities exchange services described in the next section.

Similar mechanisms are implemented in the programmatic CTI interfaces (APIs) used between CTI software components. In addition to being used by CTI client implementations and applications, these mechanisms are used by API-specific adapter software associated with CTI components that don't support standard CTI protocols directly or through mappers.

6.8.2. Capabilities Exchange

Once two components have agreed on a CTI version to use, the computing domain must learn about the switching domain before it can begin to observe or manipulate resources within it. The first thing the computing domain must do is find out what the general capabilities of the switching domain are. This includes finding out what capabilities exchange services are supported.

Switching domain implementations that support CTI Plug & Play implement the *get switching domain capabilities* service. This service is a request for information; it returns the desired capabilities information in the positive acknowledgment to the service request. The positive acknowledgment reports such things as:

- The name of the switching domain implementation. This name identifies the vendor and model of the switching domain being used. This name is used for determining what, if any, vendor specific extensions might be applicable.

- A default or suggested device within the switching domain that the switching domain believes is associated with the computing domain. If provided, this saves the computing domain from having to ask a human user what device it should monitor and manipulate by default.

- Which device identifier formats are supported.

- Whether the switching domain supports external incoming or external outgoing calls.

- If and how the forwarding feature is supported.

- If *dynamic feature availability* (see next section) is supported.

- The time indicated by the switching domain's internal clock.

- Maximum sizes of certain variable-sized parameters.

- What services and events are supported and of these what optional parameters are supported.

Once this information has been digested, the computing domain then typically tries to find out what devices are visible in the switching domain. The *get switching domain devices* service allows it to find all the visible devices in the switching domain and determine the type and device identifier for each.

Once the computing domain has determined what device(s) it is interested in observing and/or controlling, it uses the *get logical device information* and *get physical device information* services to learn about each device of interest.

The *get physical device information* service provides such information as:

- Product name of the physical device

- Size of the display (if present)

- Number of buttons (if any)

- Number of lamps (if any)

- Number of ring patterns supported

- Physical element-related services and events supported by a given physical element, and the optional parameters supported for each

- Logical elements associated through the device configuration for which the given physical element is the base

The *get logical device information* service provides information that includes the:

- Maximum number of callback requests supported for the device

- Maximum auto-answer value supported

- Maximum number of connections supported

- Maximum number of held calls supported

- Maximum number of forwarding rules supported

- Maximum number of devices that can be conferenced together

- Media services supported by the logical device or available in conjunction with the logical device

- Supported ways to prepare to transfer or conference a call

- Logical element related services and events supported by a given logical element, the optional parameters supported for each, and the required initial states associated with call control services

- Physical elements associated through the device configuration for which the given logical element is the base

- Type, behavior, and addressability of the logical element's appearances

A key step in the computing domain's preparation sequence for working with the switching domain involves figuring out what services are supported and, for services involving call control, when these services may be requested for a given device. The information provided by the *get logical device information* service, relative to the services supported and associated initial states, is used to build up this picture.

6.8.3. Dynamic Feature Availability

Dynamic feature availability is a feature that may or may not be supported by a given switching domain. If it is supported (as indicated through capabilities exchange), every event that involves call control includes a dynamic feature availability parameter that indicates what services are applicable to a given connection, given its current state.

6.9. Status Reporting

Status reporting services are those services used to support the observational portion of the CTI interface's role. If a given computing domain does nothing but observe a switching domain, it uses this subset of services.

6.9.1. System Status

System status services are bidirectional services that are used by each domain to inform the other domain of their current operational status. The *system status* service request contains a parameter that indicates the status of the domain issuing the request. The values that this parameter may take are:

- *Initializing*

- *Enabled*

- *Normal*

- *Message lost*

- *Disabled*

- *Overload imminent*

- *Overload reached*

- *Overload relieved*

If the value is anything other than *normal*, the domain receiving the service request must take appropriate action. System status service requests indicating normal status may be sent periodically by either domain as a *heartbeat* to indicate that it is "alive and well." The positive acknowledgment returned in response by the other domain confirms that it is also "alive."

6.9.2. Snapshot

Snapshot device and *snapshot call* services, as the names imply, allow a snapshot of all information associated with a particular device or call to be obtained.

The snapshot services typically are used when a computing domain is just initializing its view of the switching domain. The *snapshot device* service returns a positive acknowledgment message listing all of the connections associated with a particular device and the state of each. If dynamic feature availability is supported, then the services that can be applied to each connection is reported also.

The *snapshot call* service returns all of the basic call associated information (called device identifier, calling device identifier, associated called device identifier, associated calling device identifier, correlator data) and information regarding each device associated with the call. The per-device information includes the corresponding device identifier, the state of the corresponding connection, and any media services associated with a given connection. If dynamic feature availability is supported, then the services that can be applied to each connection are reported also.

6.9.3. Monitoring

Monitoring refers to the act of requesting that CTI event messages be generated for any changes in state, status, or settings relevant to a particular call or device.

The family of monitoring services includes:

- Monitor Start

  The *monitor start* service requests that monitoring be started for a particular call or device. The parameters for this service include a *monitor filter* that specifies which events, if any, are not desired and should not be sent. This is also referred to as an *event mask* or *message mask* by some.

- Change Monitor Filter

  If the computing domain needs to change the filter (or mask) originally specified, the *change monitor filter* service can be used instead of stopping a monitor and starting a new one.

- Monitor Stop

  The *monitor stop* service is a bidirectional service. If issued by the computing domain, it is a request for the switching domain to clear an existing monitor. If issued by the switching domain, it is a notification that the monitor has been cleared and a request for the computing domain to take appropriate action.

When the *monitor start* service succeeds in starting a monitor, the positive acknowledgment that it returns includes a monitor cross-reference identifier parameter that uniquely identifies the monitor that was established. All events generated as a result of this monitor are identified with a monitor cross-reference identifier parameter of the same value.

Monitor Objects

Monitor object refers to the item in the switching domain to which a given monitor applies. Monitors can be started on either devices or calls. *Device-object monitors* will exist as long as a given device exists within the switching domain, or until they are stopped by either domain. *Call-object monitors* are stopped when a given call ceases to exist, or if they are stopped earlier.

When a particular call is being monitored (either through call-object monitoring or because it is associated with a device that is being device-object monitored), all events relevant to all connections involving visible devices within the switching domain are reported. Information on connections involving devices not visible within, or outside of, the switching domain may be only partially available or go completely unreported.

Monitor Type

Monitor type determines how calls are monitored when they leave a device. Once a *call-type monitor* has begun reporting events for a particular call for any reason, it continues to report events for that call until the call ceases to exist. In contrast, *device-type monitors* only deliver events relevant to a particular call for as long as they remain associated with a particular device. As soon as it leaves the specified device, monitoring of the call ceases.

The following is the behavior resulting from the four combinations of monitor object and monitor type:

- Device-object, device-type

    This monitor is started on a particular device and reports all events relevant to this device and to calls associated with this device. It continues reporting events relevant to the device for as long as the device exists, but stops reporting events relevant to a given call when that call leaves the device.

- Device-object, call-type

    This monitor is started on a particular device and reports all events relevant to this device and to calls associated with this device. It continues reporting events relevant to the device for as long as the device exists and continues reporting events relevant to a given call for as long as that call exists—even after it leaves the device.

- Call-object, device-type

  This monitor is started on a particular call and continues to report all events relevant to the given call until the call leaves a specified device.

- Call-object, call-type

  This monitor is started on a particular call and continues to report all events relevant to the given call for as long as that call exists.

6.9.4. Device Maintenance

Device maintenance events are a family of events that the switching domain can send to the computing domain to indicate that the functionality associated with a particular device is changing for some reason. These events typically are used to indicate that a device is being manipulated through the OA&M interface or has been physically removed or replaced.

These events include:

- Out of Service

  The *out of service* event indicates that a device has been taken out of service and that events for it may or may not be generated. Service requests cannot be issued for it.

- Device Capabilities Changed

  The *device capabilities changed* event indicates that a device's capabilities have changed and that capabilities exchange service requests should be reissued for the device to determine its new set of capabilities.

- Back In Service

  The *back in service* event indicates that a device is back in service.

6.9.5. Normalized Behavior

Once a monitor has been started for a particular device or call, the switching domain begins sending every applicable event relevant to that entity. A given device or call may be acted upon by service requests from the computing domain or other components in a CTI system, by manual interaction with a telephone set, or by the switching domain itself. The sequence of events that result when a service is invoked for any reason is referred to as the event *flow* for that service. If a standard CTI protocol is being used, the event flow is *normalized*. This means that the behavior seen by any computing domain in the form of an event flow is the same for a given service,[6-16] regardless of how the service was invoked.

If the behavior of a given switching domain is not normalized, the computing domain will be able to track the state, status, or setting of any given telephony resource but will not necessarily be able to interpret any activity that is observed.

6.10. Routing Services

Routing services are among the most powerful of the services provided by a CTI interface. These services permit the computing domain to override the default call routing as determined by the switching domain's call processing resources.

If a computing domain wishes to use routing services, it first uses the capabilities exchange services to determine what devices in a particular switching domain, if any, support routing services. The computing domain then registers[6-17] to use routing services on the device(s) of interest.

6-16. Normalization — Normalization does not apply to vendor specific extensions, which are by definition not standardized in any fashion.

6-17. Routing registration — Certain switching domain implementations do not require registration. They simply assume that the switching domain will want to use routing services. Certain other switching domain implementations allow registering for all the devices in the switching domain at once.

Whenever a call arrives at a device for which routing services are active, a *routing dialog* is initiated by the switching domain. This dialog involves the following steps:

- The switching domain sends the computing domain a *route request* service request, asking it to specify a new destination for a particular call. The *route request* may include a time limit parameter that specifies how much time the computing domain has to make a decision on the routing of the call.

- The computing domain may respond by issuing the *route reject* service to reject the call, the *route end* service to end the dialog (and let the switching domain use its default routing), or *route select* to indicate where the call should be routed.

- If the *route reject* service was specified by the computing domain, the switching domain responds with a *route end* to indicate that the routing dialog has been concluded.

- If the *route select* service was specified by the computing domain and the switching domain found the new destination acceptable, it may optionally issue a *route used* request followed by a *route end* to indicate that the routing dialog is complete. The switching domain might instead respond with a *re-route* service request to indicate that the call could not be routed to the requested destination and the computing domain should repeat the attempt.

Examples of three complete routing dialogs are illustrated in Figure 6-33.

A typical use of routing services is in the distribution of incoming calls to specialists within a team. For example, a travel agency maintains a database of all its clients and each client is assigned to a particular travel agent. When a particular client calls the travel agency, the call is delivered to a hunt group device that ordinarily would just route the call to the next available travel agent. By using routing services, however, the agency's computer is able to override this default

Figure 6-33. Routing dialog examples

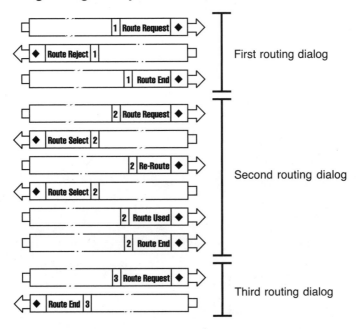

routing. It uses the callerID information associated with the call to identify the customer in its database; the call is then routed to that client's assigned travel agent.

6.11. Media Stream Binding

Computers have long been able to deal with telecommunications data, exchanging faxes, playing and recording sound, and generally working with multimedia data. Excellent technologies, including multimedia architectures and programmatic interfaces, for the manipulation of these data types already exist.

CTI does not supersede, replace, or even encompass these things. Instead, CTI allows existing multimedia technologies to take advantage of a new source of media streams—the telephone network. Switching domain implementations that are capable of providing access to media streams implement support for media services in their

CTI interfaces. In fact, media services are the cornerstone of a very large portion of all CT solutions. (Media services concepts are the subject of Chapter 7.)

The media-related services provided by a CTI interface are therefore directed at media binding, or referencing the media stream associated with a call in the switching domain, and not media manipulation. Media manipulation is done through corresponding media services interfaces.

6.11.1. Media Stream Binding Concepts

A *media stream identifier* allows an association to be established between a given call and the media services available from a particular *media service instance* that is, or can be, associated with the call through a logical device element that supports media binding. (Refer to Chapter 3, section 3.7.4 and Chapter 7, section 7.1.5 for more on media service instances.)

Media stream binding is accomplished through the *attach media service* service request, which is responsible for binding an appropriate media service instance to the call (if available) and returning a media stream identifier. The *detach media service* service request clears the association (and reverses any actions taken by the corresponding *attach media service* service request).

The desired media service instance may or may not be associated with one of the devices already participating in the specified call.[6-18] For example, if the media service instance exists within a modem peripheral attached to a line corresponding to a particular logical device already associated with a given call, then it already has access to the call's media stream. The same is true if the media service instance exists within the switch itself and can be associated with the call transparently. On the other hand, if the media service instance

---

6-18. Multiple media service instances — Depending on the implementation, if there are multiple media service instances available, the computing domain may be able to choose which one is preferred.

exists in a media server,[6-19] an appropriate logical device associated with the server must be added to the call. (See Chapter 11 for more on the actual configuration of CTI systems that utilize media servers.)

If gaining access to the media service instance requires call control activity, the call may be associated with the logical device providing the media services either by joining the device to the call (conferencing), or by diverting the call to the device (transferring) according to the computing domain's request and depending on the switching domain implementation. These operations, if required, are performed by the switching domain and may be implemented using the *transfer call*, *single step transfer call*, *conference call*, or *single step conference call* service. If any call control activities take place as a result of the *attach media service* service request or the *detach media service* service request, appropriate events are provided to indicate what is taking place.

6.11.2. Media Stream Binding Model

The mechanisms for utilizing a given media service using a media stream identifier are determined by the definition of the media service itself and generally are industry or de facto standards. The flow of messages between the computing domain and the switching domain implementations relative to a particular media service are referred to as *media service sessions*.

Figure 6-34 illustrates the basic model for media stream access. When the CTI client implementation requests the use of a particular media service, it does so using the CTI protocol flowing through the CTI stream. The logical CTI server binds the media service requested to the call specified and returns the appropriate media stream identifier (through the CTI stream). The media stream identifier then can be used by software running on the logical CTI client to access the media service stream appropriately. This includes determining how the

---

6-19. Media servers — Examples of media servers include stand-alone voice mail systems, voice processing servers, voice response units (VRUs), interactive voice response units (IVRs), fax-back servers, and modem pools.

communication path for the media service stream is to be established. In most cases the communication path used for the CTI stream is related to the one used by the media service stream in some way (e.g., the same LAN, the same multiplexed RS-232 link, etc.).

Figure 6-34. Media stream binding model

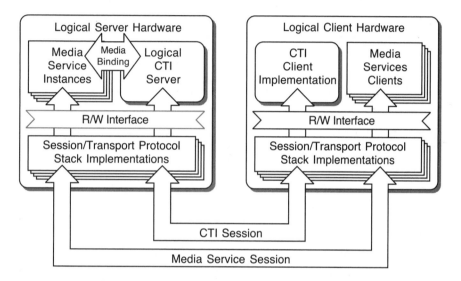

6.11.3. Tone Detection

Tone detection not only is a special form of media access that is intrinsic to the switching domain, but it is also frequently used to detect in-band signaling conventionally associated with the use of media access on voice calls.

The *start telephony tones collection* service requests that the switching domain monitor a given call, or the next call to arrive at a specified device, for telephony tones. If one or more tones are detected, the switching domain generates a *telephony tones detected* event to indicate that the tones were detected. The *stop telephony tones collection* service indicates that an outstanding tone collection request should be canceled.

If a fax calling tone (fax CNG) or modem calling tone (modem CNG) is detected, it indicates that a fax machine or modem on the call would like to begin exchanging modulated data. The computing domain can then use the *media attach service* to request that the appropriate type of modem media service instance be attached to the call. An example showing how this technique is used to receive a fax is shown in Figure 6-35.

Figure 6-35. Receiving a fax

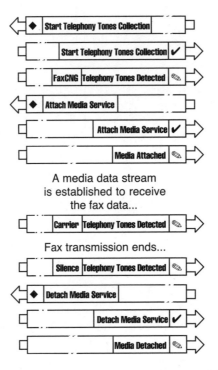

The *start DTMF tones collection* and *stop DTMF tones collection* service requests and the corresponding *DTMF tones detected* event behave in a similar fashion.

6.11.4. Tone Generation

The switching domain is capable of generating telephony and DTMF tones on behalf of the computing domain using the *generate telephony tones* and *generate digits* services respectively.[6-20]

Using these services, a fax transmission can be initiated in much the same way as the fax reception shown in the preceding example.

6.12. Vendor Specific Extensions

Support for vendor specific extensions, often referred to as the *escape mechanism*, is implemented using the *escape* service request and the *private* event. The contents and use of this service and event pair are defined uniquely by every switching domain implementation. These two messages are essentially envelopes into which vendor specific information can be placed for transmission across service boundaries.

To ensure that a computing domain does not use vendor specific extensions inadvertently, the computing domain first must register for them using the *escape register* service (which later can be canceled using the *escape cancel* or *escape abort* services).

Every service and event also carries a *private data* parameter that can be used by a switching domain to extend the functionality of a given service or event in a proprietary way, assuming that this use is negotiated by the computing domain.

The version of a switching domain's vendor specific extensions and private data parameters is independent of the version of the CTI interface being used, so a separate private data negotiation mechanism is used. This involves using the *private data version* service to indicate which of the private data versions supported by the switching domain is to be used.

6-20. Generating tones — It should be noted that whenever tones are generated, the corresponding tone detected events are issued by the switching domain even if tone collection is not active. Tone collection applies to tones from other devices on the call.

6.13. Review

In this chapter we have explored the concepts that apply to building and using a CTI interface.

Every component in a CTI system connects to its neighbor(s) through an *inter-component boundary*. As far as each component is concerned, this boundary is a *service boundary* through which it exchanges *CTI messages*. The side of the service boundary that has access to telephony resources is known as the *switching domain* and the side that is seeking to observe and control those resources is known as the *computing domain*. In the relationship between the two components, the one acting as the computing domain is a *logical client* of the one representing the switching domain, which acts as a *logical server*. *Fan-out components* allow the CTI functionality available across one service boundary to be made available to multiple logical clients.

Communication across a service boundary takes place through CTI messages, which may be conveyed through a *CTI protocol* or, if the two components are software running on the same computer, through a *programmatic interface*.

CTI messages contain parameters. *Identifiers* are parameters that make reference to specific telephony resources within the switching domain. Other parameters reflect the actual or desired states, statuses, and settings of resources. *CTI event messages* are sent by the switching domain to provide update information on individual telephony resources. *Service request messages, positive acknowledgment messages,* and *negative acknowledgment messages* are used by each domain to request services of the other, and to respond to those requests. An *atomic service* is one that is carried out in one step, and in which success or lack of success is indicated through the type of acknowledgment received. A *multi-step service* is acknowledged positively or negatively when the request itself is validated. In this case, success is indicated only through events that indicate the service completion criteria for the requested service were satisfied. Lack of success is indicated through a special service failure event.

Before any messages can be exchanged between the switching domain and the computing domain, the two first must *negotiate* a CTI protocol (if standard CTI protocols are being used) and the version of the protocol or programmatic interface that is to be used. The computing domain then must use *capabilities exchange services* to determine the telephony resources accessible within a given switching domain, and the capabilities (services and events supported, implementation options, support for media data, etc.) of all devices of interest. The *scope* of a switching domain represents the set of visible telephony resources, that is, the set of telephony resources that may be observed and controlled. Certain resources may not be visible, but their effects within a switching domain can be perceived (e.g., an invisible network interface device or pickup group device). All devices outside the switching domain are part of one or more *external networks*.

In addition to the services for controlling telephony features and services (described in Chapter 5) and services which manipulate all of the components of a physical device element, a switching domain implementation may support *status reporting services, routing services, media binding services,* and *vendor specific extensions.*

The computing domain can observe telephony resources by using the switching domain's *status reporting services*. To the extent that they are available from a given switching domain implementation, these services allow the computing domain to keep track of the operational status of the switching domain (*system status services*), to take a snapshot of a given call or device's status (*snapshot services*), and to request that any change in state, status, or setting associated with a particular device or call be reported through events (*monitoring*). *Device maintenance events* are used to indicate if the capabilities of a monitored device change in some way.

Once monitoring of a particular device has begun, the computing domain receives a *flow* of events corresponding to all of the services that affect that device. If the switching domain implementation supports a standard CTI protocol, the flow of these events is *normalized*. This means that regardless of how a service is invoked, the same event sequence will be observed.

Routing services allow a computing domain to override the default routing of a call from devices that support routing services. *Media binding services* allow the media stream of a call to be accessed through a particular *media service instance. Private data*, the *private event*, and the *escape* service request message (often referred to collectively as the *escape* mechanism) allow support for vendor specific extensions that go beyond the functionality provided for in standard CTI protocols and general-purpose programmatic CTI interfaces.

As noted in previous chapters, the concepts presented here do not necessarily reflect the implementation of a given CTI component. They represent a *CTI abstraction* that can be applied to any CTI component in order to integrate that component into a larger CTI system.

7.
Media Services
Concepts

This chapter explains media services technology. In contrast to the CTI services and associated call processing technology covered in the last few chapters, media services deals with technology for interacting with telephony media streams rather than directing them.

Media processing involves generating, consuming, or manipulating media streams associated with a telephone call. *Media services* involves making the media processing capabilities of a given telephone system accessible by allowing observation and control of media resources and their interaction with media streams. While the area of media services does not have the same broad scope and complexity associated with call control, and it can easily support very basic computer telephony solutions, the efficient sharing and management of media resources does involve some complexity. Unlike call processing, media processing relies on specialized hardware and software media resources that must be made to physically connect with telephone system media stream channels. One result of these differences has been that media processing and media services have evolved independently from call processing and CTI.

Media Services Concepts

This chapter will describe the roles played by the various telephony resources involved in media processing, present all of the key concepts related to media servers and media clients, and explain the media services abstraction that has emerged in the computer telephony industry.

7.1. Media Services Abstraction

As we saw in Chapter 6, section 6.1 abstractions are a means by which we can describe any implementation with the same terminology and framework. Just as in CTI, industry efforts to define standards for interoperable media services implementations began with the development of a universal abstraction of media services that could be applied to systems of all sizes and with a diversity of capabilities. While specific implementations may or may not include every aspect of the overall abstraction or conform directly to it, they can all be described, compared, and contrasted with the same concepts and terminology.

In the area of media services, the ECTF has been responsible for defining a media services abstraction that applies to the wide variety of standardized and proprietary APIs and protocols used by media services implementations. This specification is known as ECTF S.100[7-1].

7-1. ECTF S.100 — The ECTF S.100 specification actually defines an abstraction for media services as well as a 'C' language API that corresponds to this abstraction. This chapter is concerned only with the abstraction. The API will be discussed in Chapter 12, section 12.7.1.

7.1.1. Media Processing Model

Media processing is the technology for generating, consuming, or manipulating[7-2] media streams using media resources. Media processing involves connecting media resources with a media stream channel and directing their activity.

Media processing requires creating a direct association between a given call's media streams, a group of *media resources*, and a particular *media services client*. Media resources interact with media streams and *stored media data* based on an exchange of *media services messages* with media services clients. The media services client is effectively a party to the call and is able to extract media data from the media stream and specify what media data is to be generated. This media processing "pipeline" is illustrated in Figure 7-1.

Figure 7-1. Media processing pipeline

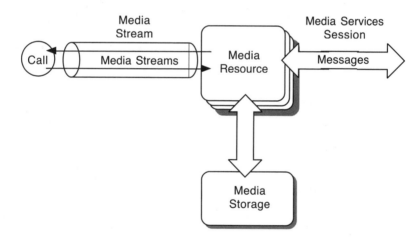

7-2. Generating and consuming — Most media resources either consume or generate media streams but some, such as mixers, filters, echo cancellers, and transcoders, manipulate the media stream by consuming it and generating a modified version simultaneously.

The collection of telephony resources directly involved in media processing include:

- Media resources associated with media access devices, or embedded in the telephone system's switching fabric and other devices;

- Switching fabric resources for conveying media streams to the media resources; and

- Media services interfaces for making media resources accessible to clients.

In addition, call processing resources may be required to associate calls with devices where the media resources are located.

7.1.2. Media Resources

Media resources, also known as *media processing resources* or *media access resources*, are responsible for actually providing the media processing services requested by clients. They are at the core of the media processing model. As shown in Figure 7-1, media resources interact with:
- One or more media streams,
- Local media storage, and
- Clients through the exchange of media services messages.

Media Stream Consumers

Media resources that consume media streams may perform one or more of the following:

- Filtering
    Media resources may filter a media stream to attenuate sound or noise in certain frequency ranges, to remove selected tones or artifacts of in-band signaling[7-3], to compensate for the

---

7-3. In-band signaling — In-band signaling refers to mixing signaling information into a media stream. This is discussed in Chapter 8, section 8.1.4

effects of passing through certain types of switching fabrics, to perform echo cancellation, or to pre-process the media stream for some other reason.

- Pattern matching

  Media resources may search the media stream for certain patterns. For example, a tone detector searches the media stream for specific tones and combinations of tones. A modem resource identifies the patterns representing specific symbols by demodulating the incoming media stream. A speech recognizer may search the media stream for phonemes in order to identify words.

- Conversion

  Media resources may convert a media stream that is being consumed into another, based on the first. For example, a media resource that performs compression would convert an uncompressed media stream into one which is compressed. Media resources can also convert a media stream into media data that can be stored or used by a media services client. For example, after searching a media stream for phonemes a speech recognizer may convert the media stream into the corresponding text.

- Storage

  Media resources may place processed media data into local media storage.

- Client delivery

  Media resources may initiate media services messages to a media services client containing media data extracted from the media stream.

A media resource that consumes a data stream for purposes of storage and/or client delivery typically does not also regenerate the media stream and is known as a *media sink*.

Media Stream Generators

Media resources that generate media streams may do so based on information supplied by a client, information stored locally, one or more media streams that are simultaneously being consumed, or some combination of these. Resources of this type may perform one or more of the following functions in combination:

- Retrieval

    Media resources may retrieve media data from a client or from local media storage to be used as the basis for generating a new media stream. For example, a *text-to-speech* (*TTS*) resource may retrieve stored text.

- Modulation

    A media resource with modem functionality will modulate a a media stream (typically a carrier wave) to encode data.

- Mixing

    A media resource may mix a number of source media streams to generate one or more new media streams. For example, a resource used to perform conferencing would generate a media stream for each party in the conference that combines all of the other media streams.

- Conversion

    Media resources may generate a new media stream by converting retrieved media data into a media stream. For example a tone generator resource would convert media data representing tone frequencies and duration into tones in the media stream. A TTS resource would convert specified text into corresponding spoken language based on instructions about the desired speech attributes.

    Media resources that transform incoming media streams may perform a conversion function such as transcoding from one compression format to another.

A media resource that generates media streams is known as a *media source*.

7.1.3. Media Services Interface

A *media service interface* is a telephony resource that creates a mechanism through which media services clients can direct media resources to interact with media streams. As with CTI interfaces (as described in Chapter 6) media service interfaces operate by generating, sending, receiving, and interpreting messages containing status information and requests for services to be performed. This is illustrated in Figure 7-2.

Figure 7-2. Media services messages

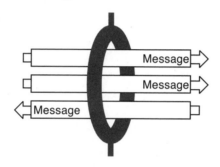

Like CTI messages, media services messages consist of:

- Service requests or command messages,

- Acknowledgments, or responses, that may indicate positive or negative outcomes to commands, and

- Event messages that provide notification of state changes and other occurrences of interest to media services clients.

The structure, content, and rules governing the flow of messages back and forth through a media services interface is defined by a media services protocol or API.

7.1.4. Media Access Devices

Media access devices, also referred to as *media attachment devices* or occasionally as *media terminal devices,* are devices (endpoints on the telephone system's switching fabric) that are added to a call in order to provide access to media resources through a media services interface. They are typically dedicated to providing access to media services.

The process of attaching a group of media resources to a call's media streams is known as *media binding.* (See Chapter 6, section 6.11 for a description of media binding.) The role of media access devices in a given media binding operation depends on the location of the media resources that are needed:

- Within telephone system switching fabric and other devices
  Accessible media resources can be embedded in the telephone system switching fabric and in devices that are active in a particular call. In this case, no additional media access devices need to be added to the call when a media services client requests that these media resources begin interacting with the call.

- Within a media access device
  Media resources may be clustered in a media access device which can be connected to a call when the corresponding media resources are needed.

- Shared by multiple media access devices
  Media resources may be located in a media server which is connected to the telephone system's switching fabric using multiple media access devices.

These three arrangements are illustrated in Figure 7-3. It should be noted that all three logical configurations for media binding are irrelevant to most media services clients as they don't affect the operation of the media services themselves. For media services clients that do track whether or not a media access device is used in the media binding operation, the second and third cases are effectively the same.

Figure 7-3. Media access devices and media binding

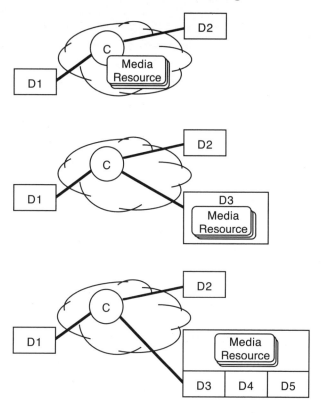

7.1.5. Media Service Instances

The results of a successful media binding operation are references to a particular *media service instance* and the call-associated resources within it. A media service instance encompasses an implementation of a media services interface, the available media resources, any required media access device, and the switching fabric and media services technology required to manage and interconnect the media resources and media streams. This is illustrated in Figure 7-4.

Figure 7-4. Media Service Instance

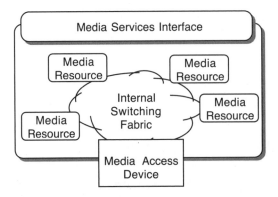

7.2. Media Services Clients

Media services clients, or *media clients* for short, are CT system components that use media resources to interact with telephone call media streams. Media clients include:

- Call processing resources,
- CTI interface implementations, and
- Media services applications.[7-4]

Media services clients form logical client-server relationships just as CTI components do (see Chapter 6, section 6.3). However, the nature of media clients is very different from CTI logical clients.

CTI clients are concerned with monitoring and controlling selected devices, calls, connections, agents, and call processing features in a given switching domain. In contrast, media services clients are concerned with using selected media resources in a given media service instance. CTI clients are typically concerned with call progress and all the connection state transitions associated with the connections of a call. On the other hand, media clients are only active when they have media streams to work with, so they are typically concerned with only active connections in the *connected* state. Where CTI clients

7-4. Media service applications — ECTF S.100 refers to media clients as applications despite the fact that it supports other types of media clients.

operate in a context that assumes many different clients may be interacting with the same resources simultaneously, media services involves media clients interacting with individual media resources in an exclusive arrangement. It should also be noted that just as CTI interfaces typically support frequently required media functionality such as tone detection and generation, media services interfaces typically support basic call control services. This comparison of CTI and media clients is summarized in Table 7-1.

Table 7-1. Comparing CTI and media clients

	CTI	Media
Interface	CTI Interface	Media Services Interface
Service boundary context	Switching Domain	Media Service Instance
Objects of interest	Devices, Calls, Connections, Agents, Features	Media Resources
Connection states of interest	All	Connected
Relationship to objects	Many to One	One to One
Cross-over	May require basic media services. e.g., tone detection, and tone generation	May require basic call control services. e.g. make call, transfer call, consultation call, clear connection

7.2.1. Client-Server Operation

Basic media services solutions often involve physically collocating both the media service instance and the media services client. This is often referred to as a *local media services application* or a *server hosted media client* and is common practice due to the properties of media client-server operation. However, in the general case, media services clients may be physically located anywhere as long as they can establish a communication session with the media service interface for the appropriate media service instance.

Supporting distributed media services clients add some additional requirements to a media services implementation, but it provides a much more compelling solution. Media clients that are location independent support CT solutions with greater flexibility, scalability, and modularity. Utilization of available media resources can be maximized by allowing media clients to share media resources.

One consequence of distributed media clients is that the abstraction of media service instances, media resources, and their services makes no assumptions about the physical location of components. Specifically:

- Media clients are not concerned with the physical location of, or media stream interconnection between, media resources.

- Media resources have access to local media storage in the media service instance.

- Media clients can delegate tasks to the media service instance to avoid potential delays resulting from slow client-server communication.

- Media clients can themselves be modularized and distributed so that media client modules can hand off media resources and associated call media streams among one another.

- Media clients are able to independently interact with multiple media service instances allowing a single media client to simultaneously interact with any number of calls using multiple threads[7-5] of the same software logic.

Support for different types of media clients and the need to integrate both CTI clients and media clients is another key factor in the definition of media services.

7-5. Software thread — In software a thread is a single sequence of steps to be executed by a processor. Using multiple threads involves simultaneously executing the same sequence multiple times with each thread being independent of the others.

7.2.2. Media Services Client Types

There are two basic types of media services clients:

- Media–only

  Media–only clients use only the services provided by a given media services interface. They are only concerned with using media resources to carry out a particular interaction for each call that is presented to them.

- Integrated

  Integrated media clients use both CTI and media services interfaces. They are concerned with both call/device control and with manipulating the media streams of a given call.

Media–only clients represent the majority of all media clients found in CT systems. This reflects the preference for highly modular, task-oriented client components.

For example, a solution that performs incoming call screening could be implemented as an integrated media client that detects undesirable calls based on call associated information, binds them to a media service instance that supports audio playback, uses media services to play a polite message, and clears the call. The same solution could be implemented using a CTI client that detects and deflects undesirable calls to a media access device along with a media–only client that plays the message and drops the call. Both approaches provide a solution with the same functionality, but the second approach offers much greater flexibility.

7.2.3. Media Services Client Operation

Media services clients operate in a call-centric fashion because media processing only takes place when active calls with allocated media stream channels are attached to the appropriate media resources.

Once it has established a session with the appropriate media services interface, a media client can operate on multiple calls using multi-threading or some other equivalent approach. Handling each call involves performing the following:

1. Bind media service instance

   Integrated media clients that want to operate on a particular call use CTI media binding services (described in Chapter 6, section 6.11) to perform this step.

   Media–only clients simply wait for a notification that a call has been presented for them to handle. If supported, a media–only client may initiate a call from the media service instance.

2. Allocate media resources

   Each media client must confirm the allocation of a group of media resources of various classes which collectively provide the media services that it requires. Before the media client can begin interacting with a call's media streams it must ensure that the media resources required are allocated and that they are appropriately interconnected.

3. Operate

   With the media streams and media resources in place the media client operates on the media streams by issuing commands and monitoring for events through the media services interface.

The media client may free resources in use, allocate new resources, and reconfigure the group of media resources in other ways, if needed.

4. Free resources and unbind

Once the media client has completed its particular task it may return control of the media resources it was using to the media service instance or it may hand them off to another media client. A media–only client could also invoke a media services function to disconnect the call or transfer it away. An integrated media client should then use the appropriate CTI service to detach the media service instance from the call. Freeing, or deallocating, the media resources makes them available to other clients or to other threads of the same client.

To summarize, using media services involves:

- Binding and unbinding calls to media service instances in order to attach and detach the corresponding media streams,

- Allocating and deallocating media resources which can interact with the call's media streams, and

- Operating on the media stream using these media resources by exchanging messages with the media services interface.

7.3. Media Resource Allocation

While simple media service instances may be dedicated to individual media service clients, in the general case they must be shared among a diversity of media services clients distributed throughout a given CT system.

7.3.1. Dedicated and Sharable Media Resources

Media resources are implemented as either hardware, software, or a combination of hardware and software. Examples include:

- Specialized analog or digital[7-6] electronic circuits;

- Signal processing software running on a DSP[7-7] coupled with appropriate hardware for funneling a media stream to and from the DSP; and

- Signal processing software running on a conventional microprocessor with access to a media stream.

The number of media resources of a certain class that are present in a given media services implementation is finite. Where resources are implemented as individual electronic circuits, the total number of resources is determined by the number of functional circuits that can be used independently. Where resources are implemented in software, the number of available resources is determined by the capacity of a given DSP or microprocessor to run a certain class of media resource software.

While a single media resource is all that is required to constitute a media service instance, most clients require a number of different media resources that can work together. For example, to implement a simple auto-attendant[7-8] one would need media player resources[7-9] for generating announcements that explain the caller's options and

7-6. Analog and Digital Media — The concepts of analog and digital media stream channels are explained in Chapter 8.

7-7. Digital Signal Processor — A digital signal processor, or DSP, is a specialized digital microprocessor designed specifically for tasks such as media processing that benefit from powerful mathematical functions and data fetching optimizations for stream-oriented processing.

7-8. Auto-attendant — An auto-attendant is a media services client that interacts with callers and routes their calls to the device they request. This is described in Chapter 5, section 5.6.3.

7-9. Player resource — A player resource is a media resource that retrieves media data from local storage and converts it to a media stream. See section 7.4.3.

tone detectors that determine their selection. If the auto-attendant were to support the option of faxing a directory to the caller it would also need a fax resource.

If the auto-attendant in this example were expected to handle up to 20 calls simultaneously, it might be appropriate to dedicate 20 groups of player and detector resources. However if only one caller in every thousand requests a fax, it would not be appropriate to dedicate 20 fax resources to this media client. As it is impossible to predict which of the 20 simultaneous calls will require the use of a fax resource, one or two fax resources would be required for this solution but they would need to be pooled, or shared. Furthermore, if the volume of calls being delivered to the auto-attendant only rarely exceeds 4 simultaneous calls, then dedicating 20 groups of resources would be an inefficient use of these resources if other media clients could be using them.

Media resource implementations, whether they be hardware or software, benefit from significant economies of scale and are much more cost effective when they are implemented in quantity. In addition, many media services solutions involve media clients that utilize different classes of resources but do not require that all needed media resources be dedicated. For both these reasons, media resources are typically pooled for shared use by multiple media services clients as shown in Figure 7-5.

Figure 7-5. Media resource sharing

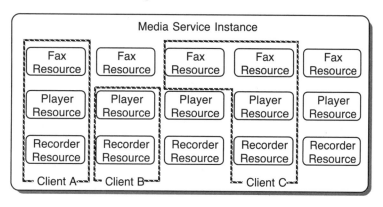

The extent to which available resources can be combined or used simultaneously is also a function of the internal media stream channels for delivering call media streams to, from, and between media resources.

7.3.2. Monolithic and Modular Media Servers

Media services are implemented as either monolithic, fixed functionality media servers or as modular media servers that can be customized to suit the requirements of a particular CT solution.

Monolithic Media Server

Monolithic media server implementations consist of an arbitrary group of media resources that are hardwired together to provide a specific media service solution. An example of a monolithic media services implementation is a stand-alone modem peripheral. The modem has certain built-in media resources that include a tone detector, a tone generator, a data modem, and a fax modem (which uses the data modem). It also has a built-in switching fabric that interconnects these resources with the hardware that connects to the telephone system's switching fabric and a media services interface that allows a computer to control it.

Figure 7-6. Monolithic media server

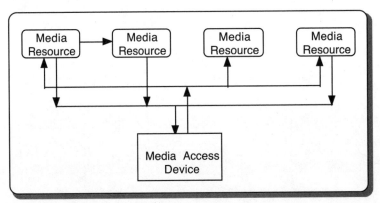

Modular Media Server

Modular media servers are built around a common switching fabric and media service software that allows a variety of different media resources to be integrated with one another. The core media service software implements the functionality necessary to support a media services interface and to manage a pool of media resources. This includes session management, local media storage, call control services, administration functions, switching fabric control, and resource management.

The ECTF framework provides specifications that standardize the interfaces between the components of a modular media server so that they can be assembled using components from a diversity of vendors and can be scaled or upgraded as required.

Figure 7-7. Modular media server

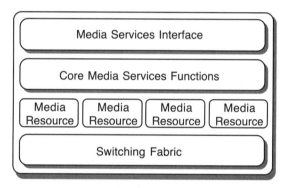

Although monolithic and modular media servers vary in the flexibility they bring to the implementation of a complete solution, both allow media clients to work with media resources in interconnected groups and both allow those groups to be shared in some fashion. When an individual media client establishes a session with the media services interface for a particular media server, the interface provides a view of available resources. This view is the media service instance.

7.3.3. Media Group Concept

A *media group*[7-10] is a subset of a media service instance's switching and media resources that are dedicated to a given media client. Groups are the basic building block for media services. They are the basis for media clients to request, allocate, and deallocate media resources and to establish media streams between them.

A media group consists of:

- A primary resource

  At the core of every media group is a *primary media group resource* which corresponds to the switching resources for establishing media stream channels between resources (intra-group), between groups (inter-group), and for the media streams associated with a call.

- Secondary resources

  Media groups have zero or more *secondary media group resources* which correspond to the media resources to be used by the media client.

Media groups are generally represented as a triangle as shown in Figure 7-8.

The media group abstraction simplifies the implementation of a media client because it models all switching through the primary resource and an associated arbitration scheme (see section 7.3.6). This means that all manipulation of media stream channels is performed implicitly by the media service instance rather than by the media client. It also delegates responsibility to the media service for managing, locating, and allocating pooled media resources that can be interconnected.

7-10. ECTF Media Groups — The ECTF S.100 specification defines the abstraction of media groups. While older media services interfaces do not use this abstraction, it provides a basis for representing media services implementations of all types.

Figure 7-8. Media group abstraction

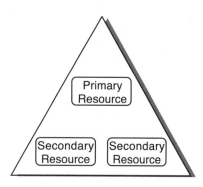

While most media clients will require only a single group per call, this abstraction still allows media clients to perform their own high-level resource management by creating and interconnecting multiple groups. Media groups can be constructed to act as compound media resources and can be interconnected to form media processing pipelines or vectors. For example, a media client could create its own "single step fax back resource" by building a group containing a player resource, a signal detector resource, and a fax sender resource. (These resource classes are described in sections 7.4.3, 7.4.1, and 7.4.7 respectively.) The resulting group would then be a self-contained compound resource for asking the caller to ready their fax machine and press a button (using the player resource), wait for a desired tone (using the signal detector resource), and send a fax (using the fax sender resource).

7.3.4. Primary Resources

Primary media group resources, or *primary resources*, are the anchor resources for media groups. Each media group has a single primary resource that represents the switching resources allocated to the group.

Primary resources are responsible for establishing the group's primary media stream channel through which it receives the group's input media stream and through which it sends the group's output media stream.

Primary resources are used to establish media stream channels for both intra-group and inter-group switching:

- Intra-group switching

   Intra-group switching involves establishing unidirectional or bidirectional media streams between the primary resource and the secondary resources within a group.

Figure 7-9. Intra-group switching

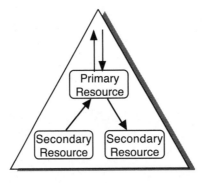

- Inter-group switching

   Inter-group switching involves establishing media stream channels between groups. A given group may connect it's primary media stream channel to a single other group and may establish media stream channels to one or more other groups as if they were secondary resources. (See section 7.3.7.)

Figure 7-10. Inter-group switching

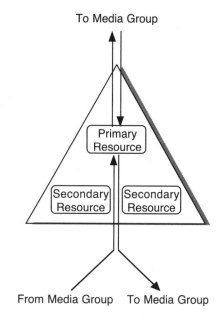

There are three types of primary resources:

- Switch port resources

    A *switch port resource*, or *SPR*, is the simplest type of primary resource. This resource accepts a media stream from the primary resource of another group and directs it to all the secondary resources in the group that consume media streams. It also accepts a media stream from one of the secondary resources that generates media streams (based on a media stream arbitration mechanism, see section 7.3.6) and conveys this output media stream to the primary resource of the other group.

- Call channel resources

  A *call channel resource,* or *CCR,* is like a switch port resource but it is only used to bind a group to calls. It receives the group's input media stream from a call and conveys the group's output media stream back to a call. Call channel resources are associated with the media access device in a given media service instance.

- Conference port resources

  A *conference port resource,* or *CPR,* implements multi-point switching by combining the media streams received from the group's secondary resources and delivering the resulting media stream to all of the secondary resources that are configured to consume media streams. A conference port resource has no primary media stream channel as it works exclusively with the media stream channels associated with secondary resources. A conference port resource may have a fixed number of unidirectional (from and towards the secondary resources) and bidirectional media stream channels available.

Groups built around call channel resources are the most common because most groups are created to interact with calls. Media clients will create groups based on switch port resources and conference port resources in order to carry out media processing using multiple groups.

7.3.5. Secondary Resources

The *secondary media group resources,* or *secondary resources,* are the media resources that are interconnected through media stream channels associated with the group's primary resource. The secondary resources are also related to one another through the group's *runtime control (RTC)* mechanism. (This is described in section 7.5.1.)

All secondary resources that consume media streams receive the same media stream from the primary resource simultaneously.

Secondary resources that generate media streams deliver their media streams to the primary resource. They are governed by an arbitration mechanism within the group that determines from which secondary resource the primary resource will receive at any given time.

Figure 7-11. Secondary resource

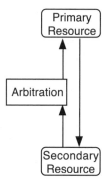

7.3.6. Media Stream Arbitration

Each group has a media stream arbitration mechanism which determines which secondary resource is permitted to transmit a media stream to the primary resource. Rules for arbitration include:

- Preemption

    If a group's arbitration mechanism works on the basis of *preemption*, a resource that is instructed to begin generating a media stream preempts any resource currently generating a media stream.

- Last-talker

    The *last-talker* rule is a variation on preemption in which a list of preempted resources is kept. When the preempting resource stops generating its media stream, the last preempted resource may continue.

- Blocking

    The *blocking* arbitration rule means that no resource can generate a media stream if another resource is already doing so.

- First-talker

    The *first-talker* rule is a variation on blocking in which a queue of resources that have requested the right to generate a media stream is maintained. When the blocking resource stops generating its media stream, the next resource in the queue may begin.

- Summing

    Conference port resources can accept media streams from multiple secondary resources simultaneously. This is permitted by the *summing* arbitration rule.

The arbitration mechanism a given media service instance supports is implementation dependent, however the media client may request a particular arbitration rule if more than one is supported.

7.3.7. Inter-group media streams

In addition to the intra-group switching already discussed, media stream channels can be directed between groups to create hybrid media processing configurations. Under the media client's direction, the media stream channels between groups can be set as bidirectional or unidirectional in either direction.

There are three ways of interconnecting two groups:
- Attach
- Bridge
- Loopback

The permissible interconnections are determined by the type of primary resource within each of the groups.

Attached

An *attached* interconnection involves attaching one group (referred to as the *secondary group*) to another (the *primary group*) as if it were a secondary resource of the primary group. Any group can be the primary, but the secondary group must contain a switch port resource.

Figure 7-12. Attached type connection

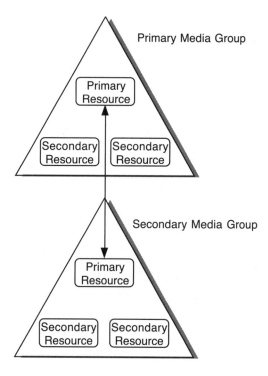

Bridged

A *bridged* interconnection involves establishing media stream channels between two groups such that each appears as a secondary resource of the other. Any two groups[7-11] can be connected in this manner.

Figure 7-13. Bridged type connection

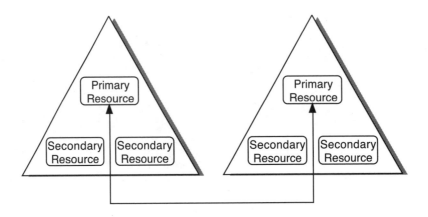

Loopback

A *loopback* interconnection involves interconnecting the primary media streams of two groups. Each group becomes the primary input for the other. Both groups must contain a switch port resource.

Multiple groups can be interconnected through any combination of bridged, attached, and loopback connections as long as there are no circular media stream paths.

7.3.8. Group Configuration

Media clients create groups by specifying a desired *group configuration*[7-12]. This consists primarily of a list of the resources that are desired in the group to be created. Each resource is described in terms

---

7-11. Bridged conferences — Two groups containing conference port resources should not be bridged together as this could lead to a feedback loop.

Figure 7-14. Loopback type connection

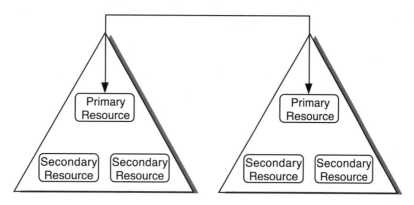

of its desired attributes. For example, a media client could request a group containing an ASR (automatic speech recognition) resource and could specify that it should support German language speakers and include support for echo cancellation.

The group configuration also specifies initial parameter settings for the group and all the individual resources.

The successful creation of the group simply indicates that all the desired resources exist and could be made available for use with the group. However, the resources are not necessarily allocated, or reserved, for the media client.

7.3.9. Static and Dynamic Resource Allocation

Before a media client can use a particular resource within a group that it owns, that resource must be allocated, or reserved and integrated into the group.

7-12. Application profiles — The ECTF S.100 specification defines application profiles that can be administered into a media service. Application profiles include group configuration information for individual media clients.

There are two resource allocation strategies:

- Static

    If a resource is *statically allocated*, it is allocated to a group as soon as the group is created and remains allocated to the group until the group is reconfigured or destroyed.

- Dynamic

    If a resource is *dynamically allocated*, it is allocated to a group (along with the associated media stream channels) only when the media client issues a command that requires the presence of the resource.

The advantage of dynamic resource allocation is that it prevents a media client from monopolizing scarce or valuable resources. It also permits a media client to specify a configuration consisting of all the resources that it might require without actually allocating them until they are needed. The disadvantage is that a required resource may not be available for allocation when it is actually required and the time to allocate a resource may create an unacceptable and unexpected delay in the media processing activity of a given group.

The resource allocation strategy of a given media service instance is implementation dependent. Implementations may:
- Support only static allocation,
- Make certain resources subject to dynamic allocation, or
- Make all resources dynamically allocated.

After creating a new group a media client should check to see if any resources were dynamically allocated. If so, it should take care to explicitly allocate the group's unallocated resources at some stage before they are actually needed.

Because dynamic allocation has the potential for non-deterministic results, it is likely that the majority of implementations will support static allocation and rely on media clients to be efficient in their allocation of resources. Media clients should not rely on dynamic allocation to make their media resource utilization efficient. They should initially only configure groups with necessary resources and then either reconfigure them or attach other groups when additional resources are required.

7.4. Media Resource Abstractions

Individual media resources are grouped into families, or *media resource class*, based on common operational characteristics. All media resources of the same class support a common set of service requests, responses, and events. However, within each class there are many variations in capability that are distinguished by media resource attributes. Different resources of the same class may belong to a particular sub-class or may have more or less functionality than others of a particular class or sub-class. When specifying the configuration for a requested media group it is important to carefully specify the desired attributes of a media resource so that an appropriate resources is added to the group. It is also important to check the attributes of all the resources in a newly configured group to adapt to the specific capabilities of the available resources.

The ECTF S.100 specification has standardized a number of common media resource classes which are described in the following sections. Additional media resources are also discussed in Chapter 3, section 3.7.

7.4.1. Signal Detector Resources

Signal detector resources are a general class of media resources that are used to detect tones, sequences of tones, or other simple patterns in a media stream. A signal detector resource has only two states, *idle* and *detecting*, as illustrated in Figure 7-15. (Refer to the sidebar "States and State Diagrams" on page 103 for an explanation of state diagrams.)

Signal detector resources in the *detecting* state run continuously and provide events to indicate all tones, or signals, detected. They may also be instructed to buffer the tones they detect, filter their buffer for patterns to ignore, and search their buffer for patterns to match.

Figure 7-15. Signal detector resource state model

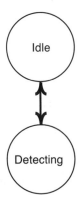

7.4.2. Signal Generator Resources

Signal generator resources are a general class of media resources that are used to generate tones in a media stream. These resources have a single state as they may generate tones at any time and support a single command that causes them to generate a tone.

7.4.3. Player Resources

Player resources generate a media stream by retrieving stored media data[7-13] in some media storage format and converting it into a corresponding media stream. Stored media data is typically pre-recorded audio but it may also consist of encoded ADSI (Analog Display Services Interface) or TDD (Telecommunications Device for the Deaf) data. Players vary according to the media formats that they support. When a media client configures a group it must specify the type, or types, of media storage formats that the player must support.

Player resources have three states as shown in Figure 7-16. In the *idle* and *paused* states the player is not generating a media stream and in the *active* state it is.

7-13. TVM File — ECTF S.100 refers to the media data used by player and recorder resources as time varying media, or TVM.

Figure 7-16. Player resource state model

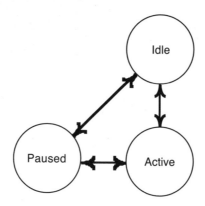

Player resources can, for example, be instructed to:
- Play a particular item,
- Queue another item for playback upon completion of a previous playback,
- Increase or decrease the speed of playback,
- Increase or decrease volume, or
- Pause or resume playback.

7.4.4. Text-to-Speech (TTS) Resources

Text-to-Speech, or *TTS*, resources are actually just a special class of player resources. They are identical to player resources except that they convert text files containing words into corresponding media streams.

Player resources that support TTS functionality have additional attributes indicating, for example, optional speech characteristics (such as gender, age, character, context, prosody, part-of-speech tagging). They also support additional commands for directing the resource to load and activate specific pronunciation dictionaries.

7.4.5. Recorder Resources

Recorder resources consume media streams by converting them to a storable format and saving them in local storage. Operation of recorders is very similar to players as they have the same state model as shown in Figure 7-17. As with players, the media client must indicate the type or types of media storage formats that the recorder is to support.

Figure 7-17. Recorder resource state model

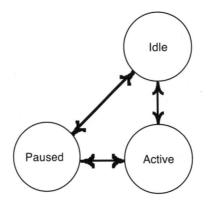

Recorders may be instructed to:
- Record a media stream, or
- Pause and resume recording.

A common requirement of media groups that record is that they must prompt with a beep. As a convenience, recorder implementations may offer the option to generate a beep before recording so that a separate signal generator resource is not required.

7.4.6. Automatic Speech Recognition (ASR) Resources

Automatic speech recognition, or *ASR*, resources are used to detect, or recognize, spoken words in a media stream. ASR resources may also be used to identify a particular speaker, verify the identity of a speaker, and identify the language that is being spoken.

Recognizing human speech is challenging. ASR resources must be initialized with a wide variety of parameters and context information. Some implementations also require or permit that they be operated in a training mode in order to allow or improve recognition.

The state model for an ASR resource in recognition mode is illustrated in Figure 7-18. Once the ASR resource has been appropriately setup and has been instructed to begin recognizing, it enters the *recognizing* state. It may send events to the media client with intermediate results while it remains in the *recognizing* state (this corresponds to the loop that starts and returns from this state) but final results are not available until the resource transitions into the *results available* state.

Figure 7-18. ASR resource state model

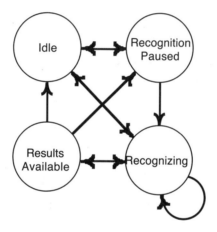

While the ASR resource is recognizing, the media client may instruct it to:

- Abort the recognition by transitioning to the *idle* state,
- Stop the recognition and deliver a final result, or
- Pause and resume recognition.

The resource itself may stop the recognition process based on timeout parameters for the total recognition period, for silence before speech is detected, and for silence after some speech is detected.

7.4.7. Fax Resources

Fax media resources are used to send and receive page images using Group 3 fax protocols. Fax resources include modem resources that are able to establish a data connection over a negotiated modem session. Image data is sent over this data connection one page at a time with handshaking according to the standardized (ITU-T T.30) fax protocols.

Fax resources that send page images retrieve stored media data[7-14] and convert it as specified by the standard fax protocols. Fax resources that receive pages do the reverse. When a media client configures a group it must specify the type, or types, of media storage formats that the fax resource must support.

Low Level Fax Resources

A *low level fax* resource is one which allows the media client to control each phase of the resource's activity. The state model for a low level fax resource is shown in Figure 7-19.

The media client is responsible for issuing commands to move the low level fax resource to each successive state in either a fax receive process or a fax send process. At each step the media client waits for events that indicate a state transition has taken place and then initiates the next step.

High Level Fax Sender Resources

To reduce the effort required on the part of a media client, a high level *fax sender* resource has been defined. The fax sender is a media resource which is layered over a low level fax resource and performs all of the low level state monitoring that media clients using a low level resource would have to perform. The simplified state model is shown in Figure 7-20.

7-14. SM File — ECTF S.100 refers to the media data used by fax resources as spatial media, or SM.

Figure 7-19. Fax low level resource state model

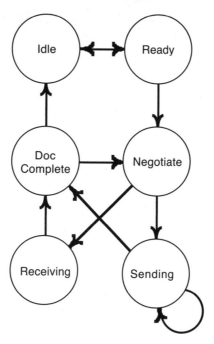

Figure 7-20. Fax sender resource state model

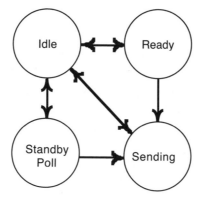

To use the fax sender resource, the media client simply initializes the resource and identifies the media data storage containing the images to be sent.

Figure 7-21. Fax receiver resource state model

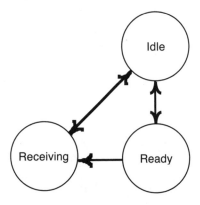

High Level Fax Receiver Resources

Like the fax sender resource, a high level *fax receiver* resource has been defined to simplify the task of receiving a fax. Once again this high level fax resource is layered over a low level fax resource. The media client simply sets up the fax receiver resource and indicates where the received media data is to be stored. The simplified state model of the fax receiver is shown in Figure 7-21.

7.5. Media Service Interfaces

In addition to providing media clients with the ability to form groups of media resources and allowing observation and control of media resources, media service interfaces provide access to other core media services functionality. This includes runtime control, local storage, media call control services, call presentation, and media group hand-off.

7.5.1. Runtime Control

Runtime control, or *RTC*, is a means for media clients to delegate certain timing critical actions to the media service instance. It is an essential feature of client-server implementations because it ensures that media processing activities can be performed in a timely fashion regardless of the connectivity between a given media client and the corresponding media service instance.

Runtime control is based on condition–action sets that are applied to a group as a whole. They allow a change in the condition of one resource in a given group to trigger an action applicable to another resource in the group.

RTC conditions are resource class-specific state, or status transitions, that can be used to trigger some action. For example, a signal detector resource will support RTC conditions which correspond to each tone that it is able to detect. A player resource will support RTC conditions corresponding to the beginning and ending of a play operation.

RTC actions are also specific to each resource class. An RTC action is a request for a resource to modify its behavior in some way. For example, most resources support an RTC action that stops the activity in progress. Resources such as players also support RTC actions such as increase volume, decrease volume, jump ahead, jump back, and pause.

By combining RTC conditions and actions into a list of *RTC sets*, a media client can specify a whole series of actions that the media service instance can carry out autonomously. For example, the media client could specify a different player RTC action (jump to start, jump back, jump forward, jump to end) to a series of different signal detector RTC conditions ("DTMF 1", "DTMF 2", "DTMF 3", "DTMF 4" respectively). In this scenario if a caller pressed "1" the player resource would immediately jump to the beginning of the media data it was playing. This eliminates the possible delay associated with the time for an event to travel from the media service instance to the media client and for the media client to respond with a request for the player resource to jump to the beginning.

Another example of RTC use involves ensuring that no caller input is missed. A media client could specify an RTC condition–action pair that instructs an ASR resource to resume recognizing when the player resource generates the condition that it has completed playing. The media client would then set the ASR resource to its *recognitionpaused* state and finally it would instruct the player resource to play a prompt to the caller. When the player completes the prompt it triggers the recognizer to begin actively recognizing speech input. In this way no speech input is lost due to a delay between the end of the prompt and the start of the recognition process.

There are two kinds of runtime control:

- Persistent RTC

  Persistent RTC involves RTC sets that the media client applies to a media group and that remain active for the life of the group or until explicitly cleared.

- Non-persistent RTC

  Non-persistent RTC involves RTC sets that are associated with a single service request and expire when the service request completes.

7.5.2. Local Data Storage and Manipulation

Media services implementations include support for local data storage because player, recorder, and fax resources retrieve and store media data, ASR resources make use of stored grammars, and TTS resources require access to stored dictionaries. Media clients must be able to locate and specify individual media data objects and other data objects required by resources.

Data objects, or files within the media service instance, are organized into a hierarchy consisting of nested *containers* as shown in Figure 7-22.

Figure 7-22. Containers and objects

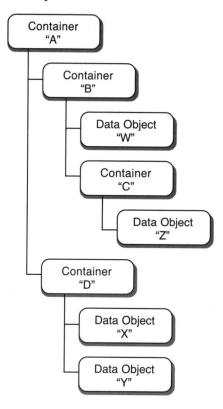

Media services interfaces permit media clients to operate on data objects and containers in much the way that a file system permits applications to operate on files and the directories that contain them. Services include:

- Creating and deleting containers and objects,
- Enumerating the objects and containers within a given container,
- Copying and renaming objects,
- Opening and closing an object, and
- Reading and writing the contents of an object.

Local storage is another essential feature of media services implementations that support media clients that are not collocated with the media service instance.

7.5.3. Media Call Control Services

Media–only clients (see section 7.2.2) are principally concerned with performing media processing using media streams from active calls. Most media–only clients simply interact with calls that are presented to them and delegate all call control functions to the media services implementation.

However, just as CTI clients occasionally require tone detection and generation services without the overhead of binding to media services, media–only clients occasionally need basic call control services without the overhead of using a CTI interface.

 Media services implementations include a CTI client component known as a *system call router*, or *SCR*, that is responsible for call control operations carried out on behalf of media clients.

Figure 7-23. Relationship between CTI and Media Services

The SCR allows the media services interface to provide basic call control services such as:

- Make call,
- Consultation call,
- Transfer call, and
- Clear connection.

The SCR is also the component within the media services implementation that is responsible for associating a particular call, and media access device if applicable, with a particular call channel resource and for supplying call associated information to media clients as parameters associated with the call channel resource.

7.5.4. Call Presentation

Media–only clients that simply wait for calls to be presented to them, perform their media processing function, and then wait for the next call, rely on the system call router to present them with new calls.

Another feature of the system call router is that it can be administered with a set of *application profiles* that describe rules for determining how calls are to be allocated to media clients. These rules are based on call associated information such as callerID/ANI, DNIS, and last redirected device as well as context information such as the day of the week and the time of day. An *application service identifier*, or *ASI*, is used to correlate application profiles with individual media clients. The application profiles also specify group configuration information that is associated with each ASI.

Media-only clients may simply use a service of the media services interface to request that they be presented with the next available call for a given ASI. When a call is presented to a media access device being monitored by the SCR the following occurs:

- The rules in the application profiles are used to determine how to handle the call. (This includes whether or not to answer it.)

- Which media client should handle the call is determined based on the application profiles and the registered ASIs.

- A call channel resource is assigned to the call.

- A media group is created around the call channel resource based on the configuration specification provided by the application profile and the media client.

The completed media group is then presented to the media client that requested it. Once the media client has completed its operations with the call it can dispose of the group by handing it back to the system call router or by handing it off to another media client.

7.5.5. Media Group Handoff

The *group handoff* mechanism used to pass new groups from the system call router to a media clients also permits media client to hand off calls to one another.

Media services solutions can be assembled in a modular fashion using multiple media clients that each handle one aspect of a larger media processing task. For example, one media client could be dedicated to answering incoming calls, reading out a main menu of options, and collecting a caller's choice. Each of the options can be implemented by other secondary media clients. When a new call arrives, the system call router builds a new group and hands it off to the media client that manages the main menu. When the caller selects a menu option, the main menu media client hands off the group to the corresponding secondary client. When the selected media client is finished with the call, it can return the group to the main menu client or directly to the system call router.

Each handoff works the same way as the original handoff from the system call router:
- The intended receiving media client is located;
- The media group configuration required by the new media client is determined;
- The media group is reconfigured (the call channel with the active call remains at the core of the group); and
- The group is handed off to the new media client.

One property of every media group is an ownership stack which keeps track of not only the current owner of a given media group, but all of the previous owners that wish to remain associated with the media group. This allows a media client to easily return a media group to a previous owner and allows previous owners to track the media group.

7.6. Review

Media services clients, or *media clients,* use *media services* to observe and control *media resources* and their interaction with telephone call media streams. Media services clients may be *media–only* or may be *integrated,* using media services and CTI functionality.

A *media service interface* is a telephony resource that creates a mechanism through which media services clients can direct media resources to interact with media streams. Media resources interact with media streams and stored media data based on an exchange of *media services messages* with media services clients.

Media resources, also known as *media processing resources* or *media access resources,* provide the media processing services requested by clients. Media resources that consume media streams may perform *filtering, pattern matching, conversion, storage,* and *delivery* of media data to a client. Examples include *signal detectors, recorders,* and *automatic speech recognition (ASR)* resources. Media resources that generate media streams may perform *retrieval* of stored media data, *modulation, mixing,* and *conversion.* Examples include signal generators, and player resources. Certain media resources, such as *fax* resources, both generate and consume media streams.

Media access devices, also known as *media attachment devices* or *media terminal devices,* may be added to a call in order to bind, or associate it, with a set of media resources located in media server.

Monolithic media server implementations consist of an arbitrary group of media resources that are hardwired together to provide a specific media service solution. *Modular media servers* are built around a common switching fabric and media service software that allows a variety of different media resources to be integrated with one another in *media groups.*

A *media group* is a set of switching and media resources that are dedicated to a given media client. Media groups are the basis for media clients to request, allocate, and deallocate media resources and to establish media streams between them. Media clients create media groups by specifying a desired *group configuration* which includes a list of the media resources required and their attributes. Once a group is configured, its resources may be *statically* or *dynamically allocated*.

Each media group has a single *primary resource* that represents the switching resources allocated to the group. Primary resources are used to establish media stream channels for both *intra-group* and *inter-group switching*.

The *secondary resources*, are the media resources within a media group that are interconnected through media stream channels associated with the group's primary resource. All secondary resources that consume media streams receive the same media stream from the primary resource simultaneously. Secondary resources that generate media streams deliver their media streams to the primary resource. Each group has a *media stream arbitration* mechanism which determines which secondary resource is permitted to transmit a media stream to the primary resource.

Runtime control, or *RTC*, is a means for media clients to delegate certain timing critical actions to the media service. Runtime control is based on condition–action sets that are applied to a group as a whole. They allow a change in the condition of one resource in a given group to trigger an action applicable to another resource in the group.

Media services implementations include a CTI client component known as a *system call router*, or *SCR*, that is responsible for call control operations carried out on behalf of media clients. When a call is presented to a media access device being monitored by the SCR it determines the appropriate media client to handle the call, configures an appropriate group and hands-off the call to the media client. Media clients may hand-off calls in a similar manner.

Media services clients generally operate in a call-centric fashion. After binding a media service instance for a call, if necessary, they allocate media resources, operate on the call's media streams using the resources, and then free the resources and unbind once finished.

8.
Switching Fabric
Implementation

Now that we have explored the concepts of call processing and media processing, we can begin exploring the layer of the telephony resource framework that holds everything else together. This chapter deals with all of the concepts, terminology, and technology involved in switching fabric implementations. Switching is at the core of the functionality provided by telephone systems. The telephony switching fabric carries out switching operations that involve conveying media stream data by allocating and deallocating media stream channels between telephone network endpoints, and collecting and communicating signaling information.

This chapter begins by explaining telephone switching concepts and terminology, implementation considerations, and design options. The balance of the chapter provides a complete review of each phase of telephony switching fabric innovation including analog circuits, digital transmission, voice over packet-based networks such as IP, and wireless telephony technologies that are used in telephone network switching fabric implementations.

8.1. Switching Resources

 A switching fabric implementation consists of a *telephony switching network* layered over one or more *transmission networks*. The transmission networks each provide one or more transmission facilities which are used by the telephony switching network implementation to establish media stream channels and signaling paths to individual devices and to external telephone networks.

Figure 8-1. Switching resources

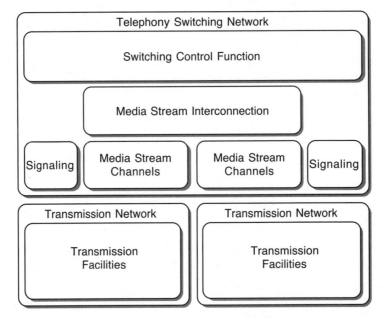

The telephony switching network also includes mechanisms for media stream interconnection which allow the media stream received on one media stream channel to be transmitted, or switched, on to another media stream channel. The switching fabric is managed by a switching control function that is responsible for carrying out switching operations and conveying signaling information to the call processing implementation.

8.1.1. Switching Control Function

The *switching control function* operates the telephony switching network. The responsibilities of the switching control function include:

- Managing availability of transmission facilities and internal resources used to provide interconnections;

- Allocating and deallocating media stream channels;

- Interconnecting media stream channels associated with each transmission facility;

- Interconnecting media stream channels for multi-point calls; and

- Conveying signaling information to and from call processing.

While all modern switching control function implementations are software-based, early telephone system implementations were completely unautomated so people—live operators—performed the control function. In the first generation of automated telephone systems the control function was mechanical and its logic was hardwired into the equipment.

The switching control function in private telephone systems is typically located at a single central point in the switching network known as the *switch*. This type of switching control function is the basis for a *centralized telephony switching network*. This type of switching network is simple, robust, and efficient.

Some telephone system implementations support a *distributed telephony switching network* (see section 8.3.6 for more information on distributed switching). In these implementations, a single centralized switching control function is typically used to control all the distributed switching network components. Some implementations are decentralized so there is no central coordination or management of the switching network. In a *decentralized telephony switching network* all decision making is decentralized and each instance of the switching control function within the switching fabric in question is a peer to all of the others and can operate independently. One benefit is that a part of the network can fail without affecting the rest.

Switching Fabric Implementation

In any switching fabric implementation where multiple switching control function components are interconnected by various transmission facilities, each switching control function must use signaling to communicate with the others in order to coordinate the allocation of capacity on the interconnecting circuits so as to minimize the overall consumption of media stream channels. Conventional enterprise telephone systems use proprietary *switching fabric control protocols*. However, the public network uses a well-defined standard protocol known as *Signaling System 7*, or *SS7*. In open VoIP implementations (see section 8.6), MGCP or MEGACO may be used to implement a distributed switching fabric and the H.323 or SIP protocols may be used between peer switching control functions. These open protocols allow system customers to buy and integrate switching fabric equipment from different vendors.

8.1.2. Media Stream Interconnection

 The role of the *media stream interconnection resource* is to interconnect media stream channels. It is therefore a logical hub out from which radiate the transmission facilities used to implement media stream channels.

Figure 8-2. Media stream interconnection is a logical hub

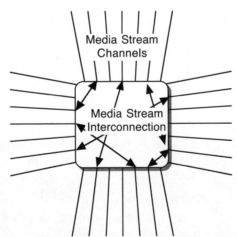

As switching resources are the foundation of the telephony resource framework, media stream interconnection resources are the foundation of the switching resources.

The implementation of a given switching network's media stream interconnection resources are governed by numerous factors. These include:

- The media stream channel implementations that are to be employed;

- The number of media stream channels that must be capable of interconnection;

- The number of media stream channels that must be simultaneously interconnected;

- The end-to-end quality of service (see section 8.2) that must be supported;

- The nature of any multi-point call support; and

- The complexity and architecture of the switching control function.

Media stream interconnection technologies include primitive automated and non-automated mechanical switches, switch matrices, switch buses, software-based voice processing and routing applications, and even interconnection proxy implementations where the interconnection mechanism is virtual. (These technologies are explained in section 8.3.)

8.1.3. Media Stream Channels

The principal payload of a transmission facility is the one or more media streams that it carries. Switching fabrics incorporate media stream channel implementations that determine how media stream data is conveyed over a particular transmission facility.

Media stream channel implementations are specific to the transmission facility technology being used. The implementation is governed by QoS requirements (see section 8.2), the capacity and properties of a particular transmission facility, and the power of the computing technology that is available.

8.1.4. Signaling

In addition to the media stream data that is conveyed over transmission facilities using media stream channels, the switching network must establish communication mechanisms for signaling information.

Signaling includes command and control information such as:

- Signals to reserve the use of a media stream channel and to indicate the presence of a new call on a media stream channel;

- Commands from telephone station equipment to execute services; and

- Physical element control and status information for each connected telephone station on lines that allow a switch to detect and control telephone station equipment.

Additionally, *call associated information*, including the following, may need to be conveyed:

- Direct Inward Dialing (DID)

- CallerID

- ANI

- DNIS

- Correlator data

- User data

Finally, in a distributed switching network the distributed elements of the switching control function must communicate with one another to coordinate their activity.

These different types of control and call associated information may be communicated in-band or out-of-band.

In-band Signaling

In-band signaling refers to a technique in which commands and information share the media stream channel for a particular call. The information is encoded as breaks in the channel, special tones or tone sequences, or modulated data. A significant disadvantage of in-band signaling is that the contents of the actual media stream can be mistaken for signaling information.[8-1]

Out-of-band Signaling

Out-of-band signaling refers to techniques where separate electrical circuits or separate signaling channels are used to keep control and status information separate from media stream information. (See sidebar "Multiplexing" on page 405).

8.1.5. Transmission Facilities and Networks

Switches, telephones, and media-based telephone system equipment connect with one another using communication links conventionally referred to as lines and trunks, or alternatively as *transmission facilities* or simply *circuits*. Circuits which allow access to the public telephone

---

8-1. In-band signaling — The best illustration of the negative consequences of in-band signaling is the so-called Captain Crunch™ whistle. A novelty whistle included in a certain brand of breakfast cereal was found to have a pitch of exactly 2600 Hz. At the time, this was precisely one of the frequencies that AT&T had chosen to use as part of its long-distance network's in-band signaling protocol. People could connect to AT&T's long-distance network, blow the whistle, and then dial free calls using tone dialing (assuming they knew the rest of the protocol). Since then, the vast majority of the world's long-distance network infrastructure has switched to out-of-band signaling.

Switching Fabric Implementation

network for customers (*subscribers*) are also known as *subscriber circuits*, *subscriber loops*, or *local loops*. In this chapter we'll explore the various forms that these circuits can take.

Telephone circuits carry all of the information necessary to convey control information, call associated information, and media stream(s) associated with a call. The *capacity* of a particular transmission facility can be measured in terms of the number of media stream channels and other signaling it has the ability to carry, given particular media stream and signaling specifications. Another measure is the overall *bandwidth*[8-2] of the facility.

Some transmission facilities, such as analog lines, correspond directly to physical layer circuits that directly connect two pieces of telephony equipment. Others, such as the ethernet-based IP networks used to connect IP telephones, provide virtual circuits over a lower-level transport mechanism through a transmission network.

There are four distinct families of transmission facility technology which correspond to successive generations of technical evolution in communications. These are:

- Analog lines;
- Digital facilities;
- Packet networks; and
- Wireless networks.

Sections 8.4, 8.5, 8.6, and 8.7 address each of these transmission facility families respectively. For each, the physical and low-layer protocols are described along with the associated media stream channel and signaling technologies used in popular switching fabric implementations based on these types of networks.

8-2. Bandwidth — Technically the term bandwidth refers to the range of frequencies a facility can carry. In common usage, it is a measure of the information-carrying capacity in thousands of bits per second (kbps) or millions of bits per second (Mbps).

8.2. Quality of Service (QoS)

The most important factor in the implementation of switching fabrics is the *quality of service* for each of the media stream channels it supports and for the end-to-end media streams it delivers through their interconnection. The characteristics of different transmission networks limit the quality of service options for media stream channel implementations.

Quality of service, or QoS[8-3], refers to the nature of a call or connection (i.e., whether it is voice or digital data), attributes relating to bandwidth or digital data rates, the number of media stream channels used, and other transmission characteristics. Quality of service is a property that is generally common to a call and all connections in that call. The quality of service applicable to a new call or connection is limited by what a given switching implementation supports. The quality of service associated with an existing call or connection determines the switching operations (and therefore the call processing services) that may be applied to it.

Quality of service is generally described as a collection of parameters involving bandwidth, isochronous delivery, latency, and jitter associated with a particular media stream channel or call.

8.2.1. Voice Bandwidth

One parameter of QoS is *bandwidth*. A voice call is one that carries media streams requiring no more than 3.1 kHz of bandwidth. The media stream channels supported by a particular transmission facility

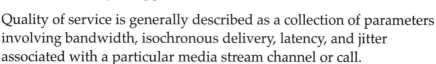

8-3. QoS and CoS — Quality of Service (QoS) should not be confused with Class of Service (CoS). Class of service involves tracking which users or which devices have access or priority use of the call processing features or services within a telephone system. CoS is thus associated with call processing, while QoS is an implementation consideration for telephony switching. Certain telephone systems that support media stream channels may actually have a class of service parameter that governs what QoS parameters must be satisfied for different calls.

may be implemented using analog or digital technology. Analog media stream channels support only voice calls (3.1 kHz of bandwidth for speech and modulated data). Transmission facilities supporting digital media stream channels allow both voice calls and digital data calls. Voice calls are encoded to pass through the digital network in a fashion such that the voice can be reconstituted in analog form as needed. In this case, an indication as to whether a particular call is a voice call or a digital data call becomes yet another piece of call associated information.

An uncompressed voice stream corresponds to digital bandwidth of 64kbps. A media stream channel capable of delivering 3.1 kHz of analog bandwidth or 64kbps of digital bandwidth is the traditional unit of measure in a telephony network and represents the traditional size of a media stream channel.

Some switching fabric implementations utilize media stream channels with less than 64kbps of bandwidth and use compression techniques to squeeze the voice call media stream into the reduced bandwidth.

8.2.2. Isochronous Streams

One unique characteristic of the telephone network is that it is capable of delivering data in an *isochronous* fashion. Only certain data networks are capable of supporting isochronous data streams.

Isochronous streams deliver bits of information at a guaranteed, constant rate. For example, your telephone network connection might deliver 64000 bits of information to your telephone every second. These bits are converted to sound, which you hear as a continuous stream of voice information.

In contrast, *asynchronous streams* deliver bits of information on an as-able basis. Data is received in random bunches based on the availability of data and transmission resources. When voice data is transmitted on an asynchronous connection, you might hear pops, clicks, and periods of silence. Figure 8-3 illustrates the difference between isochronous and asynchronous streams.

Figure 8-3. Isochronous streams

An isochronous bit stream

1010101010101010

An asynchronous bit stream

10 101   0101 01

8.2.3. Latency

Latency is another important QoS parameter. It refers to the delay associated with the delivery of media stream data between two points in the switching fabric. A propagation delay of more that 200 milliseconds is generally considered unacceptable for a voice call. Figure 8-4 illustrates the concept of latency as the sum of all the individual delays in the transmission of media stream data.

Figure 8-4. Latency

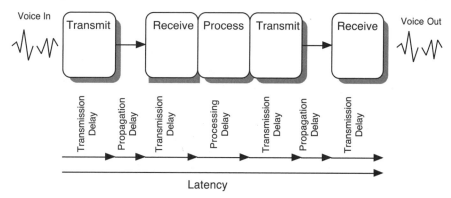

Latency for a given path through a switching fabric is the sum of the three types of delay that may be present in a given implementation:

• Propagation delay
> *Propagation delay* refers to the delay associated with the propagation of the signal over the transmission facility used to implement the media stream channel. Generally speaking,

signal propagation is at the speed of light ($3x10^8$m/s) so it is negligible over short distances. However, a call over a fiber optic cable stretched between New York and LA (a distance of approximately $4x10^6$m) will have a propagation delay of approximately 12 milliseconds. A media stream channel connecting to earth stations via a satellite in geosynchronous orbit is approximately $72x10^6$m long and thus has a propagation time of approximately 216 milliseconds.

- Transmission delay

  Transmission delay refers to the delay introduced by the encoding, framing or packetizing of the media stream. For example, if media stream data was encoded into 80 byte packets for a given media stream channel implementation capable of delivering 64000 bps and if the encode and decode time were each 2 milliseconds per packet, the total transmission delay would be 2 milliseconds for encoding, 10 milliseconds for transmission, and 2 milliseconds for decoding for a total of 14 milliseconds.

- Processing delay

  Processing delay refers to any incremental delay introduced by the switching fabric for processing each packet in a packetized media stream. Sources of processing delay include time required for interpreting and updating headers, time required for determining how to route the packet and, most significantly, any buffering required prior to the forwarding of the packet. Processing delay is quite arbitrary and is a function of both the nature of a given type of switching fabric and the specific implementation. Some routers, for example, are capable of reserving bandwidth so there is no buffering delay for congestion and of processing at *wire speed*. This means they add no incremental delay over the inherent transmission delay. Processing delay on a congested network is potentially unbounded.

8.2.4. Jitter

Jitter refers to variance in latency. The latency associated with a given media stream channel may be constant (in which case there is no jitter) or it may change from moment to moment. For example, switching fabric implementations that packetize the media stream and are capable of routing each packet through a distinct path will have non-zero jitter as the delay associated with each delivery path will vary.

Figure 8-5 illustrates the concept of jitter as the uncertainty or range of possible latency values that results from all the individual variations in the latency associated with a particular switching fabric implementation. It should be noted that switching components can be built to eliminate jitter by storing, or buffering, media stream data (increasing latency) and forwarding it at a constant rate.

Figure 8-5. Jitter

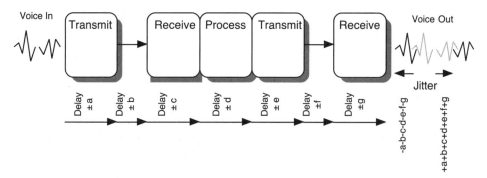

8.3. Interconnection Technologies

At the heart of the switching resources within a given telephone system is a *media stream interconnection* resource (or in larger systems, multiple media stream interconnection resources).

Switching Fabric Implementation

8.3.1. Mechanical Switching

The first generation of switching function implementations were not just mechanical, they were human-mechanical systems. The mechanical part of the system consisted of a patch panel known as a *switchboard* or *cord board*. It consisted of cords with plugs on them and jacks into which the plugs could be inserted. The cords and jacks corresponded to various individual telephone circuits. A person was responsible for detecting each request for service from a given circuit, answering it verbally (this was the principal form of signaling) and manually making the connection by plugging the appropriate cord into the appropriate jack.

The first generation of automated mechanical switching involved what was called step-by-step or progressive control. In this approach the signaling on the telephone circuit consisted of a sequence of pulses and pauses which corresponded to a telephone number of a desired party. More importantly however, this sequence of pulses directly controlled the switching operation itself.

Step-by-step switches are made up of thousands of individual switching modules, or selector units. Each of these consist of a wiper assembly which is mechanically rotated by a relay. The relay is controlled by the pulses on the line. The first digit sent down the line rotates the mechanical switch in a 1st selector unit which then connects the circuit to a 2nd selector unit. For each digit dialed another selector unit is connected to the circuit. The last digit on the last selector unit connects the circuit to the desired destination circuit.

8.3.2. Switch Matrix

The problem with the step-by-step switch (aside from the fact that it was slow, noisy, and subject to mechanical failure) was the fact that the signaling actually drove the switching process itself so that there was effectively no call processing resource in the picture. The signaling had to identify the exact sequence of switches to be activated to make the

Necessity is the Mother of Invention

Necessity is the Mother of Invention

The inventor of the step-by-step switch was Almon B. Stowger. So these switches are generally known as Stowger switches.

Stowger was an undertaker in Kansas City Missouri in 1889. When potential customers wanted to contact him at his funeral parlor they would pick up the telephone and ask the switchboard operator to connect them to the funeral parlor. Unfortunately, the telephone company appeared to be connecting most such calls to his competitor's funeral parlor. When he learned that the telephone operator was actually the wife of the other funeral parlor's owner, he decided to develop the technology to eliminate the human operator.

He patented his invention in 1891 and by 1978 23 million subscribers in the US alone were connected to Stowger switches.

desired connection. This approach certainly was not user friendly nor was it scalable. Most importantly, it didn't allow the addition of any functionality beyond the switching itself.

The next key step in the evolution of switching technology was the concept of the *switch matrix* or *crosspoint matrix*. A switch matrix is pictured in Figure 8-6. It consists of a number of inputs and a number of outputs. Each crossing point in the matrix represents a switch that can be activated to connect the corresponding input to the output. Switching based on the matrix concept are known as *space division switching* because multiple media streams share the switching fabric by being physically separated in space.

An implementation in which the number of outputs is at least as great as the number of inputs is known as a *non-blocking switch* because if all inputs are in use, an output is available for each one. An implementation with fewer outputs than inputs, a *blocking switch*, may occasionally be unable to complete a switching operation because all switch matrix outputs are in use.

Figure 8-6. Switch matrix

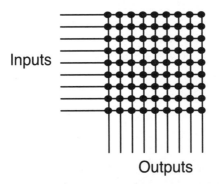

Inputs

Outputs

Implementations of the switch matrix approach allowed signaling and control of the switch matrix to be maintained as separate logical components as shown in Figure 8-7.

Figure 8-7. Switch matrix implementation

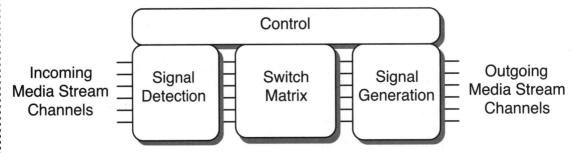

Control

Incoming Media Stream Channels

Signal Detection

Switch Matrix

Signal Generation

Outgoing Media Stream Channels

The first switching matrix implementations were electro-mechanical crossbar switches. Operationally, they were very similar to the Stowger switches they replaced in that they were operated by relays and electromagnets. Crossbar switches consist, as the name implies, of a matrix of bars which can be operated by the application of electric current to the magnets that control them. By activating the magnets associated with a given input and a given output, a switch was closed to connect the two. The crossbars would then be released in order to handle the next switching operation and the switch just set would remain closed as long as the input stream was active. Crossbar switches take up a great deal of space and are subject to reliability problems as switch contacts and other mechanical components wear. Another problem was that control only applied to the interconnection of two media streams. Disconnection was under the control of the caller and suspension of the media stream was not possible.

Multiplexing

Multiplexing is a technique for dividing the bandwidth available on a single facility into multiple channels.

One approach to multiplexing is referred to as frequency domain multiplexing, or FDM. In this approach, each channel is assigned a frequency band and its data is encoded within that band. (This is where the term bandwidth as description of channel size was derived.) For example, FDM is the technique used for multiplexing multiple TV channels on a single coaxial cable.

The multiplexing approach commonly used in telephony is time division multiplexing, or TDM, in which the bandwidth of the channel is divided into individual time slots and each channel uses a share of the time slots. Each channel being multiplexed takes its turn to insert a few bits of information into its slots at the appropriate interval, as shown below.

For example, if six channels of equal size are being multiplexed together, the 3rd, 9th, 15th, 21st, etc., time slots would be used by the third channel.

If the channels are of different capacities, some channels have more time slots than others. For example, if three channels are to be multiplexed together but the first two are both four times the size of the third, then the pattern for channel allocation would be such that of every 18 time slots, channels 1 and 2 would each have eight slots and channel 3 would have two.

Switching Fabric Implementation

The next steps in the evolution of switch matrix implementations involved first the use of reed relays (vacuum tube-like devices which used small metal reeds in a sealed nitrogen capsule under the control of an external electrical current) and ultimately the use of transistors embedded in integrated circuits. While all new matrix switch implementations now rely on transistor-based crosspoint switching, there are still many crossbar and reed relay switches in operation today around the world.

8.3.3. Switch Bus

The next innovation in switching function implementation went hand-in-hand with the implementation of TDM-based digital circuits and forms the basis for all modern digital switching equipment.

The *switch bus* approach uses *time division switching* rather than space division switching. Just as time division multiplexed transmission facilities (see section 8.5) carry multiple media stream channels by dividing a single circuit's capacity, the switch bus is a high-speed synchronous bus with capacity divided into *time slots* (see sidebar "Multiplexing").

Media stream channels are interconnected by assigning media streams to specific time slots. A given bidirectional media stream channel is instructed to insert its incoming media stream into a particular time slot and to obtain its outgoing media stream from a given time slot.

An implementation that supports at least as many time slots as the number of possible inputs is a non-blocking switch because each input can be placed on the bus and interconnected with another media stream.

The most popular architecture for implementation of switch bus technology is the switching backplane architecture in which a backplane with slots and connectors for plug-in cards forms the foundation of the implementation. The backplane connectors provide one or more buses for power and control to the cards along with one or more switch buses. Cards corresponding to various telephony devices

406

Figure 8-8. Switch bus architecture

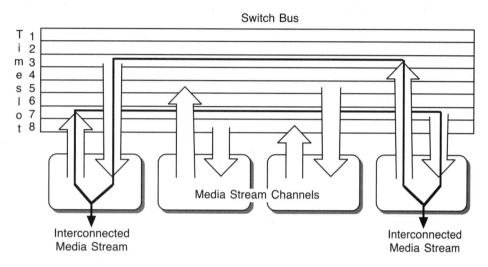

plug into the backplane. Cards that provide an interface to transmission facilities are known as *line cards*. Cards that attach other telephony resources such as media resources to the switching fabric are known as *resource cards*. Typically the switching control function is implemented on another card or occasionally on the backplane itself.

Virtually every vendor of digital switching equipment has one or more proprietary bus architectures. This requires that a customer using that vendor's backplane can only purchase cards from the same vendor. In order to allow greater modularity and provide customers with a wider range of options, a number of vendors published specifications for their switch buses. The two popular specifications became SCSA and MVIP. The ECTF has since published the H.100 and H.110 bus specifications which provide backwards compatibility to SCSA and MVIP and effectively replace both of these older specifications. See the sidebar "ECTF CT Bus: H.100 and H.110."

ECTF CT Bus: H.100 and H.110

The CT Bus is a bit-serial, byte oriented, synchronous, TDM transport bus operating at 8.192MHz. It consists of two clocks, two frame sync pulses, one backup network timing reference, and 32 independent bit-serial data streams. Voice/data transfer on the bus is accomplished by assigning one or more time-slot numbers and bus stream numbers to the sender and receiver(s). At the selected time-slot, the software-selected sender drives the bus and somewhere on the bus, the receiver(s) clock in the data bits. CT Bus has been designed with more capacity than any of the previously deployed buses, in order to support the next generation of high capacity servers. At the same time, CT Bus includes well-defined subsets so that economical low-end systems can be built involving as few as two boards.

ECTF H.100

The H.100 specification documents CT Bus Clocks and Synchronization; data bus lines, interface device requirements, data bus timing, clock skew, reset, power on, and other timing requirements; electrical specifications, mechanical specifications including the design and location of connectors, pin assignments, PCB layouts, and cable requirements; support for partial implementations and optional signals; and inter-operation with other buses.

ECTF H.110

The H.110 specification is functionally identical to the H.100 specification. However, some of the features in H.100 relating to high availability realize their full utility only in the hot swap CompactPCI environment. There are electrical differences between H.100 and H.110 due to the differences between a ribbon cable and a backplane implementation. The CT Bus is defined on 6U CompactPCI form factor printed circuit boards (PCB) and is implemented on the CompactPCI J4/P4 connector. In addition to the topics covered in the H.100 specification, the H.110 specification includes detailed hot swap requirements.

408

8.3.4. Memory-based Switching

Memory-based switching works on a principal similar to switch bus.
However, rather than using time slots on a bus, memory-based switching uses a series of buffers in computer memory. Each buffer operates
in a circular fashion where media stream information is written into
consecutive memory locations as it arrives and when the end of the
buffer is reached, writing continues at the start of the buffer. A given
bidirectional media stream channel is instructed to write its incoming
media stream into a particular buffer and to obtain its outgoing media
stream from another given buffer.

Figure 8-9. Memory-based switching

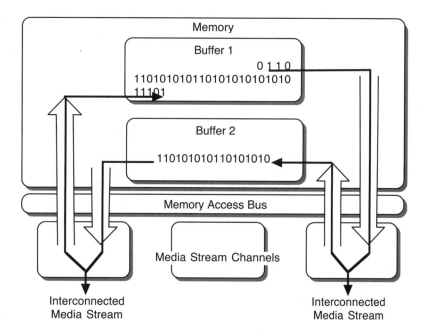

Unlike the switch bus, media stream information remains stored in the
buffer for a certain amount of time before it is read back out because it
cannot be written and read at the same time. This represents a key
source of latency in switching implementations that depend on memory-based switching. The more media stream information that is in
buffers at any given time, the more latency there is in a given system.

Integrated memory circuits and hardware architectures optimized for memory-based switching minimize contention for memory and maximize throughput.

One challenges associated with trying to reduce the amount of media stream data that is buffered is avoiding jitter. If there is too much jitter in the incoming media stream, the buffer may be empty (a buffer under run) when the attempt is made to read from it. This typically causes an increase in jitter in the outgoing media stream. On the other hand, by increasing the amount kept in the buffer (and increasing latency) the system can reduce or eliminate jitter in the output.

8.3.5. Multi-Point Call Support

In order to implement a broadcast multi-point call of the type illustrated in Figure 8-10, the switching function need only make use of the technologies we have already discussed: a crosspoint switch matrix can be set to deliver the same input to multiple inputs, or multiple outputs can be instructed to source their media stream from the same time slot on a switch bus.

Figure 8-10. Broadcast multi-point call

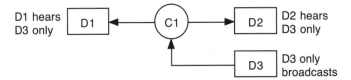

However, if the switching fabric is called upon to construct any other type of multi-point call (rather than having the multi-point call synthesized externally as described in Chapter 3, section 3.3.6) it must perform a mixing operation in which all of the media streams associated with the call are mixed together with the result being the media stream delivered to all the appropriate media stream channels. In the announcement scenario shown in Figure 8-11 the media stream channel associated with D2C2 carries a media stream which is the result of mixing the media stream received from D3 and D1 (and optionally from D2 itself).

Figure 8-11. Announcement multi-point call

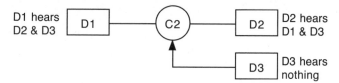

The mixing, or conferencing, process is of course implementation dependent. It could be as simple as suppressing all but the loudest input, it could involve the summing of all inputs after some equalization function, or it could involve performing a separate mix of the inputs for each device in order to eliminate any echo from the given device's media stream.

Multi-point calls are applicable to voice calls only. Audio is a unique data type in that many media streams containing audio can be simply summed together to produce a useful audio stream in which all of the original audio media data can still be discerned. The same cannot be done with digital data streams or modulated data streams. Adding these data streams together would simply corrupt the data. Synthesizing multi-point calls for digital data and large numbers of voice calls is described in the sidebar "Synthesizing Multi-Point Calls."

8.3.6. Distributed Switching Implementations

Traditionally, the switching function has been implemented as a single monolithic server known as a *switch* so it was both a logical and physical hub of a telephone system.

Newer implementations involve distributed switching architectures where the media stream interconnection resources are compartmentalized and physically distributed. The physical components are collectively still a switch but individually they are servers that communicate over a network. The benefits of this modular approach are greater flexibility, the ability to localize the impact of a failure, and the ability to locate the interconnection resources closer to endpoints to simplify wiring.

411

Synthesizing Multi-Point Calls

In a multi-point call in which there are two or more bidirectional media streams, switching resources take the media streams from all connections to the call, mix them together, and deliver the result to all of the participants that are receiving a media stream from the call.

In situations where the switching resources cannot combine two or more calls into a multi-point call (because they are digital data calls, for example), multi-point calls can be synthesized using a series of point-to-point calls in a centralized or distributed fashion.

In the centralized approach, all of the participating devices establish point-to-point calls to a conference bridge, which is a special piece of equipment that is capable of decoding or interpreting the media stream sent from every participating device and intelligently mixing the media streams together. Each participating device only sees a single other device in the point-to-point call it establishes to the conference bridge, but it receives a media stream specifically tailored to be the device's "view" of all of the other devices. A conference bridge may handle just voice data (good for cutting down the noise on multi-point calls with many people), or may handle specific protocols using digital data, such as videoconferencing calls. In fact, a sufficiently intelligent conference bridge (also called a multi-point control unit or MCU in the context of H.320 videoconferencing) can mix caller types, so that a voice call to a conference bridge on which a videoconference was being hosted would hear a special mix of the audio from the videoconference but wouldn't see the video.

The distributed approach to synthesizing multi-point calls involves each participating device placing a call to every other device or using multicasting on networks where it is supported. Multicast refers to the ability to use the network as a media stream interconnection that delivers the same media stream to multiple endpoints (or other switching resources) simultaneously in the same fashion that a timeslot is available to all the points on a given switch bus. As with the conference bridge, the transmitters prepare a specific version of the media stream for each participant to which they are connected. In the distributed case, however, it is the responsibility of each participant to do the appropriate "mixing" of the media streams being received so they can be presented as a single media stream to the person actually using the device in question.

Figure 8-12. Traditional telephone switch

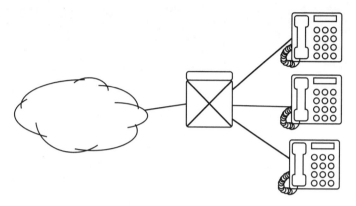

In some cases the network used to interconnect the distributed resources is a fibre optic network dedicated to carrying telephony media stream channels. This type of implementation is illustrated in Figure 8-13.

Figure 8-13. Distributed switching function

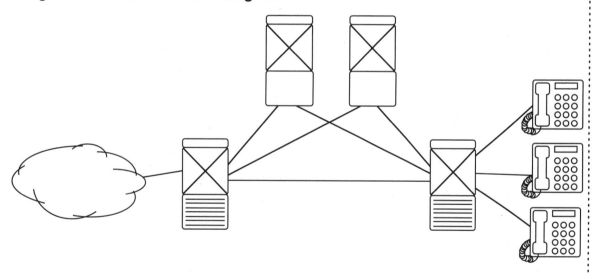

The next generation of distributed switching implementations involves placing the switching servers on a shared voice and data network such as an IP LAN. This allows, for example, a customer to build a single backbone network for their LAN and voice traffic as shown in Figure 8-14.

Figure 8-14. Converged backbone network

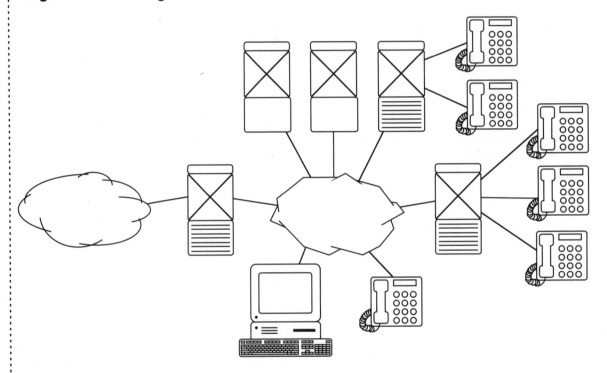

8.3.7. Media Stream Gateways

Distributing media interconnection resources around a network involves interconnecting media stream channels from endpoints and from external networks as well as between servers on the network. In this way, the network itself acts as a media interconnection resource and the interconnection resources around the edges of the network are *media stream gateways*.

8.3.8. Interconnection Resource Availability

Regardless of the implementation of the media stream interconnection mechanism or the transmission facilities involved, the resources available for allocating media stream channels through a switching fabric are finite. There are only so many time slots, matrix outputs, channels of capacity, or so much bandwidth available.

While a non-blocking switch can always interconnect media stream channels associated with directly connected devices, allocating a media stream channel for a remote device requires the use of transmission facilities. If no capacity is available on any of the circuits connecting to the remote device, the media stream channel cannot be allocated. A blocking switch must manage availability of both the time slots or outputs and the transmission facilities.

8.4. Analog Circuits

Analog circuits are the oldest form of subscriber loop technology. They can be generally categorized as:

- Copper analog circuits
  These are analog circuits in which a single pair of wires[8-4] forming an electrical circuit carries no more than one media stream channel in each direction. This represents the vast majority of subscriber loop circuits in the world today and the lowest common denominator in telephone circuit technology.

- Wireline analog carrier circuits
  Analog carrier circuits use frequency shifting to deliver more than one media stream on a given electrical circuit just as conventional radio broadcasting divides the radio spectrum into channels. Digital carrier circuits (see section 8.5) proved to be a superior technology so analog carrier circuits are not of significance.

- Radio frequency analog circuits
  RF, or wireless, circuit technology is quite different from wireline technology and is covered in section 8.7.

In this section we will focus on copper analog circuits. In fact, this category alone actually represents an extensive family of different analog circuit implementations, each one varying in terms of its support for different types of associated information and the way that this information is conveyed.

8-4. Tip and Ring — For historical reasons having to do with the tip and collar (ring) of the plug used in manual patch-panel switchboards, the two wires used are arbitrarily referred to as tip and ring, or "T" and "R." The term twisted pair refers to the fact that the pair of wires forming the circuit is twisted together to minimize electrical interference and cross-talk.

Two Pair

A two pair, or four wire, analog circuit uses one pair for transmitting in one direction and one pair for transmitting in the other. In other words, one electrical circuit only supports a unidirectional media stream so two are used to form a subscriber loop circuit capable of supporting bidirectional communication. Given how inefficient this type of circuit is, both in terms of wire and transceiver electronics, two pair analog transmission facilities are extremely rare.

Single Pair

Single pair, or two wire, analog circuits represent the dominant form of telephony line interface in the world today. Figure 8-15 shows a single pair analog circuit or *local loop*. Analog telephone equipment connects to the analog loop through a hookswitch. When the hookswitch is open (as shown) the telephone is said to be *on-hook* (in a simple mechanical phone, the receiver is in its cradle), and the electrical circuit is open so no current flows. When the hookswitch is closed (the telephone is said to be *off-hook*), electric current flows through the resulting completed circuit. The media stream from the switch and from the microphone of the telephone is then applied to the line.

Figure 8-15. Analog local loop

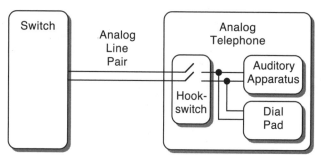

Both the telephone and the switch are able to listen to the media stream on the analog loop, using *hybrid circuits*, or just *hybrids*, that differentiate between the portions of the media stream that are being generated and received at the same point on the loop. The hybrids mix and separate the bidirectional media streams on the analog pair as shown in Figure 8-16.

Figure 8-16. Hybrid circuit

Figure 8-17 shows the same analog loop with multiple telephone stations attached or *physically bridged* onto the line. One of the chief advantages of analog loop technology is that it offers a reasonably high degree of flexibility with respect to the way that telephone sets can be added or removed. The loop can be extended and added to, up to the limits of the switch's signal strength.

Figure 8-17. Analog local loop with bridged telephone stations

Loop Start Signaling

When all the telephone stations on a given loop are on-hook, no loop current is flowing and there is no media stream—therefore there is no means of conveying in-band signaling. As a result, analog lines require additional signaling mechanisms for use while in this state.

Loop start is the most common technique used to allow a telephone station to request service from the switch on an analog local loop. With loop start signaling, the telephone set goes off-hook, which closes the electrical circuit and allows current to flow. The switch detects this flow of loop current, creates a new call with a connection in the *initiated* state, and applies dial tone to the media stream.

On the other hand, the switch indicates that it is trying to connect a call in the *ringing* mode of the *alerting* state by generating an alternating current (AC) on a local loop that is open (i.e., no telephones are off-hook, so the circuit is open). The AC voltage causes the ringers in all the telephones on the loop to signal that an attempt is being made to connect a call. When one of the telephones goes off-hook, the resulting flow of loop current indicates that the *answer call* service is to be invoked. The switch than transitions the connection from the *alerting* state to the *connected* state (stopping the AC ringing signal) and delivers the media stream(s) for the call using the media stream channel represented by the analog line. The switch assumes that it can begin using the line at any time.

Ground Start Signaling

Ground start is another technique used for requesting service on analog loops. Ground start lines are typically used for two-way combination trunks.[8-5] Two-way operation requires that use of the line be controlled in such a way that both switches do not try to use the same trunk at the same time.[8-6] In ground start signaling, each switch is assigned one of the two wires; each indicates that it wants to reserve the line for use by grounding its assigned wire (which is detected as the flow of electrical current in one of the wires, but not around the

---

8-5. Combination trunks — A combination trunk is a line between two switches that allows inbound and outbound operation. See Chapter 10, section 10.9.5

8-6. Glare — Glare occurs when the equipment at both ends of a given circuit inadvertently attempt to use that circuit for two different calls at the same time. Ground start loops prevent glare.

loop). The grounding action allows one switch to signal the other that it wants to use the line so that the second one doesn't. In every other respect, ground start lines work the same way as loop start lines.

Wink Start Signaling

Wink start is the third common technique for performing signaling on an analog line. Wink start is used primarily for DID trunks that are used only for incoming calls from a CO switch to a CPE switch, and thus don't involve supplying dial tone or entering the *initiated* state. Wink start is quite different from the other configurations of analog lines because it is the CO switch that uses the hookswitch and it is the CPE switch that detects the flow of loop current used by the CO switch to indicate the presence of an incoming call. When the CPE switch detects the loop current, it *winks* by reversing the polarity of the voltage on the loop. This signals the CO switch to provide the DID information associated with the call. When the CPE switch has received the DID digits, it winks again to indicate that it is ready to accept the actual media stream associated with the call.

8.4.1. POTS

POTS, or *plain old telephone service*, refers to the lowest common denominator of analog telephone services. Telephony features on a POTS line are restricted to making, answering, and dropping calls.

To make calls, either the loop start or ground start technique is used, depending on the type of line. This request for service corresponds to invoking the *make call* service with no digits. The digits to be dialed are conveyed on the analog local loop using either pulses or DTMF (touchtones). Pulse dialing involves breaking the loop current approximately ten times per second to count out each digit desired, with a pause of between 0.6 and 0.9 seconds between the digits. DTMF dialing simply generates the appropriate tone for each of the desired digits. All dialing on a POTS line is multi-stage, and each digit conveyed is a request for the *dial digits* service.

The telephone station conveys the command to clear the connection by putting the telephone station on-hook. When every telephone station on the local loop is on-hook, the electrical circuit is broken and the switch interprets this as a request for the *clear connection* service for the logical device in question.

In general, the behavior of a logical device element representing a POTS line corresponds to the behavior described for interdependent–shared–bridged appearances (in Chapter 4, section 4.4.3). If a given POTS line is dedicated to just one telephone station, however, and there is no physical bridging taking place, then the logical device may be modeled as having non-addressable standard appearances.

8.4.2. DTMF Feature Codes

One enhancement to POTS that is supported by most switches involves the ability to access many, if not all, of the telephony features and services it supports by issuing commands using special sequences of DTMF digits.

There is no universal standard for the assignment of these codes to telephony features and services, so each telephone company and switch vendor assigns them as needed. These sequences usually begin with the digit "*" or "#" in order to distinguish them from digit sequences to dial. This feature works by having the switch look at the first digit pressed when a new call is connected to the device in the *initiated* state. If the first digit is "*" (or "#" depending on the switch implementation), the rest of the sequence is interpreted as a command. Otherwise all tones are interpreted as digits to be dialed.

For North America, Bellcore has defined some feature codes for use by all public carriers. These include the sequence "*67" which invokes the *set callerID* service to deactivate sending of callerID information with the next call and "*82" which uses the *set caller ID* service to activate sending of callerID information with the next call.

8.4.3. Hookswitch Flash

Nearly all switches support the use of *flash hook,* or just *flash,* on their analog lines. This involves detecting short breaks in loop current during an active call as an invocation of the *consultation call* service if there are no other connections at the device, or the *conference call* or *alternate call* services if there is another connection at the device that is *alerting* or on *hold*.

If there are no other connections at the device, the flash is interpreted as a request for the *consultation call* service with no digits dialed (just as with POTS *make call*). The active connection is placed in the *hold* state and dial tone is heard on the media stream for the new call (which is associated with the device using a connection in the *initiated* state). If the switch support DTMF feature codes, then the first digits are screened to determine if a command is being issued or if a dial string is being entered. Depending on the digits pressed, a feature or service is invoked or a new call is originated.

If there is another connection at the logical device corresponding to the analog line when the hookswitch is flashed, the flash is interpreted as invoking either the *alternate call* service or the *conference call* service, depending on the state of the other connection, the switch implementation, and the class of service assigned to the device. Typically if the second connection is in the *alerting* state, the *alternate call* service is used, which places the first call on *hold* and transitions the second call to the *connected* state. If the second connection is in the *hold* state, the *conference call* service typically is the one invoked. This merges the two calls into a single call with everyone participating.

Most systems also support the *transfer call* and *single step transfer call* services through a related mechanism. If flash was used to place a connection on *hold* and to originate a new connection, dropping the call before the new call is answered is treated as a request for the *single step conference call* service. Dropping the call after it is answered is treated as a *transfer call* request.

8.4.4. CallerID

One of the most important pieces of call associated information for CTI solutions is callerID. Despite the limitations of analog lines, callerID information can be delivered and most telephone companies now include callerID among their service offerings for analog lines. CallerID information is sent by the CO switch as a burst of modulated data between the first and second ring cycles as a call is being presented. To use callerID, telephone station equipment must be equipped with appropriate electronics for intercepting and decoding the callerID information, and the phone must not be answered until after the second ring has started. (See Chapter 5, section 5.5.1 for more information on ANI and callerID.)

8.4.5. Distinctive Ringing

Some analog lines are able to indicate the original dialed number associated with a call that is being presented to a particular device using the *distinctive ringing* feature. This feature is effectively an analog line version of DNIS that involves using a different ringer pattern (the cadence of the AC signals making up a ring cycle) to correspond to each of the subscribed-to numbers that are associated with the analog line in question. The mapping between numbers and ringer patterns is specific to each and every line, so this feature is rarely taken advantage of in CTI systems. To do so requires special hardware capable of detecting different ring patterns and prior knowledge of the assignments of ringer patterns to numbers. (See Chapter 5, section 5.5.2 and Chapter 10, section 10.9.10 for information on DNIS.)

8.4.6. Call Waiting Indication

Call waiting indication is used in conjunction with the flash hook capability. It involves allowing for a second call to be presented to the logical device corresponding to an analog line while a first call is active. The presence of the second call in the *alerting* state is signaled using a special in-band signaling tone that is inserted into the media stream channel on the analog line. The second call can be answered by issuing the *alternate call* service using a hookflash, or can be ignored.

The latest variation on call waiting involves delivering callerID information for the second call using additional in-band signaling. While having this information is very useful, the disruption to the active call, resulting from the in-band signaling that otherwise blocks the active media stream, can be significant.[8-7]

8.4.7. Proprietary Second Pair Signaling

Despite the fact that analog local loops (in fact, most local loops, period) require only a single pair of wires, standard telephone cable has at least two pairs of conductors, and the standard North American telephone jack, the RJ-11, provides a connection for both pairs. Certain switches put the normally unused pair of wires to work as a separate, independent signaling channel.

These implementations support full POTS compatibility with the standard pair of wires so that any existing analog station equipment will work. Special proprietary phones compatible with the proprietary protocol on the second pair, however, are specially handled by the switch when detected. These products use the second independent (typically digital) channel in order to allow full functionality on the analog loop. Information conveyed includes all available call information, status and control information for the physical device, etc. This arrangement is illustrated in Figure 8-18.

8-7. Call waiting and in-band signaling — Call waiting illustrates a second disadvantage of in-band signaling. If the switch assumes that a human is listening and can take action, it may generate in-band tones at any time. This can disrupt modulated data connections (i.e., modem or fax transmissions) that might be present on the line. For this reason, most systems that support call waiting also support DTMF commands to disable the feature.

Figure 8-18. Proprietary second pair

8.4.8. Analog Telephone Station Equipment

Analog telephone station equipment is anything that can be connected to an analog telephone line. This includes:

- Analog telephone sets

- Fax machines

- Data modems

- Fax modems

- Low-speed video phone telephones

Individual devices typically combine multiple functions and can be found in a limitless variety of form factors.

The portions of these products that correspond to the components of a telephone set represent each item's physical element. Other portions (such as a fax machine's scanner, printer, and modem) are considered media services associated with the device. Telephone stations are discussed in more detail in section 10.5.

8.5. Digital Circuits

This section deals with transmission facilities based on wired circuits that carry digitized media streams over a synchronous network medium. (We'll be looking at packet-based virtual circuits in section 8.6 and at wireless digital circuits in section 8.7.)

8.5.1. Digitizing Voice

Voice data is digitized and encoded in order to be carried on digital circuits. This process is shown in Figure 8-19.

Figure 8-19. Voice coding process

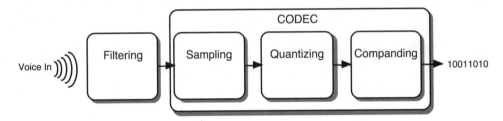

It involves four steps:

- Filtering
  Filtering involves the use of bandpass filters to limit the range of frequencies that are transmitted to being within the voice channel range. In the PSTN, a bandpass filter is used that limits the all coded audio to be in the 300Hz to 4kHz range.

- Sampling
  Sampling involves converting the smooth curve of an analog signal into a series of steps that can then be quantized as shown in Figure 8-20. The key variable in a sampler is the rate at which it samples the incoming analog signal—this corresponds to the width of the steps. The greater the sampling rate the better the approximation of the digital representation. The Nyquist criterion states that the sampling rate must be at least twice the highest frequency component

found in the signal. This means that if the signal is being filtered to make the highest possible frequency component 4kHz, the sampling rate must be at least 8kHz.

Figure 8-20. Sampling

- *Quantizing*

    Quantizing converts the sampled analog signal into a sequence of digital levels.

- *Companding*

    Companding (which is short for compressing and expanding) involves encoding the quantized voice samples into a coded sequence of numbers. The simplest coding scheme is linear in which each increment in the output corresponds to an equal step in the quantized input. Other coding schemes use non-linear formulas to deliver better sound quality. Still more use special compression techniques to reduce the amount of bandwidth required by making assumptions about the nature of the input signal.

A *CODEC* (which stands for coder-decoder) is used to perform the sampling, quantization, and companding function. CODECs can be implemented in hardware or software.

Digital carrier circuits in the PSTN use the following coding schemes:

- µLaw PCM

    PCM stands for *Pulse Code Modulation*. Standard PCM involves sampling at a rate of 8000 times per second (8kHz) where each sample is 8 bits. The highest bit represents the sign and the other 7 bits represent the absolute amplitude of the sample. The result is a 64kbps stream of digitized voice data. *µLaw PCM* (pronounced mu-law or micro-law) refers to the non-linear companding formula which divides the amplitude into 16 chords (identified by 4 bits) which are in turn divided into 8 levels (identified by 3 bits). The chords are not equally sized so that the low-amplitude chords have more granularity than the high-amplitude chords (the distribution is roughly logarithmic). The result is a digital signal that has less perceived noise on quiet or idle channels. µLaw PCM is the standard for digital carrier circuits in North America and Japan.

- ALaw PCM
  ALaw PCM is the standard for digital carrier circuits in Europe. The only difference between µLaw PCM and ALaw PCM is the companding formula. ALaw's companding formula results in less noise for quieter signals but more noise for an idle (silent) channel.

Adaptive Pulse Code Modulation, or ADPCM, is a scheme which further compresses PCM so that only 32kbps of bandwidth are required.

8.5.2. Pair Gain Concept

The original impetus behind the telephone network's shift from analog to digital circuits was not, as one might think, about adding functionality or improving quality. It was about increasing the efficiency of cable use. Metallic analog circuits required two or four wires per media stream channel. With the demand for telephone lines climbing, the amount of cable required and the number of individual wires that had to be connected and tracked was rapidly becoming unmanageable. It became clear that the only way to cope with demand was to find ways to pack multiple media stream channels on a single transmission facility. *Pair gain* refers to the result of multiplexing several media stream channels over a circuit that would previously have carried only one. Initially many different multiplexing schemes were tried, including frequency division multiplexing (resulting in analog carrier circuits). However, time division multiplexing quickly proved to be the most reliable and scalable approach. (See sidebar "Multiplexing" on page 405.)

8.5.3. Digital Signal Level Hierarchy

As we have seen, a standard voice channel represents a 64kbps stream in the world of digital carrier circuits. This is also referred to as a *DS-0* channel. (DS stands for digital signal.) The real benefit of digitized voice channels is that they can be easily multiplexed together. The ways in which DS-0 channels are multiplexed are also standardized according to the *digital signal level hierarchy* for a given region.

A few of the most popular levels of the standardized DS-level hierarchy for North America are shown in Table 8-1. The Japanese hierarchy is based on the North American hierarchy but has a number of differences as shown in the examples in Table 8-2. The European hierarchy is quite different and only DS-0 is common to the others as shown through the examples listed in Table 8-3.

DS Levels

Each level in the hierarchy is assigned a *DS level designator* which identifies the number of DS-0 channels that are being multiplexed together. For example, a North American DS-4 represents 4032 DS-0 channels for a total channel carrying bandwidth of 258.048 Mbps (64kbps x 4032).

Carrier designations identify a particular transmission facility specification corresponding to the DS level in question. The carrier designation refers to both the medium used and the way in which the channels are multiplexed. There may be multiple carrier specifications for a particular DS level. For example, North American DS-4s can be carried by a 4 wire T-4, but also by coaxial cable (T-4M), or Fiber Optic cable (FT-4).

Table 8-1. North American hierarchy

DS Level Designation	Number of DS-0 Channels	Carrier Designation	Carrier Bandwidth
DS-0	1	N/A	N/A
DS-1	24	T-1	1.544 Mbps
DS-1C	48	T-1c	3.152 Mbps

Table 8-1. North American hierarchy

DS Level Designation	Number of DS-0 Channels	Carrier Designation	Carrier Bandwidth
DS-2	96	T-2	6.312 Mbps
DS-3	672	T-3	44.736 Mbps
DS-4	4032	T-4	274.176 Mbps

Table 8-2. Japanese hierarchy

DS Level Designation	Number of DS-0 Channels	Carrier Designation	Carrier Bandwidth
DS-0	1	N/A	N/A
DS-1	24	T-1	1.544 Mbps
DS-2	96	T-2	6.312 Mbps
DS-3	480	T-3	32.064 Mbps
DS-4	1440	T-4	97.728 Mbps
DS-5	5760	T-5	400.352 Mbps

Table 8-3. European hierarchy

DS Level Designation	Number of DS-0 Channels	Carrier Designation	Carrier Bandwidth
DS-0	1	N/A	N/A
DS-1	30	E-1	2.078 Mbps
DS-2	120	E-2	8.448 Mbps
DS-3	480	E-3	34.368 Mbps
DS-4	1920	E-4	139.268 Mbps
DS-5	7680	E-5	565.148 Mbps

8.5.4. T-1 and E-1

The circuit for carrying a channel from one switch to another is called a trunk circuit, and in most cases there is a desire to maximize the number of trunks that can be delivered while using as few physical wires as possible. The way this generally is accomplished involves using a high-speed digital connection, called a span, between two points and using time division multiplexing to divide it into multiple fixed-bandwidth channels.

There are many variations of digital trunking technology available, depending on the capacity desired. The data rate of any given carrier circuit, or the number of usable DS-0 channels it provides, is standardized (see Tables 8-1, 8-2, and 8-3) but standards are different around the world. The most common forms of digital trunking facilities in use are the T-1 span (used in North America and Japan) and the E-1 span (used in Europe and elsewhere).

T-1

T-1 involves a facility, referred to as a *T-1 span*, that has a total capacity of 1.544 Mbps and supports 24 independent multiplexed channels. T-1 is typically implemented using four wires (two pairs). One pair is used to transmit data in each direction. T-1 also may be delivered through fiber-optic cable. The term *fractional T-1* refers to a normal T-1 span on which fewer than the full 24 channels are going to be used.

Figure 8-21. T-1 span

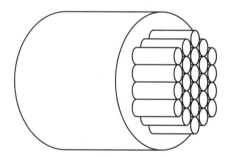

Each DS-0 channel on a T-1 span is equivalent to an analog line. In fact, T-1 was originally designed purely to concentrate the voice traffic from multiple analog lines. The 3.1 kHz audio bandwidth of an analog line is digitized and encoded into a 64 kbps media stream, which then is allocated to one of the T-1's channels. Unfortunately, because the voice media stream completely fills the DS-0 channel and because each of the T-1 channels operates independently, signaling for calls is in-band.

Robbed Bits

The in-band signaling approach used in T-1 is referred to as the *robbed-bits* method. This involves *stealing bits* from voice data and using them for encoding signaling information. Specifically, the standard T-1 method involves stealing every 48th bit (in each channel) for signaling. The signaling information is present at all times, but the lost bit results in an imperceptible loss of audio quality.

A limitation of the robbed-bits signaling method is that it limits the transmission of digital data using one of the DS-0 channels to 56 kbps in order to keep the digital data and the in-band signaling apart.

The signaling bits are used to implement very simple signaling protocols that are based on the operation of the analog lines that the multiplexed channels supersede. For example, when a particular channel is not being used, the signaling bits are zeros in both directions. When the CO switch wants to present a call on a given channel, it changes the bits to be all ones. The CPE switch interprets the transition to all ones as a ringing indication, and it changes the bits it is sending to ones also in order to indicate that it wants to answer the call.

E-1

The *E-1 span* has a total capacity of 2.048 Mbps, which represents 32 DS-0 channels. E-1 uses out-of-band signaling, however, so 30 bearer channels are available for media streams and 2 channels are used for signaling and control. Because all signaling is out-of-band, each DS-0 supports digital data at 64 kbps. Like T-1, E-1 is typically available through either wire pairs or fiber optics.

8.5.5. ISDN-BRI

Integrated services digital network, or *ISDN,* refers to the standards defined for providing subscriber access to the full capabilities of the digital portion of the telephone network while still interoperating with the analog/voice portion. There is a series of different line interface

Figure 8-22. E-1 span

standards that represent different levels of ISDN capability and capacity. The basic version is called the ISDN *Basic Rate Interface,* or *BRI.*

Basic rate ISDN is a digital (rather than analog) subscriber loop technology. BRI was conceived as the evolutionary path for analog lines. It was designed to be compatible with most existing analog subscriber loop wiring, with the intention that ISDN basic rate service would be a simple upgrade for people using analog lines and wanting to move to digital technology.

With ISDN, all the signaling, control, and media stream information and other call associated information is sent in digital form. BRI uses multiplexing to provide three channels:

- One *signaling* channel referred to as the "D" channel. This channel is a 16 kbps (bidirectional) channel; and

- Two *bearer* channels, referred to as "B" channels. Each B channel is a 64 kbps (bidirectional) channel.

An ISDN BRI line can be visualized as shown in Figure 8-23.

The B channels are media stream channels. The D channel is used for all other information exchanged between telephone station equipment and the switch. This means that all signaling is out-of-band. The protocols used for information exchanged on the D channel are defined as part of the international ISDN standard. The D channel is capable of conveying all of the call associated information and

Figure 8-23. ISDN BRI

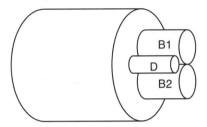

requests for most of the popular telephony features and services. In fact, in most locations callerID service was available on ISDN BRI long before it was available on analog lines.

User to User Information Elements (UUIE)

The ISDN signaling protocol defines a mechanism called User to User Information Elements (UUIE) that allow the exchange of user data between the endpoints in a call. This is a special feature in which a special class of call associated information is actually delivered from one endpoint to the other endpoints associated with a call in much the same way as a call's media stream.

The user data conveyed between ISDN endpoints may be used for any purpose. One way to take advantage of it is to deliver electronic business card information including, for example, digitally signed credit information. The Versit defined V-Card format which is now used extensively in directory services applications and in e-mail was originally developed for this purpose.

Versit has also defined a standardized means of encoding both correlator data and user data into the UUIE. This is referred to as Versit UUIE and is documented in Appendix A of Volume 3 of the Versit CTI Encyclopedia. The Versit UUIE encoding is used to allow separate telephone systems connected by an ISDN call to appear as a single system for purposes of their CTI interfaces.ECMA has adopted this same mechanism in CSTA phase III.

Both bearer channels are capable of carrying both voice calls (including modulated data such as modem and fax) and digital data calls.[8-8] The combination of uses subscribed to for a specific channel is referred to as the channel's *bearer capabilities*. Depending on how the switch and line are configured for use, restrictions may be applied to the use of each B channel for voice or data calls. This is often done, for example, to prevent both channels from being used for data, leaving no channel for a voice call if one had to be made. The bearer channels are typically used independently so that, for example, one may be used for a voice call while another is being used to connect to the Internet using a digital data call. It is also possible to use both B channels for a single digital data call (128 kbps) using a standard known as BONDING, which in reality still establishes two independent calls to another ISDN interface, but uses protocols to synchronize the contents of both calls to appear as a single digital data media stream in each direction.

Installation and Configuration

Technically speaking, ISDN BRI actually refers to a whole family of individual line interfaces that work together. Attaching telephone stations (or *terminal equipment* as it is known in ISDN) to a BRI line is not as easy as attaching equipment to an analog line. Figure 8-24 shows a generic ISDN BRI installation.

Figure 8-24. ISDN BRI interface points and equipment

8-8. **Packet switching** — Both bearer and signaling channels can also be used for carrying X.25 packet-switched data.

The BRI line directly from a central office switch is known as the "U" interface. The U interface is capable of encoding[8-9] all of the ISDN BRI digital information on a single pair of wires that is compatible with most of the analog subscriber loops currently in use.

The ISDN standard also defines an "S" and a "T" interface. These are actually the same interface electrically, but the letter indicates what type of equipment is exposing the interface. The ISDN standard defines a piece of equipment that converts the U interface to the T interface as a *Network Termination–Type 1*, or *NT1*. If the line is to be connected to a switch that in turn has ISDN basic rate lines, those lines are considered to be S lines and the switch, for purposes of the ISDN standard, is considered a *Network Termination –Type 2*, or *NT2*.

Terminal equipment is attached to the S/T interface. (S and T are equivalent from the perspective of terminal equipment.) The S/T interface is a four-wire interface (two pairs of wires) and is the interface provided on generic ISDN telephone station devices. A big advantage of ISDN is that terminal equipment can be used anywhere in the world because this interface is standardized internationally.

A *terminal adapter* is a piece of ISDN terminal equipment that allows one or more pieces of non-ISDN equipment to be adapted to the ISDN line. The number of adaptations possible is unbounded and thus is not standardized. The "R" interface generically represents all of these non-ISDN connections.

8-9. 2B1Q — The encoding used for the U interface is not internationally defined. Each country defines the U interface to be used on subscriber lines provided by local exchange carriers. In North America the standard for the U interface is referred to as 2B1Q (ANSI standard T1.206); this standard was not established, however, until after vendors had already begun shipping switches that used a different U interface encoding referred to as AMI (Alternate Mark Inversion). AMI is the encoding used on T-1 and E-1 circuits and while AMI is no longer used for new BRI lines, older equipment and lines may still be in use.

Power

One important difference between ISDN and analog lines is that analog lines utilize loop current from the switch, so that all the power needed by the telephone station can be obtained from the line itself.

With ISDN lines, the line (U interface) only carries low-voltage signaling. This means that all power for operating the various devices that allow for a functional ISDN line (NT1, etc.) must be powered independently. If there is a power failure and the NT1, NT2, and other ISDN station equipment lose power, the ISDN line cannot be used. For this reason it is generally recommended that, unless a given ISDN line is intended only as a spare line or for recreational purposes, associated equipment should be powered through an *uninterruptable power supply* (*UPS*).

Multi-Point Installations

The S/T interface supports the ability for up to eight pieces of terminal equipment to connect to the same S/T interface, in much the same way that multiple analog telephones can be connected to the same analog line. This is shown in Figure 8-25.

Figure 8-25. ISDN multi-point

ISDN Telephone Station Equipment

ISDN telephone sets connect directly to the S/T interface. They take direct advantage of the D channel for receiving notification of connection attempts, placing calls, requesting telephony features and services, and receiving call associated information such as callerID. They can access either B channel for voice media stream data (assuming that the line is configured to use both B channels for voice), but can only listen and talk on one channel at any instant.

While other types of stand-alone ISDN station equipment do exist, such as videoconferencing devices, most ISDN station equipment is in the form of terminal adapters that integrate with other communications products.

Terminal Adapters

Terminal adapters, or *TAs*, typically are stand-alone peripherals or add-in cards for personal computers. They also may appear in the form of cards inside other forms of communications equipment or inside an ISDN telephone set.

Terminal adapters intended for use with personal computers (either as add-in cards or as external peripherals) typically are used to establish digital data calls for connecting to the Internet, exchanging large documents, and doing videoconferencing.

Some terminal adapters can handle both voice and digital data calls by using built-in modems.[8-10] They can also exchange faxes and data with analog fax and data modem equipment in the voice telephone network.

Analog Telephone Station Equipment

A very popular R interface for a terminal adapter is an analog telephone line circuit. Terminal adapters may include one or two analog line interfaces that correspond to one or either B channel. This

---

8-10. ISDN modems — Terminal adapters that have built-in modems are generally called ISDN modems. People sometimes misuse this term to refer to terminal adapters that have the external appearance of traditional analog modem peripherals but don't actually have modem functionality. This is not correct and can be confusing.

allows existing analog telephone station equipment (telephones, fax machines, etc.) to be attached to an ISDN line. This is shown in Figure 8-26.

Figure 8-26. Analog stations on an ISDN line

Built-in NT1s

Many ISDN terminal adapters and other station equipment incorporate the NT1, so that they plug directly into the U interface. This is done to avoid the expense and complexity of having two separate devices. There is a trade-off to be made, unfortunately, because only one such device can be attached to a given U interface, so one can have, at most, one of these types of devices per line. In addition, unless the device also supports a T interface (from the built-in NT1), other S/T equipment cannot be used and multi-point operation is not supported.

Modeling ISDN

There are many different ways that ISDN can be modeled. Different switch implementors have pursued different approaches.

One approach is to model the ISDN line as a single logical device and to model each ISDN bearer channel as an independent media stream channel that is available for use by an appearance. Another approach is to model each ISDN bearer channel as a distinct logical device. In either case, appearances may be addressable or non-addressable.

The telephone stations on an ISDN line correspond to physical device elements. If there is only one, it may be modeled as being part of the same device as the logical element. Otherwise, it is modeled as part of a *multiple logical elements* device configuration or a *bridged* device configuration.

Sub-addressing can be used, if supported by the switch and the external network, to indicate which station (i.e., physical element) is to hear ringing with respect to an *alerting* call.

8.5.6. Proprietary Digital Subscriber Loops

Proprietary digital loops are digital lines from switches that are proprietary to a particular switch vendor. There are too many proprietary implementations to enumerate, as each switch vendor typically has one or many that they have invented. Proprietary telephone station equipment is compatible only with the specific variety of proprietary digital subscriber loop technology for which it was built.

Despite all of these incompatible designs, the underlying concepts are usually all very comparable to ISDN, if not actually based on ISDN. Typically, most proprietary digital loops have a signaling channel and one, two, or more bearer channels (although the additional bearer channels usually are either never used or used only for digital data). This is illustrated in Figure 8-27.

Figure 8-27. Proprietary digital loops

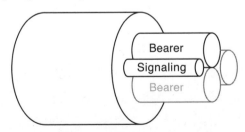

The key differences between implementations involve the range of functionality that can be accessed and controlled (in either direction) by the signaling channel and the bandwidth associated with the bearer channels. (If supported, this is usually 56 kbps or 64 kbps in order to take advantage of standard voice media stream encodings.) For example, one vendor's signaling protocol may provide complete functionality for control and observation over the components of the physical telephone set while another's may support full access to every imaginable telephony feature and service.

8.5.7. ISDN Primary Rate Interface (PRI)

ISDN Primary Rate Interface (PRI) provides ISDN signaling and services using T-1 and E-1 spans. Where T-1 is applicable, PRI consists of 23 ISDN B channels and one 64 kbps ISDN D channel for signaling. Where E-1 is applicable, PRI consists of 30 ISDN B channels and one 64 kbps ISDN D channel.

PRI offers two very significant advantages over basic T-1. The first is the use of a D channel for signaling. This means that each of the other channels in the T-1 span is free of in-band signaling and each is capable of carrying the full 64 kbps B channel media stream.

The second key advantage of PRI, which applies to both T-1 and E-1 spans, is that ISDN signaling provides a much richer feature set than the primitive, bit-oriented protocols that are used by basic T-1 and E-1. This additional functionality means, among other things, that individual B channels can be managed dynamically—in contrast to raw T-1/E-1 channels, which must have their use (incoming, outgoing, voice, digital data, etc.) predefined.

8.5.8. DSL

DSL stands for *digital subscriber loop*. DSL is a family of technologies used to deliver one or more digital channels over a wired connection. It is often written *xDSL* to indicate that it refers to a family of specifications which all include "DSL" in their names. While DSL is relatively new as a family of technologies, individual DSL technologies form the basis for conventional T-1, E-1, and BRI circuits.

DSL technology makes use of modems that are similar to the modems used on analog local loops. The key difference is that modems designed for use on analog lines are limited to a bandwidth of 3.1 kHz (corresponding to a single DS-0 voice channel) because they establish connections through the voice network. In contrast, DSL modems attempt to use as much of the frequency bandwidth available on the copper circuit as possible because they are establishing connections with a modem on the other end of a physical circuit.

Figure 8-28. DSL spectrum vs. analog modem

0   3.1kHz   80-120kHz

Subscriber Line Bandwidth (Frequency)

DSL technologies vary according to which parts of the spectrum they use, how much energy they dissipate, how they allocate the bandwidth on a give circuit between upstream channels and downstream channels, and the encoding used. These variations determine the maximum distance between the DSL modems at each end of the circuit, tolerance for bad wiring (including bad junctions, splices, and other sources of signal echo), the susceptibility to electromagnetic interference, the number of wire pairs required, and the number and size of the communication channels that are supported.

Some forms of DSL provide unidirectional channels. The most popular of these is ADSL which stands for asymmetric digital subscriber loop. Asymmetric channels are seen as a vehicle for delivering TV programming or other media-intensive data streams to homes as an alternative to cable TV. ADSL also features the ability to share the same circuit as a POTS loop because it doesn't use bandwidth in the 0-3.1 kHz spectrum used by POTS.

Table 8-1. xDSL specifications[a]

DSL Type	Total upstream bandwidth	Total downstream bandwidth	Maximum Distance	Other properties
IDSL	160 kbps	160 kbps	26,000 ft.	Same as BRI
ADSL	1.544 Mbps	16 kbps	18,000 ft.	Avoids 4kHz voice band
HDSL	1.544 Mbps	1.544 Mbps	12,000 ft.	Requires 2 wire pairs
SDSL	1.544 Mbps	1.544 Mbps	10,000 ft.	
VDSL	12.96 Mbps	1.544 Mbps	4,500 ft.	Avoids 4kHz voice band

a.**European DSL** — The specifications shown are for US DSL standards. European DSL specification established by ETSI for HDSL, SDSL, and VDSL vary slightly from their US counterparts.

DSL is a likely contender as the future of wired subscriber loop technology.

Voice over DSL (VoDSL)

Voice over DSL refers to products and services carrying a telephony media stream over a DSL circuit in which the vendor does not specify the protocol used. Many VoDSL implementations may channelize the DSL link and thus they represent conventional TDM circuits. However other VoDSL circuits use a packet transport such as IP and thus are voice-over-packet implementations.

8.5.9. Sonet

Sonet, or *synchronous optical network*, refers to a family of internationally defined fibre optic transmission standards that are comparable to the regionally defined digital carrier specifications (described in section 8.5.3). The T-series and E-series digital carriers were initially defined for electrical circuits over copper wires and were extended with designations for implementations of the corresponding digital signal levels delivered over fibre-optic cables. Unfortunately, these designations remained specific to the US, Europe, and Japan. In contrast, Sonet was designed as a means of creating a world-wide standard that supports the multiplexing of any of the regional DS levels together in arbitrary combinations.

Sonet defines the digital hierarchy shown in Table 8-4. The optical fibre carriers are designated with an *OC* (for *optical carrier*). The equivalent signals when transported electronically (typically within a switching function) are given *STS* (for Synchronous Transport Signals) designations.

Table 8-4. Sonet digital hierarchy

Optical Signal Designation	Electrical Signal Designation	Carrier Bandwidth	Equivalent DS-0 channels
OC-1	STS-1	51.84 Mbps	672 DS-0
OC-3	STS-3	155.52 Mbps	2016 DS-0
OC-9	STS-9	466.56 Mbps	6048 DS-0
OC-12	STS-12	622.08 Mbps	8064 DS-0
OC-18	STS-18	933.12 Mbps	12096 DS-0
OC-24	STS-24	1244.16 Mbps	16128 DS-0

Sonet multiplexing is based on multiples of the OC-1 rather than the DS-0. DS carriers are multiplexed into OC-1 carriers using well defined Sonet protocols and then multiplexed into higher order OC carriers as needed.

8.5.10. ATM

Asynchronous Transfer Mode (ATM) refers to the next generation of digital networking fabric, which ultimately will eclipse the use of spans with channels of fixed bandwidth. ATM involves a form of time division multiplexing (see sidebar "Multiplexing" on page 405) in which the assignment of channels to time slots is managed dynamically. When a channel is established using ATM, the data rate and other QoS factors are used to determine channel assignments on the fly. Real-time isochronous traffic is guaranteed a consistent allocation of time slots, and asynchronous data is put into slots on an as-available basis. Each time slot in ATM actually is a *cell* of 53 bytes that carries with it the information needed to appropriately route it to its ultimate destination.

ATM networking can be implemented over a variety of different circuit technologies including DSL and Sonet.

The ability for ATM to supersede both existing telephony *and* telecommunications networks—by efficiently mixing all forms of telephony and telecommunications traffic without compromising the isochronous nature of telephony media streams—means that it represents a unifying technology in the world of wide area communications and networking.

8.5.11. B-ISDN

Broadband ISDN, or *B-ISDN*, is to ATM what primary-rate ISDN (PRI) is to the digital network fabric based on fixed-size DS-0 channels. B-ISDN refers to the set of specifications for performing signaling and for interfacing to the ATM portion of the digital telephone network.

8.5.12. Cable TV Networks

Another subscriber loop technology that may emerge as the next dominant approach involves using a portion of the bandwidth available in cable TV networks.

Cable TV networks now connect a significant portion of homes and businesses throughout the world. This network has a great deal of capacity as it snakes through a neighborhood. While most of this capacity is used to carry TV signals in a single direction, a portion of this capacity can be allocated for carrying two-way media streams. By placing the appropriate transmitters and receivers at each home or business cable TV junction, this network can be used for both voice and digital data connections to the PSTN. This is illustrated in Figure 8-29.

Figure 8-29. Cable TV network providing subscriber loop

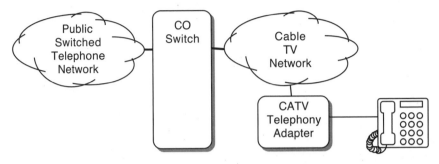

8.6. Packet-based Virtual Circuits

Another rapidly emerging type of switching fabric technology for the transmission of telephony media streams involve implementing telephony switching functionality layered over packet networks originally designed for delivery of data, such as the Internet and privately managed intranets. While Voice-over-IP is the best known technology of this kind, other voice-over-packet implementations are also quite popular.

Conventional telephony switching fabrics rely on underlying networks that allow pathways of guaranteed bandwidth to be allocated. This is historically referred to as *circuit switching* because in the days of analog telephony an end-to-end electrical circuit was formed connecting telephone sets at either end. With digital switching, time division multiplexing is used to divide the bandwidth across a particular network link or switching bus into channels. An end-to-end connection is established by interconnecting all of the channels to form a single media stream path of guaranteed bandwidth.

All the telephone circuit technologies we have examined so far have involved synchronous circuit-switched channels. This means that when a media stream channel has been established between two points, each leg of the call is handled by a transmission facility that dedicates a particular portion of its bandwidth, or channel, to conveying the data stream. The result is referred to as a switched circuit and the key attribute of such a circuit is that it guarantees QoS.

A switched circuit guarantees full isochronous delivery of media streams with minimal latency, no jitter, and no data loss. However, this quality comes at what is arguably a high cost: it involves reserving dedicated channels for each media stream channel associated with a call as long as a media stream channel is allocated even if no one is speaking. In a typical conversation only one person is speaking at a given time so that half of the dedicated bandwidth is idle. Furthermore, 10-20% of the bandwidth associated with a speaker is actually silence (between words, pauses, etc.). With worldwide data traffic growing faster than worldwide voice traffic (in terms of bandwidth requirements), data networking infrastructure is overtaking traditional circuit switched network infrastructure. It has been argued that maintaining separate voice and data networks is uneconomical, on any scale, if the data network can be made to support a voice switching fabric.

In contrast to circuit-switched telephony switching fabric implementations, packetized voice technology establishes media stream channels over networks where the available bandwidth is pooled. Any device on the network can send and receive individual packets in varying quantities and sizes. In this way the bandwidth of the transmission facility is shared and used as efficiently as possible.

Figure 8-30. Voice-over-packet clouds

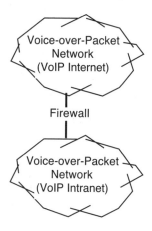

A telephony media stream in this environment is delivered as a sequence of packets from one point on the network to another. This is referred to as a virtual circuit because there is no dedicated bandwidth. The collection of endpoints that can be connected in this fashion represents a voice-over-packet telephony switching network and is represented graphically as a cloud (as shown in Figure 8-30), much the same way that the other telephony switching networks are represented. Also, as shown in the diagram, the scope of a given voice-over-packet telephony switching network may be limited to just the Internet or just an intranet within an organization, depending on the implementation of the firewall between the two.

8.6.1. Packetization vs. TDM

In a TDM media stream channel, media stream data is delivered as a constant stream of bits. The corresponding analog voice signal can be regenerated from this stream —whether it contains silence or actual voice data—and delivered to a speaker in real-time so it can be heard by a human being.

Figure 8-31. Circuit switched network

A switching fabric that implements packetized media streams divides the media stream into arbitrary chunks known as packets. These packets are transported through the packet network in a fashion dictated by the QoS rules (if any) supported by the data network in question. Typically there is no guarantee that packets will take the same route to their destination or that they will arrive in the same order they were sent. If a packet is directed through a congested portion of the network, it may be delayed or lost. At the termination of the packet media stream channel, the receiver must reconstitute the media stream with the packets it receives. If a packet is lost there is typically no time to request for the packet to be resent so it is replaced with silence or interpolated values. The resulting media stream is often called *pseudo isochronous* because it isn't guaranteed to be 100% isochronous.

Figure 8-32. Packet network

8.6.2. Voice Compression

A key to making packet-based switching fabrics more cost effective is the use of voice compression and encoding technology to minimize the bandwidth associated with a given media stream. The *vocoder* (voice coder) for a given implementation is an algorithm designed to squeeze the media stream data into significantly less than 64kbps. This is accomplished by eliminating silence and using knowledge of normal speech patterns to approximate the original media stream. There is a direct trade-off between the reduction of bandwidth required and the faithfulness of the resulting media stream to the original sound. Vocoders assume that the media stream is human speech. This means that a given vocoder may compress and decompress the voice of someone speaking English in such a way that there is very little distortion. However, the same vocoder might severely distort a media stream consisting of music. In any case, the process of compression adds to the latency associated with packet-based switching fabrics.

8.6.3. Voice over Frame Relay (VoFR)

Frame relay technology was one of the first wide-area data networks to be used to carry packetized voice in commercial applications. An enterprise with an existing frame relay infrastructure in place can realize significant savings routing long distance calls through this network rather than through the PSTN.

In May 1997, the Frame Relay Forum published FRF.11. This specification defines how *Voice over Frame Relay, VoFR,* should be implemented in order to provide interoperability between VoFR implementations from different vendors.

8.6.4. Voice over ATM (VoATM)

ATM is capable of carrying a media stream with guaranteed QoS because it is based on synchronous transmission. However, it is also capable of carrying packetized voice by taking advantage of its asynchronous capabilities. If one is willing to make the quality trade-offs, packetized voice is significantly more bandwidth efficient than constant bit rate voice in an ATM network.

The ATM Forum's AAL2 specification defines the ATM protocol for delivering packetized *voice over ATM,* or *VoATM.* Like the FRF.11 specification for frame relay, it is intended to allow various vendors to offer customers interoperable VoATM functionality.

8.6.5. Voice over IP (VoIP)

The most popular packetized voice technology is Voice over IP. *Voice over IP,* or *VoIP,* refers broadly to the technologies used to deliver telephony media streams using IP (Internet Protocol) networking technology.

The advantages of VoIP stem from the fact that IP has become a universal network infrastructure. IP is available and universally deployed in both LANs and WANs including the public Internet. IP WANs are frequently implemented by layering IP over TDM spans, frame relay, or ATM which in turn are most often layered over DSL[8-11]. IP access is also available over cable TV networks and wireless environments. So VoIP allows a single infrastructure to work in many networking environments.

The VoIP-based switching fabrics that form the basis for IP telephone systems represent a key milestone in the evolution of computer telephony solutions because they eliminate the barriers to creating computer telephony products in all categories. Until recently, much of the openness in computer telephony was limited to the periphery of conventional telephone systems because at their core was a proprietary switching fabric. Call control, media services, and administration products could be integrated with a legacy telephone system such as a PBX but the PBX itself was not necessarily open.

As IP-based switching fabrics are typically implemented using CT technology, they have the potential to be open and thus allow the complete decomposition of telephone systems while simultaneously merging telephony and data networks. Call control, media services, and administrative functions are distributed across an IP network as are the clients that are able to access them. The VoIP-based switching fabric utilizes this same network to connect the various endpoints of the telephone system including network interface devices, media access devices, and telephone stations.

Competing vendors are implementing IP telephony products in a myriad of different ways. Some vendors are building best-of-class products in a single category. Others are building products that span a number of areas and close the interfaces between these components

8-11. VoDSL — Voice over DSL refers to products and services carrying a telephony media stream over a DSL circuit using an unspecified network protocol stack. Some VoDSL implementations may channelize the DSL link use a packet transport such as IP.

VoIP and IP Telephony

IP telephony is a field of computer telephony that involves building telephone systems using the Internet Protocol (IP) stack and industry standard networking technology to implement the core telephony switching fabric over which call control, media services, and administration are layered. On the other hand, Voice over IP, or VoIP, refers broadly to the technologies used to establish and deliver telephony media streams using IP networking technology.

An iPBX is an implementation of a complete IP telephone system. See Chapter 10, section 10.8 for a discussion of iPBXs and their components.

while exposing appropriate external interfaces. Still more are building suites of modular products that allow customers to mix and match with competing products.

The use of mainstream computer technology (in this case IP networking protocol stacks and network equipment) allows the implementation of this radical new type of telephony switching fabric in a modular fashion. One of the most attractive features of VoIP is that implementations typically reflect the ideal model shown in Figure 8-1. In older circuit technologies the distinction between the transmission network and the modules making up the telephony switching network layers are not easy to isolate, but for VoIP they correspond to specific software modules that vendors must implement.

There is no industry consensus on all of the protocols to be used to implement a complete VoIP system. Even within specification setting organizations such as the IETF (Internet Engineering Task Force) there are competing approaches. However, most VoIP products are based on one or more of the available specifications so while products are not universally interoperable, there is a reasonable level of interoperability among the products of vendors backing particular protocol sets.

8.6.6. IP Media Stream Channel Protocols

Packet-based telephony switching fabrics use an underlying transmission network that shares bandwidth by delivering data in variably sized bursts called packets. With VoIP, media stream information is broken into packets and routed through an IP network. Challenges associated with VoIP involve ensuring that no media stream packets are delayed or lost as they travel across the IP network as these would lower the quality of the resulting sound output. The IP networking industry has been investing heavily in technologies which address these QoS issues.

RTP

There appears to be broad industry consensus over the protocols used for carrying media stream channels over IP. The IETF has defined *RTP*, the *real-time transport protocol*, which is layered on top of *UDP*, the *user datagram protocol*, which is widely acknowledged as the transport layer for IP telephony media streams. RTP takes advantage of any underlying IP networking QoS schemes such as the IETF-defined *DiffServ*, *differentiated services protocol*, and *RSVP*, *resource reservation protocol*, schemes.

Voice Compression and Encoding

The ITU-T (the International Telecommunication Union's Telecommunications Standardization study group) has defined a series of different voice compression and encoding schemes which have been assigned designations of the form G.xxx. The ones in use for IP telephony include G.711, G.722, G.723, G.728, and G.729. G.711 corresponds to the uncompressed PCM audio found in a DS-0 media stream channel. Once again there is little doubt that these speech coders, along with future G.xxx series specifications, will form the basis for IP telephony media streams.

8.6.7. IP Switching Fabric Endpoint Signaling Protocols

Another aspect of telephony switching fabrics involves conveying endpoint signaling information. Analog telephone technology delivers this signaling information through tones or by breaking the electrical circuit (a *pulse*). The alternative to this stimulus-based signaling is message-based signaling in which messages are sent using a separate data transport. Because the move from stimulus to message-based signaling took much longer than the move from analog to digital networks, both signaling methods are used over TDM networks—either stimulus signaling is passed through the same channel as the media stream (in-band signaling) or a separate channel is set aside for signaling messages. In IP-based switching fabrics, signaling is delivered as additional packet data over the IP network. This has the advantage of being more efficient than in traditional circuit switched networks and also has the advantage that signaling messages need not be delivered to the same network node as the associated media streams. VoIP implementations may utilize "signaling proxy servers" that handle all signaling on behalf of a number of different endpoints.

The area of endpoint signaling protocols is one area of VoIP specifications where there are two competing sets of well defined interfaces.

H.323

The ITU-T has published the *H.323* family of standards as the collection of specifications applicable to IP Telephony switching fabrics. It is defined using a model consisting of *terminals, gateways, gatekeepers, multi-point control units* or *MCUs*. H.323 defines the functional roles of each of these components and defines the protocols used for certain interactions between them. In addition to specifying the use of RTP and G-series voice coders for media stream channels, H.323 includes a collection of protocols used for signaling between endpoints and between endpoints and the gatekeeper function. The H.323 protocols are rooted in the assumption that the endpoints have autonomy. This requires that the endpoints support a rich set of signaling protocols.

H.323 defines five logical components that make up the VoIP switching fabric:

- IP transport

  The IP transport is the IP LAN or WAN implementation including the cabling, routers, and other infrastructure that deliver IP packets from one place to another. A VoIP switching fabric does not use a switch matrix or switch bus to connect all of the inputs and outputs, instead it uses the IP network which already connects all of the potential media stream sources and destinations.

- Terminal

  The *VoIP terminal* is an endpoint in the VoIP switching fabric.

- Gatekeeper

  The *VoIP gatekeeper* provides the role of switching function control in the VoIP switching fabric.

- Gateway

  VoIP gateways are special VoIP endpoints which allow the H.323 switching fabric to interface with other switching fabrics.

- Multi-point Control Unit (MCU)

  The VoIP MCU is the resource that provides dedicated support for multi-point calls. Unlike traditional switch matrix and switch bus switching functions, an IP network is capable of performing multi-casting where the VoIP terminal (which is a logical function of the switching fabric associated with an endpoint) is actually instructed to send identical media stream data to multiple destinations and to locally perform a mixing (conferencing function) on all the incoming media streams associated with the multi-cast destinations. Multicasting in this fashion requires every endpoint to have significant built-in processing resources. To more efficiently use processing resources and bandwidth, a dedicated conferencing function called an MCU can be embedded in the VoIP switching fabric to transparently provide conferencing of multiple streams just as traditional switching functions would do.

All of these logical components can be implemented on distinct IP network nodes or can be grouped. Typically a gatekeeper, gateway, and MCU will be installed on a single server with additional gateways and MCUs added to the network as required in the form of additional servers.

Figure 8-33. H.323 network

SIP

The IETF has also developed a set of protocols that are being applied in the implementation of IP telephony switching fabrics. They are *SIP*, the *Session Initiation Protocol*, and *SDP*, the *Session Description Protocol*, a simple set of protocols that allow endpoints to locate one another, negotiate media stream properties, and establish media streams.

Figure 8-34. SIP network

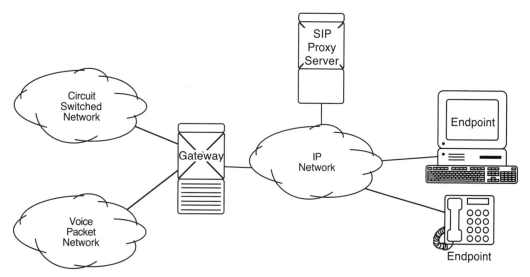

SIP defines the VoIP switching fabric in terms of logical endpoints that act on behalf of users and invite one another to establish a media stream. It also defines additional server components that allow endpoints to locate the endpoints corresponding to specific users. Redirect servers provide endpoints with alternative locations and proxy servers search for the desired user on behalf of the requesting endpoint.

SIP's functionality is comparable to H.323 but the protocol is arguably more efficient in terms of the number of transactions required to set up a call and much easier to work with as it is text-based and quite similar to HTTP.

8.6.8. IP Switching Fabric Control Protocols

Switching fabric control protocols are another form of signaling used by the switching control function to control switching fabric components. In the case of an IP telephony switching fabric, these components are distributed across an IP network and consist of gateways and other media stream endpoints.

MGCP

MGCP, the *Media Gateway Control Protocol*, was developed to allow the implementation of VoIP networks in which the switching control function could be centralized and developed independently of media gateways responsible for interconnecting VoIP media streams with other telephony circuits and with other VoIP media streams.

MGCP defines a master-slave relationship between a *media gateway controller* and a *telephony gateway*:

- In MGCP, a media gateway controller corresponds to the switching control function in the switching fabric implementation. It is responsible for managing all of the switching resources, determining how a media stream should be established through the switching fabric, and for directing media gateways to perform the necessary interconnections.

- In MGCP, a telephony gateway is a single instance of an interconnection function within a distributed VoIP network. A telephony gateway is able to interconnect media stream channels and relay corresponding signaling information between the VoIP network and interconnected telephony circuits of other types.

MGCP is available as an IETF informational RFC and has been widely implemented.

Figure 8-35. MGCP architecture

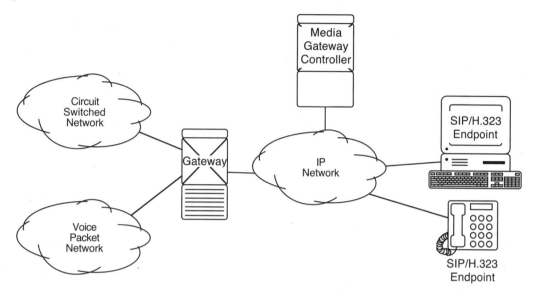

MEGACO / H.248

The *MEGACO* protocol, also known as *H.248*, is a specification under joint development[8-12] by the IETF and ITU-T. It expands on MGCP by specifying a robust framework for switching fabric control and provides extensibility for new types of gateway capabilities through the definition of new MEGACO packages.

It is expected that this specification will ultimately be adopted as the basis for switching fabric control in VoIP and all open switching fabric implementations. MEGACO will allow switching fabric control implementations to manage media and signaling gateways, station servers, and individual stations. (See Chapter 10 for more on the nature of each of these VoIP system products.)

8-12. MEGAGO status — At the time of writing, MEGACO/H.248 had not yet been published as a standard.

8.7. Wireless Circuits

Wireless telephony circuits cut the cord between telephone stations and other equipment in a telephone network. They do this by using *radio frequency* or *infrared* communication technology to eliminate wires, however the signalling and media stream protocols used in various wireless systems span the analog, digital, and packet approaches that we've already explored.

8.7.1. Wireless Telephony

Wireless telephony links come in a huge variety of different forms. Examples of common wireless telephony facilities include:

- 1.6–1.8 / 49.8–49.9 MHz cordless telephones
- 900 MHz cordless
- CT2
- Personal Handyphone (PHS)
- Digital Enhanced Cordless Telecommunications (DECT)
- 802.11 wireless IP telephones
- Radio telephones
- Specialized Mobile Radio (SMR)
- Cellular (AMPS, GSM, TDMA, CDMA)
- Aircraft public telephones
- Rail public telephones
- Local Multipoint Distribution Service (LMDS)
- Multi-channel Multipoint Distribution System (MMDS)
- Microwave links
- Satellite links

Despite the diversity of wireless options, the basic concepts of a wireless telephony circuit are very simple and are illustrated in Figure 8-36.

Figure 8-36. Wireless circuit

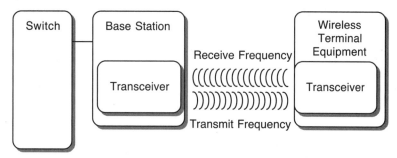

Wireless technologies establish links without cables. The general model for wireless communication involves a wireless device that uses radio frequency or light frequency links to connect to a base station which, in turn, provides an access point to a particular wired or wireless network.

8.7.2. Air Interface

Air interface refers to the protocol stack, including the physical layer, associated with a given wireless implementation independent of the actual frequency or frequencies over which a given transmitter is programmed to operate.

8.7.3. Infrared Wireless

One approach to implementing a wireless link is to use a pair of transceivers that operate in the *infrared*, or *IR*, spectrum. The disadvantage of infrared is that it requires careful alignment of the two transmitter and receivers and the link will fail if something blocks or deflects the light beam between the transceivers. The advantage of infrared is that it is harder to intercept an IR beam than it is to intercept radio frequency transmissions and is thus more secure.

8.7.4. Radio Frequency Wireless

Given the limitations of IR links, *radio frequency,* or *RF,* links represent the vast majority of wireless options. Radio frequency provides the ability to travel much longer distances and to pass through obstacles. However, aside from issues of eavesdropping, the ability for RF waves to propagate in all directions represents a significant issue when it comes to the sharing of RF spectrum. Unlike deployments of wireline networking technologies, the laws of physics limit the number of wireless transmitters that can operate in a given space at the same time. The density that can be achieved for a given system operating in a given area is a function of the transmission power of the transmitters, the total bandwidth available, and the bandwidth required per link. Bandwidth is a public resource that is licensed by government regulators (for example, the FCC in the USA) and is carefully rationed among many competing technologies and application areas. Wireless networks rely on maximizing density and optimizing the use of available bandwidth by:

- dividing coverage areas into cells with a base station in the center of each cell, and
- limiting the power of the transmissions within each cell to minimize the propagation of signals into other cells.

Another challenge in the use of radio frequencies is that different parts of the RF spectrum have different characteristics when it comes to passing through obstacles ranging from buildings to raindrops.

Air interface technologies vary in terms of their applicability to different cell sizes (from a few feet to many miles), in their applicability to a given part of the RF spectrum, and in terms of how they cope with interference from other RF transmissions.

8.7.5. Wireless Link Attributes

While the list above represents a sampling of the many wireless offerings available, a comprehensive list would be too long to be practical.

Wireless network implementations are typically referenced in terms of:

- Air Interface technology (e.g. CDMA), or

- Frequency range allocated (e.g. 824-894 MHz), or

- Service name (e.g. Cellular, PCS).

This leads to some confusion because a wide variety of air interfaces can be used to deliver a service (such as cellular) in a given frequency range.

In addition, there are a number of other very important wireless link attributes that can be used to classify and compare available wireless offerings. Given the nature of this technology, all of these attribute categories overlap with one another to some extent, but each represents an important distinguishing characteristic between one scheme and another.

Shared vs. Dedicated

Wireless devices may be configured with dedicated spectrum so that, like a wired connection, there is effectively a dedicated link available to the device at all times. For example, two microwave transmitters configured to communicate with one another will use an assigned, fixed frequency. In most cases, however, wireless devices are placed into systems that share available bandwidth among all the users of a given base station. When all the available spectrum associated with a given base station is in use, all other devices are blocked.

Fixed vs. Roaming

The end device in a wireless link may be fixed or roaming. Fixed wireless devices are rooted to one location, like a wireline device, but use a wireless link. A popular use of fixed wireless links is in wireless local telephone loops in rural areas and other situations where laying cable is cost-prohibitive. Examples of wireless technologies for fixed links include LMDS, MMDS, microwave, and infrared. Where a fixed wireless link is appropriate, the technologies used typically require that the transceivers be carefully pointed at one another with a direct line of sight.

Roaming wireless devices are used by people on the move (on foot or in a vehicle). These devices actually switch between different wireless base stations as they move and must support being handed off between one base station and another.

FDM vs. TDM vs. Spread Spectrum

Just as wireline telephony circuits options have evolved from analog to channelized digital facilities to packet-based wireless air interfaces have evolved through an analogous set of technologies.

Wireless *frequency division multiplexing*, FDM, involves dividing available frequencies into channels that can be assigned to a single analog media stream. Analog cordless telephones and AMPS cellular phones are examples of wireless schemes that use FDM.

Wireless *time division multiplexing*, TDM, involves multiplexing a series of digital media streams into a single frequency range just as wireline TDM technology does. GSM and TDMA are air interfaces that implement TDM.

Spread spectrum technology pools the entire time/frequency channel among all of the links using a specific block of RF spectrum. CDMA is an air interface that uses spread spectrum technology.

Packet vs. Circuit

A given air interface may be designed to permit transceivers to send in variably sized packets or as a continuous stream corresponding to a circuit. *IEEE 802.11* is a wireless IP network scheme that provides an untethered alternative to a wired ethernet network for VoIP.

Frequency Band

The use and allocation of RF bandwidth is dictated by different government agencies around the world. The ability to use a particular telephone in a particular location is not only a function of the technology it uses but the frequency band in which that technology is allowed to operate in the particular location in question.

Frequency also determines the applicability of particular technologies to different applications. High frequency RF transmissions (short wavelengths) travel in straight lines and are more susceptible to being blocked by everything from rain drops to buildings. LMDS and microwave, for example use very high frequencies and thus are typically applicable to line of sight applications.

Service Name

The marketing and regulatory names assigned to particular services are often quite confusing as they do not always relate to the corresponding air-interface or frequency allocation in obvious ways. The following are examples of commonly used service names.

- Cellular
    While the term *cellular* refers to any wireless network organized into cells to support the reuse of frequencies within a given service area, it is most frequently used to refer to a particular service offered by wireless carriers. In North America this refers to the services offered by carriers in the cellular spectrum. Likewise around the world, the definition of "cellular" service is generally driven by spectrum usage and the applicable spectrum licensing agency.

- PCS

  In the US, despite the fact that the same digital air interface technologies (and thus the same services) can be offered in both the cellular spectrum and the PCS spectrum, the name *PCS* is given to those that are offered in the PCS band.

- DCS-1800

  DCS-1800 was the name given to the second generation of pan-european digital cellular service. However, unlike the US, where the service name PCS is used to differentiate from cellular, DCS-1800 is used only to distinguish phones capable of operating in the new bands allocated for DCS-1800 but the service is still generally referred to as Cellular or GSN.

- SMR

  Specialized mobile radio, or *SMR*, is yet another set of FCC licensed frequency bands. A wide variety of carriers offer different services in the SMR bands, and some offer a mobile radio service that is identical in most ways to the functionality promised by digital cellular (telephony, paging, mobile fax and digital data, etc.) but adds the benefits normally associated with two-way radio (the previously dominant technology in the SMR band).

8.8. Review

Switching resources are responsible for carrying out services that create, clear, and manipulate the states of connections, and for allocating and deallocating media stream channels as needed.

Media stream channels are paths of communication that can be established to convey a *media stream* of either voice or digital data. A voice media stream consists of a signal requiring no more than 3.1 kHz of bandwidth which may be transmitted through an analog connection or as a digitized and optionally compressed stream. Through modulation, data can be carried over a voice channel, allowing the delivery of data over both analog and digital transmission facilities. Other attributes of media stream channels are the level of *Quality of Service* (*QoS*) that they support.

A *switching fabric* implementation consist of a a *telephony switching network* implementation layered over one or more *transmission networks*. The telephony switching network consists of a *switching control function* which is responsible for managing the switching fabric, *media stream interconnection resources* that are responsible for interconnecting media stream channels, and components that are capable of supporting the delivery of both signaling information and media streams over each of the available transmission networks.

Switching control functions may be *centralized, distributed,* or *decentralized.* Similarly, media stream interconnection resources can be distributed across a network in the form of *media stream gateways.*

Media stream channels and signaling are carried over *transmission facilities,* generally called *circuits* or *lines,* that include (but are not limited to) *analog, ISDN Basic Rate Interface* or *Primary Rate Interface* (*BRI* or *PRI*), *T-1/E-1 span, B-ISDN/ATM, voice over DSL* (*VoDSL*), packetized *voice over frame relay* (*VoFR*), packetized *voice over ATM* (*VoATM*), and *voice over IP* (*VoIP*). These facilities all vary in the way in which they manage signaling (in-band or out-of-band), the number of channels they support and the properties of those channels.

Switching Fabric Implementation

Simple *analog circuits* use either two wire pairs (send and transmit) or a single pair of wires using a *hybrid circuit*. A single media stream channel is supported on each circuit. Signaling is performed in-band using tones and AC voltage applied over the circuit.

Digital carrier circuits use *multiplexing* technology to combine multiple signaling and/or media stream channels onto a single set of cables or a single wireless link. While early digital carriers continued to use in-band signaling, modern implementations use *Signaling System 7* (*SS7*) or proprietary out-of-band messages.

Voice-over-packet technologies provide a family of switching fabric options which layer telephony over packet-oriented data networks. Examples include some *VoDSL* implementations, *voice over frame relay* (*VoFR*), *voice over ATM* (*VoATM*), and *voice over IP* (*VoIP*). The most popular of these is VoIP because IP networking applies to both LANs and WANs and is already ubiquitous. VoIP involves digitizing then compressing an audio stream and delivering it over an IP network as a series of packets. *H.323* and *SIP* are alternate *endpoint signaling protocols* used in VoIP switching fabrics. *MGCP* and *MEGACO/H.248* are *switching fabric control protocols* that are used to control distributed switching fabric resources such as media stream gateways, station servers, and telephone stations.

Wireless circuits may be categorized as *fixed* or *roaming*. Fixed wireless links involve some type of a fixed relationship between two *transceivers*. Mobile (roaming) wireless links involve a mobile transceiver that is able to connect with appropriate nearby transceivers, known as *base stations* and then switch to a new base station as its location changes. Wireless circuits are also categorized as being either *shared* or *dedicated*. *Dedicated wireless links* are those that dedicate a wireless channel between a given pair of transceivers even when it is not in use. *Shared wireless links* involve the sharing of wireless channels which means that one transceiver may not be able to establish a link if another transceiver is already using the channel.

9.
Administration

Administration functionality is an essential but often overlooked aspect of telephone systems. This is particularly true in the field of computer telephony where the rush to build new products and develop CTI and media services features has taken the focus away from the more mundane issues of administration. However, with the growing maturity of computer telephony technology and products, companies throughout the computer telephony value chain are making the development of administration products, and the interoperability specifications and standards that govern them, a priority. Responding to both the needs of system buyers, carriers, and product vendors, the ECTF has emerged as a leader in the effort to define frameworks and specifications for various aspects of administration. However, work is still in the early stages.

This chapter will provide an introduction to the key issues in administration, identify the significant technology trends, and provide references to the existing specifications in this field.

Administration

Administration is a field of CT in its own right because the design and implementation of software, hardware, systems, and interfaces for administering a telephone system are largely independent of the implementation details associated with call processing, media services, and switching.

Telephone system administration is concerned with five areas:

- Fault monitoring

    Fault monitoring involves monitoring the status of telephone system components to locate, identify, and possibly correct, failures and sources of potential failure.

- Configuration

    Configuration involves setting options and customizing a given telephone system for use in a particular setting and maintaining settings over time.

- Performance management

    Performance management involves collecting statistical data about the performance of the various components of a telephone system including its human operators in order to maximize the utilization of resources and ensure that system availability, responsiveness, and throughput are within acceptable margins.

- Security Management

    Security management deals with limiting access to the services provided by the telephone system as well as the information stored in, and generated by, the system.

- Accounting

    Accounting involves measuring telephone system usage for billing or budgeting purposes. It may also track account information and authorize system use based on available credit.

9.1. Fault Monitoring

Fault monitoring involves not only detecting faults but also monitoring a system in order to identify issues before they become problems and probing a system to trouble-shoot problems when they are detected. Fault monitoring tools interrogate all the components within a system about every aspect of their physical and logical configuration and status.

SNMP, the *simple network management protocol,* was defined by the IETF for remotely detecting, diagnosing, and correcting network problems. SNMP has been universally adopted as the standard protocol for monitoring and profiling system components distributed across a network. Deploying SNMP involves SNMP agent software, SNMP console software, and an information framework known as a MIB, or *management information base*:

- SNMP agent software must be installed on each system component. The SNMP agent is responsible for collecting information about that system component.

- SNMP console software use the SNMP protocol to communicate with SNMP agents and request selected information about a given system component.

- The SNMP agent tracks and delivers information based on a MIB. The MIB specifies how each piece of information is to be identified, how it is to be organized, and whether it can be modified. The MIB effectively defines the vocabulary that the SNMP agent and console will use.

- If a given piece of information is modifiable, the SNMP console may be authorized to modify it, thereby updating a setting within a given system component.

The applicability of SNMP to managing a particular system is a function of the MIBs used by the SNMP agents installed on the system's components.

9.1.1. ECTF M.500

The ECTF has published a MIB for use in computer telephony components which is known as ECTF M.500. It complements other standard MIBs which are used for management of the lower level network nodes. The ECTF M.500 MIB[9-1] covers the following areas:

- Servers
   The server services portion of the MIB covers information necessary to manage and monitor a media server. This includes determining if the server is running or not and determining what services and capabilities are supported. The ability to stop and start the server is also provided.

- Resources
   The resources portion of the MIB covers the types, number, and capabilities of the media resources available in a particular system component.

- Clients
   The client services portion of the MIB covers information necessary to manage and monitor a client component. This includes what servers the client is connected to and the applications it is running.

- Hardware
   The hardware portion of the MIB covers information about a given telephone system component's hardware-based resources.

SNMP and the ECTF MIB allows system administrators to use SNMP consoles running anywhere on a network to access current information about each system component. Robotic consoles can monitor specific pieces of information and raise alarms under specific conditions. Administrators can diagnose problems by remotely inspecting the resources of all the relevant components.

9-1. M.500 Version 1 — Version 1 of M.500 covers the areas listed. Future versions may cover additional areas.

9.2. Configuration

As we have already seen, the real promise of computer telephony is increasingly customizable systems that can be adapted to the unique individual needs of their owners. However, every option, feature, or parameter in a telephone system represents a piece of data that the telephone system must check before carrying out a relevant operation. Even the most basic of telephone systems must be programmed with a dialing plan and with information about active devices and their telephone numbers.

Most telephone systems have a significant database of configuration information. Most configuration data can be categorized as related to:

- Devices,

- Users, or

- System Features.

Device Configuration

Device configuration determines what devices are in or out of service and how they should behave. This includes the assignment of device identifiers to specific hardware ports, the specification of logical element attributes, and device configuration settings (see Chapter 4, section 4.5). Device-specific attributes such as the list of devices that are distributed to by a group device and the distribution function used by an ACD device are other examples of device configuration information.

Physical device configuration information involves assigning the meanings and functions of lamps and buttons on each telephone station.

A system administrator must also establish the default forwarding settings for each device to determine how the forwarding feature should behave.

Media access devices, media servers, and their associated media resources are also configured with information that governs how they will operate. The system call router associated with each media server must be configured and application profiles may be established to establish the media group configuration preferences of media clients.

User Configuration

User configuration information involves a database of system users and the assignment of devices, accounts, authorization codes, and class of service (CoS) settings.

System Feature Configuration

System feature configuration settings include information on system-wide features such as callback timers and the behavior of individual call control features.

9.2.1. Off-line, On-line, Start-up, Shut-down

An essential feature of configuration services is the ability to start-up and shut-down individual telephone system components and the ability to take individual devices and system resources off-line and put them back on-line.

Deactivating a device or other resource so that it is no longer being used by the system and thus no longer available to the system's users is referred to as taking it *off-line*. Typically one must take a resource off-line before it can be replaced, upgraded, or reconfigured. Activating a resource is known as putting it *on-line*. This returns it to the pool of available resources.

System components such as media servers, call processing servers, and gateways that process many calls must be shut-down for maintenance or upgrades.

The key to supporting these configuration services is the ability to do so in stages so that there is no disruption of service. When a request is made to shut-down a server, the server must first stop accepting new calls but should still continue processing existing calls. When the server has no calls, the shutdown process may continue by disconnecting all logical clients, shutting down subordinate server processes, and ultimately shutting down the server itself.

The ECTF M.001 specification defines a framework for implementing these operations.

9.2.2. Provisioning

Provisioning refers to adding a new device to a system. This includes creating new entries in the telephone system's configuration database for the device and establishing all the associated device and user related settings for the new device.

9.2.3. Moves, Adds, and Changes (MACs)

The most common configuration tasks are known as *MACs* which stands for *moves, adds, and changes*. Whenever an employee is hired, changes locations, or leaves a company the telephone system's configuration information must be immediately updated to reflect the change.

9.2.4. Command Line Interfaces

Conventional configuration management tools have been limited to rudimentary *command line interfaces*. These are typically implemented using serial ports and text-based terminals or telephone stations with text displays. Command line interfaces require administrators to learn the unique and specialized commands for a given telephone system and this limits the availability of staff that can perform these tasks, increases the likelihood of error, and increases the expense associated with performing telephone system administration.

9.2.5. Browser-based Interfaces

A key trend in the development of administrative interfaces for telephone system components is a move to mainstream web browsers to provide a graphical user interface for system administration that can be accessed from any computer on a company's IP network. Telephone system vendors are implementing their administrative interface as embedded web servers that allow administrators to log on with a secure connection and to perform all configuration activities from taking devices off-line to configuring the lamps and buttons on telephone stations through a web interface. This is another example of using off-the-shelf computer technology to simplify administration and to make administrative tools platform independent.

9.2.6. Directory Services

Another key trend in the evolution of telephone system configuration tools is the use of directory services technology.

Directory services refers to a special type of database service that is designed specifically for consolidating directory information about entities such as people, groups, and network resources. The directory databases are designed to store information that identifies and locates the entities as well as any desirable attributes of the entities. Directory systems are designed to allow administrators to quickly and easily add and remove entities as well as to update the attributes of one or more entities in a single step. Directory services implementations are designed to secure information in the directory and simultaneously to make it easily available to every component on a network.

The increasing trend towards user-specific customization of telephone systems and the priority placed on efficient MACs has made user configuration a priority. A likely result is that telephone system vendors will increasingly use directory servers as the repository for all configuration information Storing user configuration information as attributes in mainstream directory services systems

Directory services are likely to play a key role in CT-based telephone system configuration management because:

- CT is driving a trend towards user-specific customization that makes the user portion of a system's configuration database the largest. By using a mainstream directory services implementation to store telephone system configuration data, only a single list of user names exist in the directory. This eliminates the need for two separate databases that must be kept synchronized.

- System buyer's are demanding more robust, easy to use, mainstream administrative tools. By treating telephone system resources as entities in a directory system and storing the associated configuration information as attributes, mainstream directory services applications can be used to manage telephone system configuration information. This relieves vendors from having to develop proprietary administrative tools.

- MACs remain the biggest administrative challenge for telephone system administrators. By consolidating telephone system administration with the administration of other network services, this workload is greatly diminished.

- The trend towards distributed telephone system architectures, particularly IP telephone systems, requires that configuration information be distributed across the network as well. Once again, by utilizing available directory services technologies the telephone system vendor does not have to develop their own distributed database technology.

Until standardized directory schemas for telephone system configuration information are developed, software that accesses this information will remain proprietary. However the use of standard directory access protocols such as LDAP, the lightweight directory access protocol, published by the IETF make this information more accessible and ultimately standardized schemas will result in many new software tools for administration.

9.2.7. ECTF M.100

The ECTF M.100 specification defines an API and associated protocol for implementing administrative applications. The M.100 administrative interface allows an application to start and stop a server; provision, remove, and check the status of resources; take resources in and out of service and run diagnostics on them; and manage application profiles.

9.3. Performance Management

Performance management involves collecting statistical information about the operation of a given telephone system. This allows a system owner to understand how the system (and its human operators) are performing. With this knowledge the system can be incrementally modified to improve performance.

The information upon which statistics are based must be gathered from all over a distributed system. A system may provide a built-in function for delivering performance management statistics or the information from a number of different CT interfaces may be used to generate these statistics.

A significant challenge in the area of performance management has been the lack of standard definitions and formulas for telephony performance data. This problem is much worse in CT systems that are made up of components from many different vendors because performance data from all of these components must be combined in order to formulate a complete picture.

9.3.1. ECTF R.100

The *ECTF R.100* specification is the first in a series of specifications planned by the ECTF to address the issue of standardized reporting of statistical telephone system data.

ECTF R.100 specifies the terminology, practices, source data, timers, counters, and formulas that are to be used in monitoring call processing performance including the operation of ACD, ACD Group, and Agents. Call centers, or contact centers, are particularly concerned with this kind of statistical data because it allows system owners to understand not only how the system itself is performing but how effective the overall solution is.

9.4. Security

There are three aspects of telephone system security:

- Ensuring that telephone system services (and the resources that support them) may only be used by people that are authorized to do so.

- Ensuring that access to configuration information is limited to authorized individuals.

- Ensuring that information carried within, or generated by, the telephone system is only accessible to approved clients.

Implementing security in each of these areas involves using one or more of the following security technologies:

- Authentication
  The act of verifying the identity of the entity you are communicating with (one or both directions).

- Authorization
  Controlling access to resources based on the identity discerned by authentication.

- Integrity Protection
  Determining if data has been altered.

- Privacy
  Keeping unauthorized individuals from accessing information they are not authorized to see.

While computer telephony promises many benefits over conventional telephony implementations because of its modularity, flexibility, and customizability, these same attributes make CT systems susceptible to breaches of security that were previously unkown in monolithic telephone systems with few, if any, external interfaces. As a result, these system could rely largely on physically securing the system components. On the other hand, distributed CT components must communicate with one another over a network and without the appropriate security in place, anyone with access to the network could interfere with its operation or steal information. Examples of security breaches in these systems include:

- CTI software can access an unsecured CTI interface and use this to monitor all calls made to or from a particular device. Even without tapping into the associated media streams, a great deal of business information can be stolen by using this knowledge. For example, if the president of a given company is observed making many calls to a head-hunting firm or to the president of a competing firm one might infer that a merger might be in the company's future. The same data could also be captured if a legitimate client was monitoring these calls (such as software on the president's computer) and the associated messages were captured by an eavesdropper.

- An unsecured CTI server allows client software to initiate calls from arbitrary telephone stations. Someone could use this mechanism to place a call from a speakerphone in a conference room to their own telephone. Unless the occupants of the conference room noticed the speakerphone in-use indicator, they would be unaware that their conversation was being listened to by a remote telephone. This breach of security represents an easy way to effectively bug a given organization's offices in an undetectable fashion using the existing telephones.

- An unsecured media server allows client software to access or alter messages stored on the media server. In this fashion someone could instruct a media server to playback recorded voicemail and even to delete or substitute a new or different recording.

- An unsecured media server would allow an application to allocate all available resources of a given type thereby denying all other applications access to these resources and denying callers from using the corresponding services.

- An unsecured media stream in a VoIP switching fabric allows someone with the appropriate software running on a machine connected to the IP network to eavesdrop on conversations taking place in the system.

Those building and deploying CT systems must employ the appropriate security technologies to ensure that CT solutions are secured from the threats described earlier.

Authentication

Authentication involves using technologies that verify the identity of a particular user or client. The strongest forms of authentication involve using a password ("something you know"), presentation of smartcard or some physical token ("something you have"), and a piece of biometric data such as a voiceprint or fingerprint ("something about you").Most systems rely on just a password. In any case, authentication information must be transmitted between client and logical server in a secure fashion, otherwise this data may be intercepted and used by someone impersonating the valid user.

Authentication is an essential feature of all logical servers within a CT system. Authentication services are typically coupled to the implementation of directory servers.

Authorization

Authorization involves a database of information which specifies the privileges which are to be afforded to a particular authenticated client. A key entry in the authorization database is the identity or identities of those permitted to update the authorization database.

Authorization works in conjunction with authentication. It may or may not be required given the logical server in question.

Integrity Protection

Integrity protection refers to technology that allows an electronic seal to be placed on data. The seal, or digital signature, allows the recipient of some data to verify the origin of the data. This involves public key encryption technology which ensures that only the unique holder of a secret private key can generate a signature that may be decrypted with a public key.

CT systems may use integrity protection technology to guard against intruders inserting themselves into an already authenticated session. If all the messages associated with a given session are signed, an intruder would be unable to generate a fictitious message.

Privacy

Privacy involves encrypting information so that eavesdroppers are unable to intercept and decode information travelling across a network and are unable to access stored information.

Encryption is used in CT systems to protect both messages and media streams from eavesdroppers on a network. IPSec is the emerging standard for encrypting all information communicated on an IP network. As CT systems are increasingly based on IP networks, IPSec is likely to emerge as the primary privacy technology used.

9.5. Accounting

CT system accounting functions provide one or more of the following features:

- Call detail records that indicate what user (or device) made use of which service(s) or resource(s) and for how long.

- Billing information that combines call detail records with pricing data. This may be used for recovering or apportioning operating costs (in private networks) or for billing subscribers (in public networks).

- Credit tracking and reporting that uses billing information to debit a users account and cut off service when the account balance reaches a certain level. This is used for pre-paid services.

Most systems provide *call detail recording,* or *CDR,* which refers to the ability to generate call detail records. Conventional telephone systems traditionally deliver CDR information through a serial port. A printer or a computer that records these records can then be attached to the serial port. CT systems, particularly IP telephone systems, typically use the IP network to convey this information. Unfortunately, there is no clear standard covering the content, structure, and mechanisms for delivering this information.

9.6. Review

CT administration technology includes *fault monitoring, configuration, performance management, security,* and *accounting.* It is one area of computer telephony that is not yet very mature.

Fault monitoring involves monitoring a system in order to identify issues before they become problems, detecting failures, and probing a system to determine the cause of a problem. *SNMP,* the *simple network management protocol,* is the most emerging standard for monitoring CT system components. The *ECTF M.500* specification defines an SNMP *MIB* (*management information base*) for computer telephony.

The biggest administrative chore in most systems is configuration. This includes *device configuration, user configuration,* and *system feature configuration.* Features in this area of administration include starting up and shutting down system components, setting individual telephone system resources to on-line and off-line states, and provisioning new devices. *Moves, adds, and changes (MACs)* refers to updating a telephone system's configuration to reflect the addition, loss, or relocation of a system user. Performing MACs may involve updating both user configuration information as well as device configuration information. Interfaces for performing configuration tasks range from primitive command line interfaces to browser-based interfaces using HTML and other world wide web technologies.

Directory services technology with interfaces such as LDAP are also likely to be critical aspects of future telephone system implementations. The *ECTF M.100* specification defines an API for certain administrative operations in a CT system.

Performance management involves collecting statistical information that provide insights into how a telephone system and its human operators are performing. The most significant challenge in this area is defining standard terminology, formulas, and data collection practices. The *ECTF R.100* specification establishes standardized terminology, practices, source data, timers, counters, and formulas that are to be used in monitoring system performance in areas related to call processing.

Security is a critical aspect of CT system design. Securing a CT system involves administering it to prevent unauthorized individuals from using it, accessing the information it manages and generates, and accessing configuration information. Security features involve mechanisms for *authentication* of individuals, databases to track the *authorization*, or privileges, granted to a particular individual, as well as technology for *privacy* and *integrity protection* that respectively prevent capturing or altering CT system information.

CT system accounting functions may provide *call detail recording*, or *CDR*, for detailed information on system use, billing information that combines CDR with pricing information, and credit tracking that also maintains credit balance information to support pre-paid services.

10.
Telephony Equipment and Services

We turn now to looking at how actual telephony products and services are implemented and how they implement the functionality we have explored. The concrete examples provided in this chapter breathe more life into the concepts and abstractions presented earlier.

Each individual telephony product is a set of telephony resources that goes into a telephone system. Telephone systems are generally subsystems of a larger telephone system, and virtually all telephone systems are ultimately connected to, and thus are part of, the world-wide telephone network (the PSTN).

This chapter is a buyer's guide of sorts. It deals with the wide array of telephony equipment and services available to anyone building or updating a telephone system. We'll look at the various product categories and the factors to consider when comparing alternatives. Telephony product categories include conventional CPE switches and switch peripherals including, telephone stations and peripherals, media servers and gateways, and IP telephone systems. The chapter concludes with a discussion of the service offerings from traditional carriers and a new generation of telephony service providers.

10.1. Assembling a Telephone System

Assembling or upgrading a telephone system, whether it be for your company, for a client, or for your own home or business, involves combining a number of different telephony products and services to satisfy your unique requirements.

Figure 10-1 depicts a complete telephone system. At the center of the system is a CPE telephone switch which forms the hub of this private telephone network. In addition to the telephone stations and telephone station peripherals which the switch provides service to, there are a variety of computer telephony client and server products and other peripheral devices that enhance the functionality provided by the switch. The telephone system is connected to one or more external networks either directly or through gateway products.

External Networks

If you are assembling a new telephone system, or are assessing your existing system, one of the most important steps is to determine what services you want to obtain from public carriers or other telephony service providers and what services you want to implement using hardware and software that you deploy yourself.

The functionality of your telephone system (whether it is as small as just a single telephone or as large as a collection of switches forming a global private network) will be a combination of the features and services built into your telephony equipment and the telephony features and services that you subscribe to.

At a minimum you'll need a service provider for access to local calls and a carrier for long distance calls. However, shopping for telephone service is no longer just a question of calling your local telephone company. Choices include competitive local exchange carriers (CLECs), cable TV operators, utility companies, Internet service providers (ISPs), and more. Most of these telephony service providers offer features and services that go well beyond basic access to the PSTN.

Figure 10-1. Telephone System

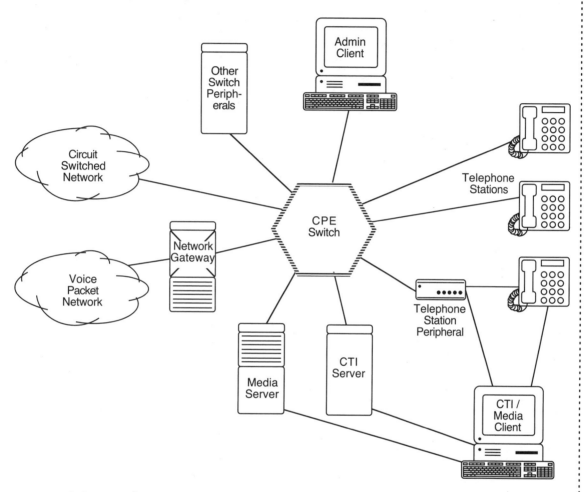

Selecting the appropriate mix of products and services is a basic "buy versus rent" question. You will need to weigh the cost of acquiring and maintaining telephone system equipment against the cost and reduced flexibility of subscribing to comparable services. You should investigate all the subscription options available before deciding on a particular implementation strategy and keep these trade-offs in mind throughout your evaluation and planning process.

CPE Switch

The trunks or lines that you already have and those that are available to you from various telephone companies (carriers) will, in part, dictate what types of telephony products are applicable to your needs. Assuming that you decide not to completely outsource all switching functionality, you'll need to choose a switch (or collection of switches). Your options in this category of customer premises equipment include:

- Front End Switch
- Key System
- Monolithic PBX
- Modular PBX
- iPBX

The switch that you select will drive all the other service and equipment choices you make.

Gateways and IADs

If you want to connect your switch to external networks that your switch doesn't support, such as an IP Telephony service on the Internet or your own private wide area network, then you'll need to install an appropriate gateways. Gateways may be packaged in an *integrated access device* (*IAD*) which delivers not only telephony media streams but other services such as cableTV and internet service.

Media Servers

Media servers are the basis for deploying customized telephone system solutions that involve automated interaction with callers. Unless your switch came with a built-in set of accessible media resources, you'll need a media server to support applications ranging from customer self-service inquiries to unified messaging.

Other Switch Peripherals and Options

Other products to integrate with your switch range from uninteruptable power supplies to dedicated voicemail systems to administrative tools.

Telephone Stations

No telephone system purchase decision would be complete without deciding on telephone stations. Options for telephone stations are constrained by the choice of switch (or vice-versa). However, the selection of telephone station features and options to choose from is vast. However it should be noted that the best phone interface of all is the right piece of CTI software running on a personal computer nearby. (See Chapter 12.)

Telephone Station Peripherals

Last but not least, there are a wide variety of add-ons available that connect to telephone stations or connect to extension circuits from the switch.

10.2. Conventional Telephone Switches

CPE switching equipment can take many different forms. Historically, switching products have been divided into four principal categories based on different underlying technologies and the resulting functionality. These four types are:

- Front-end switch

- Key system unit (KSU)

- Private branch exchange (PBX)

- Application-specific switches (e.g., ACD switch, etc.)

As all switch implementations evolved and came to be based on digital technology, features previously associated with only one type of device became easy to implement in any product, and the differences began to blur. Most new switch implementations are capable of virtually any feature, so the categorization of switch products has become increasingly arbitrary. Over time, the distinctions between switch types will continue to blur. Any switch may include any or all of the telephony resources described in Chapter 3.

The size of a switch is fundamentally measured in terms of two independent characteristics: its capacity for connecting lines (trunks and extensions) and its capacity for handling calls. A *non-blocking switch* is one that has enough capacity for calls that every line can be used simultaneously. A switch can vary in line capacity from just two lines to thousands of lines.

The other important characteristic that differentiates switch designs is their ability to be upgraded, and the limits to that upgradability. At the simplest level, a switch is simply a box, or *cabinet*, with a power supply and printed circuit boards. Switches that cannot be upgraded in any way have a fixed number of ports to which trunk and extension lines can be attached. A switch that can be upgraded may allow for the addition of entire auxiliary PBX cabinets, or for the addition of individual printed circuit boards to one or more *shelves* within the PBX cabinet. Add-in circuit boards may provide additional telephony resources of any type. Typically the cards will provide a set of additional trunk ports or extension ports of a given line interface type. Other examples of possible add-in functionality include media access devices (ranging from new DTMF detectors to a voice mail system), an ACD device function, an upgrade to switching resources (i.e., support for more conference calls), or a CTI interface.

10.2.1. Front-End Switches

A *front-end switch* is a very simple type of switch that sits between a series of CO lines (its trunks or network interface devices) and a corresponding number of telephone stations (its extensions or station devices). No switching takes place between extensions. The switch is capable only of establishing calls to connect (and disconnect) each extension to (and from) its designated trunk. A front-end switch may be modeled as illustrated in Figure 10-2. In the illustration, network interface devices are labeled with an "N," logical device elements are labeled with an "L," and physical device elements are labeled with a "P." Potential calls are shown in gray.

Figure 10-2. Front-end switch model

The main purpose for a front-end switch is providing a more sophisticated user interface to otherwise very primitive subscriber lines from another switch. This is accomplished by making the extension lines proprietary digital (or proprietary second pair) so that more sophisticated proprietary telephone stations can be used, or by providing a CTI interface on the front-end switch. The call processing function in the front-end switch accepts commands from the proprietary stations or the CTI interface and translates them into appropriate commands for the CO switch.

This type of switch is a very attractive way of providing Centrex services to a particular office or location. An appropriate front-end switch can be connected to a T-1 line from the CO switch, so that the telephone company need only provide four wires to an entire location of 24 or fewer users. Instead of accessing Centrex functionality through DTMF sequences that have to be memorized, Centrex telephone users at the location take advantage of sophisticated telephone stations, computer interfaces, or a combination of the two. (CTI configurations like this one are presented in Chapter 11.)

10.2.2. Key Systems

Key system units, or *KSUs,* are switches that are very similar to front-end switches, except that each logical station device has bridged appearances instead of the standard appearances seen in a front-end switch. Key systems use special telephone stations that are able to indicate the logical device with which they wish to interact. The typical mechanism used is a button, or "key," on the telephone set that a user can press. This is how the name "key system" was derived. A key system can be modeled as illustrated in Figure 10-3.

Figure 10-3. Key system model

The bridged appearances used in a key system are shared–bridged by default, but a system that supports privacy implements exclusive bridging.

Some key systems have a special button (corresponding to a logical device) on the phone that is associated with an intercom or with the receptionist's telephone station. One example of this is shown in Figure 10-4. In this illustration, the receptionist's physical device is PR and it has access to logical devices L4, LR, L1, and L2. If the receptionist presses the button corresponding to L4, he is connected to the public address system or intercom and can announce something like, "Joe, the call on line 2 is for you." If anyone using P1 or P2 presses the button corresponding to L0, rather than connecting to a network interface device, the system places a call to LR, which then may be answered by the receptionist.

Figure 10-4. Example key system with attendant button and PA system access

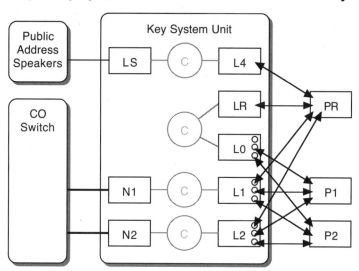

Like the front-end switch, the key system implements few, if any, telephony features and services itself. Instead, it treats its network interface devices as proxies and instructs the switch providing the corresponding line to carry out any requested services. In this way, the key system adds the functionality of bridging (and, depending on the implementation, feature-rich telephone stations, and CTI interfaces) to whatever telephony features and services are supported by the switch that it is front-ending.

Squared System

A *squared key system* is one in which every telephone station has access to every logical device; typically the button for each of the logical devices is in the same place on every physical device. The key system shown in Figure 10-3 is squared. The term *squared* reflects the fact that, in this type of system, the maximum number of possible interactions that has to be supported via bridging is the square of the maximum number of physical devices supported. (It is assumed that there will never be more logical devices than physical devices.)

Hybrid

A *hybrid key system*, or *hybrid switch*, is one that supports devices which are both bridged and not bridged. If the devices with standard appearances can establish calls to or through devices other than a single designated network interface device, the switch is effectively a PBX that supports bridging.

Virtual Key System

In a *virtual key system*, there is no central KSU cabinet. Instead, pieces of the KSU functionality are implemented among each of the special telephone stations that make up the system. Each telephone station has a connection to every other telephone station and to all of the trunks. Call processing and switching functionality is distributed among all of the telephone stations. Despite the fact that there is no physical KSU cabinet, this type of key system is still modeled as shown in Figure 10-3.

10.2.3. PBXs

A *Private branch exchanges*, or *PBXs*[10-1] is a general-purpose switch. It typically implements all telephony features and services internally, may connect any device to any other, and may support both standard and bridged appearances. A PBX may be modeled as illustrated in Figure 10-5.

Distributed PBX

A *distributed PBX* is one that is implemented as a series of distinct PBX cabinets that may be located at some distance from one another. The implementation of call processing functionality and switching

10-1. PABX and CBX — PBXs are also called private automatic branch exchanges (PABXs) or computerized branch exchanges (CBXs) by some vendors in order to contrast them with switching systems that require humans to perform the actual switching role (as was true with early key systems and cord boards). As manual systems are now all but obsolete, this clarification is not required; PBX is the preferred term for all products in this category.

Figure 10-5. PBX model

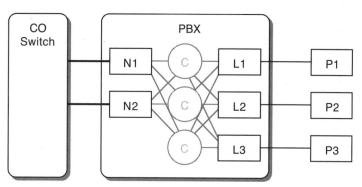

functionality is distributed among the pieces of the PBX; these pieces are constantly in communication in order to coordinate their activities. Some vendors support a feature in which, if a number of their PBXs are connected to one another in a private network using tie lines, the PBXs can establish digital data calls among themselves and use these to begin operating as a single distributed PBX.

Personal PBX

A *personal PBX* is to a traditional PBX what a personal computer is to a mainframe computer. The personal PBX is small, if not tiny by PBX standards. It has much less capacity than the average PBX, but is designed to be a peripheral in an individual's home, office, or small business. Unlike the traditional PBX, it is designed with an emphasis on requiring little, if any, administration or maintenance while delivering the complete range of telephony functionality found on a normal PBX.

Personal PBXs are designed to be used in environments that typically are equipped with a personal computer, so a full-function CTI interface is an important part of these products.

10.2.4. Application-Specific Switches

Application-specific switches are those that are not intended for general-purpose use in the same fashion as PBXs and KSUs, but instead are designed for some very specific and limited application.

The most common example of application-specific switches are stand-alone ACDs. These are basically PBXs that are built around a highly functional ACD or ACD group device. The network interface devices (trunks) for these switches are often connected to an organization's PBX, rather than to a central office switch. While many general-purpose PBXs include ACD or ACD group functionality, or have it as an option, other vendors specialize in ACD functionality. An application-specific switch is one way they can package their ACD functionality. (Another way is through a CTI interface.) One advantage—or disadvantage, depending on your point of view—is that application-specific switches may utilize proprietary telephone stations with displays that provide application-specific information.

10.3. Switch Peripherals and Add-Ons

A number of dedicated peripherals are available for most switches. These peripherals generally are dedicated products that provide service through one of the switch's interface ports, by attaching as a station device, or a combination of the two.

10.3.1. OA&M Interfaces

A switch's *operations, administration, and maintenance,* or *OA&M interface,* provides the functionality necessary for setting up and administering it. The OA&M interface may also provide diagnostic or status information and allow telephony resources to be taken in and out of service.

Standard functions of an OA&M interface include:

- Assigning device addresses;

- Assigning bridged appearance relationships;

- Defining and assigning classes of service;

- Assigning devices to hunt, pick, and ACD groups;

- Setting up dial plan management rules, including prefixes, LCR rules, and associated network interface devices;

- Defining network interface device associations for incoming calls on non-DID trunks;

- Fixing system defaults for forwarding; and

- Configuring media service devices.

Once a switch has been set up, the primary use of the OA&M interface is for *moves, adds, and changes,* or *MACs*. This refers to the day-to-day need to cope with people moving from office to office, and joining or leaving an organization. The assignment of telephone numbers to lines must be modified every time one of these events takes place.

An OA&M interface implementation not only permits administration of these features, but also exposes a mechanism by which a human user can interact with it. Three common approaches are providing a console interface, providing support for the direct attachment of an ASCII terminal, and using a LAN-based connection to a separate computer.

Console

A commonly implemented OA&M mechanism for small switches involves using telephone stations.

If the OA&M options are sufficiently simple, an ordinary touchtone phone can be used. In this case, a special sequence of DTMF tones is interpreted as a command to place the telephone station into OA&M

console mode; then more DTMF tones can be used to set features. While this type of implementation is inexpensive, it is also error-prone and difficult to use.

Another variation on this approach involves special console telephone stations with many buttons and a large display. When this telephone station is placed in OA&M mode, instructions and feedback appear on the set's display and buttons are used to select options and make settings.

Terminal Interface

The most common access mechanism for OA&M implemented in PBXs is the OA&M interface serial port. This is a serial port (or set of serial ports) that allows a terminal to be connected, either directly or remotely through a modem. Once the terminal is connected, the OA&M interface can be manipulated as if it were any other type of text-oriented computer to which a terminal had been attached. Commands are entered on the terminal keyboard and feedback appears on the screen. Some vendors implement a very primitive user interface based on a simple stream of characters that are read in and printed out. Other vendors provide a slightly easier-to-use interface that employs the whole surface of the terminal display, making status easier to read and understand.

Network Interface

The third approach, which is likely to become the dominant one eventually, involves using a LAN connection, This will allow application software on a computer to connect to the switch for purposes of monitoring and configuring it.[10-2]

10-2. OA&M protocol — Though a specific OA&M protocol for switches does not exist at the time of writing, this is a likely target for industry groups now that standard CTI protocols have been developed. Many vendors have already begun to build custom extensions to SNMP, the TCP/IP-based management protocol that is the de facto standard for network management and others have begun supporting a web browser interface using HTTP and HTML. See Chapter 9, section 9.2 for more information on these technology options.

10.3.2. Telemanagement Systems

A *telemanagement system*, sometimes referred to as a *call accounting system*, is a computer-based system designed to help manage and account for the use and assignment of a switch and associated facilities. Telemanagement systems connect to the switch's accounting interface to obtain information pertaining to the usage of telephony resources.

The switch's accounting interface is typically connected to the telemanagement system using an RS-232 serial port. The information delivered using this serial port is often referred to as *call detail recording* (*CDR*) or *station message detail recording* (*SMDR*) information. For each call originated, the accounting interface generates a record of information detailing the starting and ending date and time, plus all the call associated information including the calling device, the called device, the network interface device used, the account number and access code used, etc. This information streams out of the switch through the appropriate serial port. Some organizations don't use this information and leave the port unconnected; others attach a line printer so that this information is captured and later can be analyzed manually. In general, however, most organizations want the ability to make full use of this information, so they install a telemanagement system.

Telemanagement software typically uses the CDR information received from the switch, along with all the rate information for the carrier associated with each network interface device. The telemanagement system estimates the cost of each call and can provide reports that can be used in various ways, including:

- Identifying toll fraud activity;
- Billing clients;
- Billing against projects or departments;
- Validating telephone bills from carriers; and
- Adding a telephone charge to a hotel guest's bill.

Telemanagement systems also are used to conduct traffic studies. A *traffic study* is a statistical analysis of call data to determine whether telephony resources and systems are being used in a cost-effective or optimal fashion. For example, a traffic study might determine that even in peak periods only 75 percent of the trunks on a switch are ever used. With this information, the organization can save money by decommissioning the unneeded trunks. Another example might be the discovery that most calls being delivered to a call center are not arriving with ANI information. In this case, the call center's software can be adjusted to deal with ANI-free calls more optimally, and the carrier can be contacted to see if the problem is in their network.

10.3.3. Voice Mail

A *voice mail system* is a switch peripheral that uses media access to record and play back messages. Voice mail systems can take many different forms, but they typically have the following basic features:

- Call associated information, such as the identity of the called device, is used to determine the intended voice mail recipient.

- If call associated information is unavailable, the voice mail system may interact with a caller to obtain the identity of the desired voice mail recipient.

- Based on the identity of the recipient, a custom greeting is played. This might be simply the recipient's name incorporated into a single, system-wide greeting. It also could be a prerecorded greeting left by the recipient, or one of a set of prerecorded recipient greetings that is selected based on a rule such as the time of day.

- Messages may be tagged with a priority (urgent/not urgent), a status (new/listened-to/saved), and time and date information. When the messages are retrieved, they are accessed based on this tagging.

- Tagging can be used to determine how a recipient should be notified of messages. For example, a system might flash a special "message waiting" lamp on a telephone set to indicate new messages, and might send a pager message to indicate the arrival of new/urgent messages.

- Message tagging may be used as a basis for deleting certain messages automatically. For example, all messages older than two weeks may be deleted automatically, or all messages that have been listened to (but not tagged as saved) may be discarded.

- Recipients can retrieve messages by calling a special number and using DTMF commands to authenticate themselves and to interact with the voice mail system. Typical commands include next message, previous message, delete message, save message, rewind, fast forward, and pause. When messages are played back, all or some of the tag information may be communicated as well.

Voice mail systems may be attached to a switch in many different ways. Some voice mail systems are designed to be added directly into a switch, and all logical devices used by the voice mail system are strictly media access devices. Other voice mail systems use lines from the switch, and their logical devices are actually station devices that have associated media services.

10.3.4. Universal Mailbox

A *universal mailbox system* is a voice mail system that typically has the following enhancements:

- Accepts faxes in addition to voice messages;

- Is integrated with an electronic mail system in some fashion; and

- Can be accessed electronically through a computer interface.

A universal mailbox system generally allows a recipient to view a list of all of the items in his or her mailbox, and allows individual messages to be retrieved through a computer interface.

Universal mailbox system implementations generally have a connection to a local area network for computer access, in addition to their connections to the switch.

10.3.5. UPS

One of the most important switch peripherals is an *uninteruptable power supply* (*UPS*) that will ensure that the telephone switch will continue to operate in the event of power failures or other power anomalies. Aside from the fact that a switch often is a significant asset, it is generally one of the most relied-upon systems for any organization. A UPS will protect the switch from power sags and spikes that have the potential to do serious damage. In the event of a power failure, the batteries (or generator) associated with the UPS will ensure that the telephone system continues to operate. In prolonged power outage, the switch can be shut down in a graceful fashion automatically.

10.3.6. Cross Connect

Cross-connecting two analog circuits is the process of forming an electrical connection between two circuits. To form a complete end-to-end circuit between a telephone station and a switch, one must cross connect the pair of wires going to a telephone station with the pair of wires coming from the switch. Cross connection of circuits now refers to the permanent, or semi-permanent, interconnection of media stream channels.

A *cross connect* is a piece of equipment used to organize and establish connections between a set of telephone circuits and the inputs to a switch. A basic cross connect for twisted pair cabling is a piece of equipment that organizes the cables feeding into the cross connect and provides places to electrically join, or *punch down*, pairs of wires. *Digital cross connects* are like digital switches that allow the static interconnection of specific TDM channels. *Optical cross connects* provide a patch panel to cross connect fibre optic cables.

10.3.7. Wireless Access Controller

A *wireless access controller* is a switch peripheral that is used to convert wired extensions into wireless extensions. The wireless access controller acts as an interface between the switch and a network of wireless base stations. It allows a set of corresponding wireless telephones to roam throughout the area covered by the base stations and to place and receive calls as if they were wired extensions.

10.4. Telephone Stations

Telephone stations are the telephony products that are connected to switches using line interfaces corresponding to the switch's station devices. Telephone station design is limited only by the imagination of the product designer. Thousands of different telephone designs have been developed over the last century. A number of representative examples are presented in this section to demonstrate how easily all of these products can be modeled. These examples are by no means exhaustive. Expect to find any combination or permutation of the products described here.

In Chapter 4, section 4.1 we saw how the physical element of a station device is modeled, and in Chapter 4, section 4.5 we saw how physical device elements may be associated with one or more logical device elements through device configurations. This section puts these concepts to work as it describes a range of typical telephone station designs.

10.4.1. Single-Line Telephone

The device configuration for a simple, single-line phone can be modeled as illustrated in Figure 10-6. This example represents a basic POTS telephone set, with only the most basic features, on a dedicated line. In this case the device configuration consists of a single device with a physical element and a logical element. (Both elements have the same label.) The logical device element has non-addressable standard appearances.

Figure 10-6. Dedicated POTS line station example

Figure 10-7. Simple POTS telephone set

The telephone station itself might look like the one pictured in Figure 10-7. The physical element consists of:

- A ringer

- A hookswitch that is controlled locally

- An auditory apparatus in the form of a standard handset with a single, fixed-gain microphone and fixed-volume speaker

- A dial pad with twelve buttons

10.4.2. Bridged Line

When multiple analog telephone sets are bridged physically onto the same analog line, the result is sometimes referred to as a *bridged line* or *party line*. The logical device corresponding to the line behaves as described for interdependent–shared–bridged appearance (see Chapter 4, section 4.4.3). This device configuration for a POTS telephone station is shown in Figure 10-8.

Figure 10-8. Bridged POTS line station example

The physical element used in this case is the same as the one for the dedicated POTS line. The difference here is that the logical device element is simultaneously associated with other physical device elements as a part of their device configurations.

10.4.3. Key Telephones

A *key telephone* is a special phone designed to be used with a particular key system (as described in section 10.2.2). Simple key telephones, like the one used in this example, are essentially the same as the physical elements used in the last two examples, except that they have extra buttons (so-called *line buttons*) to allow access to multiple logical devices using bridged appearances. The device configuration for a key telephone station is illustrated in Figure 10-9 and the appearance of a corresponding telephone set is depicted in Figure 10-10.

Figure 10-9. Key phone station example

Key telephones typically use one or more lamps to indicate the presence and status of a call associated with a given button. For example, in one implementation the following indications might be used:

- Lamp mode: *off*
  There is no associated call (*null* connection state).

- Lamp mode: *steady*
  There is a connection in the *connected* or *initiated* state.

- Lamp mode: *wink*
  There is a connection in the *hold* state.

- Lamp mode: *flutter*
  There is a connection in the *alerting* state.

Figure 10-10. Simple key telephone set

• Lamp mode: *broken flutter*
There is a connection in the *queued* or *fail* state.

10.4.4. Multiple Line Telephones

Multiple line telephone stations that make use of dedicated—not bridged—logical devices come in two varieties: analog and digital.

An analog multiple line telephone station connects directly to multiple lines, so each of the associated media stream channels are simultaneously dedicated to the telephone station, but only one can be used by the telephone station at a time.

The digital version is often called a *multiple DN telephone* (for *multiple directory number*). In this case the telephone station is connected only to a single line, but the media stream channels supported by the line are associated with whichever logical device element is interacting with the physical device element at a given instant.

The telephone station itself generally appears the same as in the key telephone shown in Figure 10-10. The logical device elements in the device configuration, however, have non-addressable standard

Figure 10-11. Multiple line station example

appearances. Figure 10-11 illustrates a multiple line device configuration in which the logical device labeled "555-1234" is part of the base device for the device configuration.

10.4.5. Multiple Appearance Telephones

Multiple appearance telephones are found on digital lines. They are like multiple line stations. Rather than having buttons representing different logical devices, however, they have buttons representing different addressable appearances associated with the same logical device. The functionality for the telephone user is therefore very similar, despite the fact that the implementation is quite different. The appearance behavior may be basic or selected. If it is basic, a new call

Figure 10-12. Multiple appearance station example

will be presented in the *alerting* state at each of the appearance buttons, and any of the buttons may be pressed to answer it using the corresponding appearance. If the appearance behavior type is *selected*, new calls will appear only at one of the appearances. A *multiple appearance* device configuration is shown in Figure 10-12 and the corresponding telephone station is shown in Figure 10-13.

Figure 10-13. Multiple appearance telephone set

Appearance buttons and corresponding LCD lamps

Note that the telephone set in this example uses lamps in yet another way. In this case there are three distinct lamps for each appearance button. These lamps are in the form of symbols (implemented using LCDs for each button) that can be turned on and off. These are used as follows:

- No symbol showing
    There is no associated call (*null* connection state).

- Speaker symbol
    There is a connection in the *connected* or *initiated* state.

- Bell symbol
    There is a connection in the *alerting* or *queued* state.

- Stop sign
    There is a connection in the *hold* state.

509

10.4.6. Assistant's Telephone

In many workplace environments, an assistant handles calls on behalf of his or her manager. The logical device associated with the manager's telephone number therefore must be bridged, while the assistant's need not be. This is one case of a *hybrid* device configuration. An example consisting of an assistant's telephone with two managers and multiple appearances for the assistant's logical device is shown in Figures 10-14 and 10-15.

Figure 10-14. Assistant's telephone set

Figure 10-15. Assistant's hybrid station example

This example illustrates yet another way that lamps may be used in a given design. In this case, each appearance button is associated with two lamps, one red and one green. This implementation might use these lamps as follows:

- Green lamp mode: *off*, Red lamp mode: *off*
    There is no associated call (*null* connection state).

- Green lamp mode: *steady*, Red lamp mode: *off*
    There is a connection in the *connected* or *initiated* state.

- Green lamp mode: *flutter*, Red lamp mode: *off*
    There is a connection in the *alerting* state.

- Green lamp mode: *off*, Red lamp mode: *steady*
    There is a connection in the *hold* state.

- Green lamp mode: *off*, Red lamp mode: *flutter*
    There is a connection in the *queued* or *fail* state.

10.4.7. Attendant Console

The attendant for a PBX switch is responsible for answering and redirecting external incoming calls that are not automatically directed to a desired extension through DID, DISA, or an automated attendant. In the *attendant console* example shown in Figure 10-16,

Figure 10-16. Attendant console station example

the station's logical device element has addressable selected–standard appearances that correspond to each of the six network interface devices (trunks) that the attendant is responsible for answering. Two additional appearances are used for internal calls. All external incoming calls associated with one of these trunks are presented to the appropriate corresponding appearance. The attendant can then answer the call, find out the desired destination, and redirect it.

One rendition of the corresponding attendant console telephone set is depicted in Figure 10-17. In this example, the attendant console has four types of buttons:

- Dial pad buttons
- Function buttons ("hold," "transfer," "drop," "park")
- Appearance (trunk) buttons
- Single step transfer (extension) buttons

Figure 10-17. Attendant console telephone set

The trunk buttons are the call appearance buttons associated with the trunks. Extension buttons are special-function buttons that perform a *single step transfer call* service to a preprogrammed extension number. In normal operation, the attendant simply presses a trunk button to answer an alerting call and then presses the extension button to transfer it to the requested destination. The dial pad and function buttons provide additional flexibility, such as selecting reach extensions that are not listed. Pressing the "transfer" button in this example performs a *consultation call* service, with *transfer* as the consult purpose. The desired extension number is then entered on the dial pad, followed by a press of the "drop" button. This completes the transfer operation by requesting the *transfer call* service followed by the *clear connection* service. Similarly, the "park" button invokes the *park call* service and allows the attendant to park a new call at an otherwise busy extension.

It is useful for the attendant to be able to see what extensions are busy before attempting to transfer a call. Therefore every trunk and extension button is provided with an "in-use" lamp, in addition to the lamp used for each trunk button to indicate an actual connection for the attendant console in the *alerting* state.

10.4.8. Desk Sets

Desk sets refer to telephone stations that are attached to a wire (tethered to a desk) and are used by someone on a day-to-day basis at his or her desk. They may have an analog telephone line interface, an ISDN telephone line interface, or a proprietary digital line interface.

Speaker Phone

A *speaker phone* typically has two auditory apparatuses. The first is the standard handset; the second consists of a speaker and a microphone built into the base of the telephone station. This is illustrated in Figure 10-18. Each auditory apparatus has its own hookswitch.[10-3]

Figure 10-18. Speaker phone

Speaker-only Phone

A *speaker-only phone* is a cost-reduced version of a speaker phone where the microphone in the base is omitted; the media stream associated with a call can be heard, but only silence is sent in the other direction.

Wireless Handset

Another variation on desk set design involves wireless handsets. The handset for a telephone station may be connected to the base using a wireless connection rather than the traditional coiled cable. The wireless handset could be provided in addition to or instead of a base handset.

10-3. Internal hookswitch — In many speaker phone implementations, the actual hookswitches are internal to the telephone station and are controlled in such a way that only one is active at a time. The cradle hookswitch control and the speaker hookswitch control are used as inputs to reset the actual hook-switches.

Headsets

Those who spend prolonged periods using a telephone generally take advantage of yet another type of auditory apparatus: the *headset*. Headset designs range from the bulky, dual-cup model with a big microphone boom that drops down in front of the speaker's mouth, to a tiny unit that combines a speaker and microphone into a single ear piece that rests in the ear. Those in the first category are good in noisy environments; those in the latter are good wherever they are practical. Headsets generally attach to desk sets in place of or in tandem with a handset. Desk sets that are designed with headsets in mind use a hookswitch button on the telephone set, rather than a cradle hook switch. If both a handset and headset are attached to the handset connector, a switch is provided to switch between the two.

Headset-only telephone stations are used in environments, such as call centers, where jobs are telephone-centric. If CTI is used in such an environment, the computer interface provides all of the call control; the telephone station itself is used only for the auditory apparatus (here a headset-only unit), or for dealing with a call already in progress in case the computer fails for some reason. Such a device is pictured in Figure 10-19.

Figure 10-19. Headset telephone set

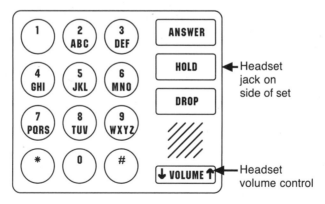

515

Wireless Headsets

Wireless headsets are a variation on headset design involving a wireless link[10-4] between the headset and the telephone station to allow more mobility for the headset user. In any office environment, this allows individuals to leave their offices for any reason and not have to "detach" from their telephones. It also means that they can be notified of incoming calls at their telephones while away from their desks. Wireless headsets are very useful for call center agents who must research items on behalf of callers and must roam away from the desk. For example, if a caller to a catalog clothing company requests a jacket the same color as the one he bought several years ago, the agent handling his call might have to walk to a bookshelf with all of the archived catalogs. With a wireless headset, the agent can walk to the shelf and look up the old jacket while continuing to talk to the caller. Without the wireless headset, the agent would have to put the caller on hold, take off the headset, check the old catalog, and then reverse the process.

Display

A *display* is a matrix of characters that can be displayed by a telephone station. The actual matrix managed internally by the telephone can be larger than the visible portion, provided that buttons on the station allow for scrolling in some fashion. The display may show characters in any character set, using LCD or LED bitmaps or by forming the character in each position using a multi-segment approach.

Though the two most common uses of a display are to allow on-hook dialing and to show callerID information, there is no limit to the type of call associated and device associated information that can be presented. For example, some telephone sets use the display to implement a type of help function that steps a person through the use of a particular feature.

10-4. Wireless headset technology — Spread-spectrum radio frequency technology is preferred because it is least susceptible to interference from other wireless headset users in close proximity.

Figure 10-20. Display phone

The display area should not be confused with other areas of a telephone set's surface that may use LCD as a technique for implementing lamps. In Figure 10-20 there is only one display area, which is one line high and 17 characters across, but it also uses a portion of the same LCD as a message-waiting indicator and a lamp that indicates that forwarding is active.

10.4.9. Wireless Telephones

Wireless telephone stations are those that connect to a switch through a wireless "line" rather than a cable of some sort (see Chapter 8, section 8.7). Wireless stations may connect to a public switch, or to a PBX, or in some implementations may connect to a public switch automatically if a designated PBX is not accessible.

Wireless telephone stations tend to have many of the same features that apply to desk sets, but are constrained by the fact that they must be as small as possible.

Display

A display on a wireless telephone station is another essential element. The need for a display is driven primarily by the need to support on-hook dialing. The telephone set's user can enter a number, backspace to correct mistakes, etc., and only press the send button (which requests a *make call* service) when the number is correct. As with desk sets, displays are also used for presenting callerID information (where available) and other implementation specific help information.

Speed-Dial Buttons

Wireless phones tend to have many (100 or more) speed-dial or "rep" buttons (i.e., stored numbers). Because the units are small, however, there isn't room to have each of these buttons on the surface of the phone. Instead, these speed-dial buttons are virtual, that is, they exist only inside the phone. Users of the wireless telephone set press a sequence of physical buttons in order to trigger a "press" of the virtual speed-dial button of interest. A CTI interface to the same phone would allow using the *set button information*, *get button information*, and *press button* services, respectively, to program a given button with a speed-dial number, retrieve the number, and dial it.

Lamps

Examples of lamps used on cellular phones include those for indicating battery levels, the presence of an active connection, travel outside the local service area, and the lack of any available switch providing service.

Ringer

The ringer in a cellular phone may be auditory or vibratory, or there may be one for each. (Indicating ringing through vibration is decidedly not a feature found on desk sets.)

Auditory Apparatuses

Wireless telephone stations are typically in the form of a handset, so the built-in auditory apparatus is a handset. There usually is a volume control for the associated speaker.

In addition, most new wireless telephones support an earphone jack designed for the direct connection of an all-in-one headset. This is very useful in nearly every context where a cellular phone may be used, but it is particularly important in vehicles because it allows the cellular phone to be used while keeping both hands free for driving.

A speaker phone auditory apparatus is an alternative to the headset for use in vehicles. It involves installing a speaker and a microphone in appropriate locations in the car.

10.4.10. Multi-Function Telephone Stations

Many telephone station products also have other built-in functionality. These *multi-function telephones* have both a telephone station portion and a separate portion that might have related or unrelated capabilities. Examples include:

- Clock-radio telephones

- Pay phones with attached data terminals

- Fax phones

- Video phones

- Combination phone and answering machine devices

In the cases where the other portion of the product's functionality relates to media access, the media service instance is treated as part of the given device and is associated with its logical element. The media service resources in these devices share the same appearance as the physical device. For example, a fax phone consists of both a physical element (the telephone station portion) and a logical element that can, when requested, transparently associate the fax media service instance with calls. When it does so, the auditory apparatus on the associated physical element may or may not be muted during any subsequent fax transmission. In this example, the device configuration is the same as for the POTS telephone stations described earlier (see Figure 10-6).

10.5. Telephone Station Peripherals

Telephone station peripherals are devices that can be connected to a telephone line in addition to a telephone station (through physical bridging or tandem connection), or instead of a telephone station. As with telephone stations themselves, the possibilities in this category are endless; the examples presented here are merely representative.

10.5.1. CallerID Displays

CallerID displays are small units with LCD displays that can be bridged onto a telephone line (typically an analog telephone line). They display the callerID information associated with each incoming call (assuming that callerID service is subscribed to). This type of product is an alternative to replacing an entire telephone set with one that has a display.

10.5.2. Call Blockers and Call Announcers

Call blockers and *call announcers* are variations on callerID displays. While they might or might not have a display for showing callerID, they are connected to the phone line coming into an office or home in tandem with the telephone(s), and they use the callerID information associated with a call to selectively block and/or customize the announcement of calls. A call blocker can be programmed with certain numbers that are to be blocked. If calls are presented with one of these numbers, no telephones attached to the blocker will be rung. The blocker might or might not play a message to the caller. Similarly, a call announcer may use the callerID information to ring the phone in some customized fashion. These devices are functionally equivalent to a telephone company's call blocking and custom ringing services.

10.5.3. Media Access Products

Media access products are products that attach to computer equipment in some fashion in order to get access to one or more media stream channels on a telephone line.

External Modems

External modems are products that can be attached externally to a computer and typically are connected through a serial cable or serial bus. External modems allow for forms of media access including modulated data, modulated fax data, and sampled audio data. An external modem may be attached to one or more telephone lines.

Data Units and Terminal Adapters

Data units and *terminal adapters* are the digital data equivalent to modems on a telephone line. They allow computer systems to get access to the digital data media streams carried by a particular telephone line.

Media Interface Cards

Media interface cards are add-in cards designed to be placed inside a computer system. They attach to one or more telephone lines and implement any type of media access, ranging from speech recognition to modem functionality.

Software Modems

Software modems implement fax and data modulation as a modem would, but do so in software running on a computer system. The software modem works with media streams that are delivered to and received from a telephone line using one of the pieces of hardware described above, or through a simple piece of hardware that looks like a modem but acts only as a digitizer and transducer of the analog signal on an analog line.

10.6. Media Servers and Server Components

Media servers are pieces of equipment that can be deployed in a telephone system to provide media services to other telephone system components and media services applications.

Media servers consist of:

- A media services interface,

- A collection of media resources that can be accessed through the media services interface,

- One or more media access devices that are used to attach the media server to the telephone system, and

- An internal switching fabric for establishing media stream channels between resources.

See Chapter 7 for media services concepts.

10.6.1. Monolithic Servers

A *monolithic media server* is a product with a fixed set of media resources and associated functionality. Monolithic media servers are typically built to support a particular application and are built in a monolithic form for cost effectiveness.

Typical examples of monolithic media servers include:

- Computer modem peripherals

- Stand-alone voicemail systems

- Stand-alone auto-attendants

- Fax servers

- Call recording servers

Monolithic servers represent hardware designed for a particular media resource, such as a modem, or around a specific application, such as a fax server.

The advantage of monolithic servers is that they are generally ready to go out of the box and are often inexpensive.

The disadvantage is that they offer little or no flexibility to add new functionality. This means that a customer using media servers of this type may need to acquire a new media server for every new application. This is inefficient and requires that the system owner must manage and maintain a large number of servers.

10.6.2. Open Servers

The alternative to monolithic servers are *open media servers* that allow system owners to assemble media servers from scratch to meet their unique requirements. With an open server, the customer specifies the physical properties of the server, the number and classes of media resources to be installed, the flexibility required for interconnecting these resources dynamically, and the number and type of media access devices.

The advantages of open servers are not only that they can be configured to meet the initial requirements of a system owner, but also that they can be easily reconfigured or upgraded to include new resources. This flexibility keeps the number of media servers needed in a network to a minimum and simplifies management and maintenance.

10.6.3. ECTF Reference Framework

The *ECTF reference framework* represents the industry consensus for standards-based computer telephony solutions. It consists of both an overall architecture and a comprehensive set of specifications that define the behavior of various system components and the interfaces between components.

The ECTF framework provides a model for open media server construction. The ECTF specifications define the following elements of a media server:

- At the heart of a media server is a piece of server software that implements media services functionality and is accessible to media clients through both a media services protocol and/or programmatic interfaces. The server is known as an *ECTF S.100 server*.

- Media services provided by the S.100 server can be accessed over a network using a protocol known as *ECTF S.200*

- Media services can also be access using a 'C' language interface known as S.100, and a Java language interface known as *ECTF S.410*.

- Media resources may be hardware-based or software-based and these interface to the media server software through a well defined service provider interface known as *ECTF S.300*.

- Hardware-based resources including media access device implementations are interconnected through a standardized TDM bus. The ECTF defines both H.100 based on the PCI bus and H.110 based on the compact PCI bus. (See sidebar "ECTF CT Bus: H.100 and H.110" on page 408).

The relevant portions of the ECTF framework are illustrated in Figure 10-21.

Figure 10-21. ECTF Reference Framework

10.6.4. Chassis

An open media server can be assembled entirely from off the shelf parts supplied by different vendors. The first step in building an open media server is choosing the *chassis* in which to build it.

At the low-end, this could be any computer with a PCI backplane into which H.100 cards can be installed. H.100 is limited by the PCI bus which requires that the computer be switched off to install and remove PCI cards.

Systems that require high availability will typically be built around compact PCI backplanes that support hot plugging of cards. Makers of compact PCI chassis offer a range of products for high availability environments that include redundant power supplies, redundant fans, rack mounting, cable routing, etc.

The chassis may or may not include a main processor and operating system. If not, this must be added to host the media services software and, optionally, software-based resources and local media clients.

Chassis vary in terms of the number of slots they have. It's important to have a good estimate of how many will be required so that the chassis is of sufficient size.

10.6.5. Card-based Resources

Card-based resources, also known as *resource cards* and *hardware-based resources*, include both media access devices for connecting to the telephone system and hardware-based media resources that perform media processing using specialized hardware and/or DSPs. These resources are packaged on H.100 or H.110 cards that are inserted into the slots on the chassis. While the media processing functionality of these cards is based on the hardware, service provider software installed on the main processor with the CT server must accompany these products.

Hardware resources may be packaged as single-function products. Examples include boards such as a 16 port analog interface or a board that supports Spanish speech recognition on 4 media streams (e.g. has 4 ASR resources). Hardware boards may also be multi-function with built-in switching among a variety of on board resources that can all be accessed using the system's H.100 or H.110 bus. For example, a single card could have a T-1 interface, 24 signal generators, 24 signal detectors, 24 players, and 24 recorders.

10.6.6. Host-based Resources

Host-based resources, also referred to as *software-based resources*, are resource implemented entirely in software that run on the host processor. These use the host processor's available memory to implement a memory-based switching fabric and rely on a high performance system microprocessor to allow them to process media streams in real-time. The memory-based switching used by software resources must be interconnected to the hardware-based switching fabric (if any) for full functionality.

10.7. Telephony Gateways and IADs

Telephony gateways are used to interconnect disparate switching fabric implementations. Gateways that interconnect switching fabrics which use in-band signaling must perform the interconnection of media streams as well as convey signaling information. A *media stream gateway* is a product that provides interconnection of media streams along with associated signaling. A *signaling gateway,* is a gateway that translates and forwards signaling information.

Gateways are functionally somewhere between a full featured switch and a cross connect. Like a switch they perform dynamic interconnection of call media stream channels, but they are actually just an interconnection function of a larger switching fabric and are under the control of a switching control function. In an H.323-based switching fabric implementation, the switching control function that manages the gateway is associated with H.323's gatekeeper component. The gatekeeper locate the desired endpoints of a call, validates the CoS and QoS for the call, and deciding how to establish the media stream channels required and manages any gateways involved. A general term for this type of role is *proxy server* because the gatekeeper is able to act as a proxy for the endpoints in the call.

Gateways are generally implemented as a suite of software combined with any necessary media processing and communication hardware packaged on a dedicated machine or integrated with other server functions on a general purpose server. In fact, the hardware requirements of a gateway are the same for that of a media server (discussed in section 10.6). A media server represents the optimal platform for deploying a media gateway as it has all the necessary media processing resources and is typically implemented on a high availability hardware platform.

Integrated Access Devices

Integrated access devices, or *IADs*, are products that connect to an external network that provides a number of different services in addition to telephony. An IAD for connectivity to a Cable TV network

527

provides Cable TV services, secure Internet access, as well as telephony services. An IAD for broadband wireless or DSL may provide just Internet and telephony services.

VoIP Gateways

A *VoIP gateway* is a gateway between one or more VoIP switching fabric implementations and one or more conventional switched telephone line interfaces as described in Chapter 8. Typically, the switched telephony circuits are used as trunks and the VoIP media stream channels may be extensions or trunks, depending on how the gateway is used. As a switching fabric component, the gateway can interconnect media stream channels for calls originating on either side, and can establish conference calls that connect multiple devices on either side.

Internet Telephony

A common application for VoIP gateways is connecting the PSTN to a public Internet-based VoIP switching fabric. This allows calls placed on the PSTN to be carried across the Internet or vice versa, thereby avoiding long distance tolls. This is known as *Internet telephony*. These calls can also originate or terminate in a CPE switch if a VoIP gateway is used to interconnect the private network. This is illustrated in Figure 10-22.

Inbound Calls

One of the best uses of Internet telephony is allowing Internet users who are browsing World Wide Web information to talk to a human if they have questions, can't find what they're looking for, or want to place an order for something. This is illustrated in Figure 10-23.

This is an example of good Internet telephony use, because many residential Internet users dedicate all their available connectivity to make their connection to the Internet and would have to disconnect in order to place the phone call. In addition, because the Internet

Figure 10-22. VoIP gateways

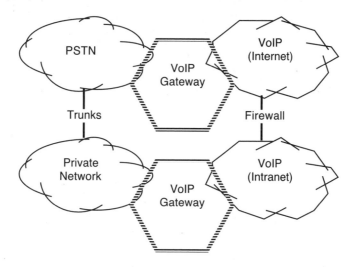

Figure 10-23. Using Internet telephony for inbound calls

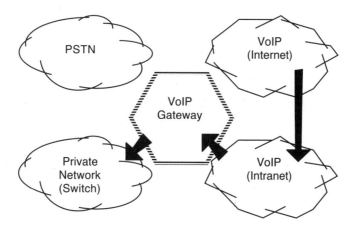

telephony call is being placed to a well-known location that is expecting such calls (i.e., the organization managing the Web site), the issues of spontaneity don't apply.

In this scenario, the Internet telephony gateway is effectively a trunk as far as the rest of the called organization's telephone system is concerned. The difference is that correlator data associated with the incoming call may be used to link the call to the Web site context. The VoIP gateway allows existing call center agents to handle Internet telephony calls in addition to their normal calls, which use existing telephone station equipment.

Internet Tie Lines, FX Lines, and OPX Lines

Another application of Internet telephony through a VoIP gateway involves alternative implementations of tie lines, FX lines, and OPX lines[10-5] in situations where the reduced quality and reliability of Internet telephony are not important. These uses are illustrated in Figures 10-24, 10-25, and 10-26.

Figure 10-24. Using Internet telephony for tie lines

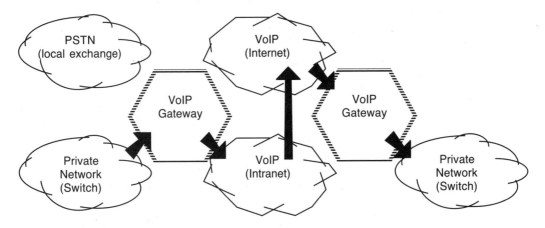

A tie line between two CPE switches is accomplished by routing calls between VoIP gateways attached to each.

An FX line can be implemented by working with an alternate wire line provider in the foreign exchange area that can provide access to the local switch in that location through available trunks on its VoIP gateway.

10-5. Tie lines, FX lines, and OPX lines — These are services normally provided by a public carrier. They are defined in sections 10.9.7, 10.9.8, and 10.9.9.

Figure 10-25. Using Internet telephony for FX lines

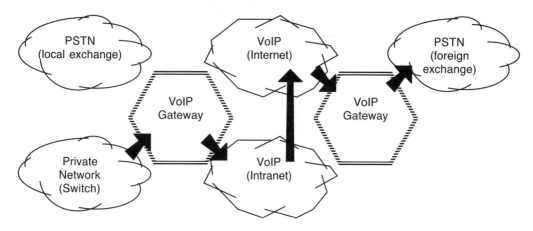

Figure 10-26. Using Internet telephony for OPX lines

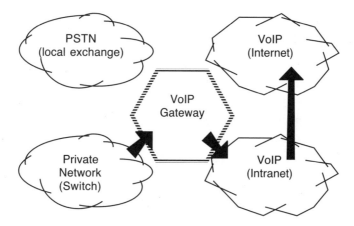

The Internet telephony approach to an OPX line is used when, for example, an employee who telecommutes is using all available connectivity for Internet access. This person will be at a fixed network address for an extended period of time, so Internet telephony calls may be directed to that person's machine on the Internet.

10.8. iPBX

An *iPBX* is an IP telephone system that takes the place of a traditional PBX. It consists of software components that provide call processing, switching fabric control, administration, and media services along with hardware components that implement the iPBX endpoints, including voice network interfaces and telephone stations. Unlike traditional telephone systems, iPBXs are highly modular. Where standard interfaces are used between modules, products from different vendors can be integrated to build a customized iPBX solution.

An iPBX allows a company to deploy a single IP network infrastructure that delivers both voice and data connectivity to every desktop. This consolidation reduces cabling costs and simplifies administration and maintenance. An iPBX provides great flexibility and, most importantly, customizability. In implementing an iPBX-based telephony solution, a business is able to assemble a unique mix of software components and configuration parameters that would not be feasible with a traditional PBX.

IP telephony is able to deliver on CT's promise of modular telephone systems because each component of the telephone system can be implemented as a separate module that communicates with all of the other modules over an IP network using standardized protocols. Figure 10-27 depicts an iPBX—an IP telephone system implemented in a private network—that is decomposed into distinct components.

10.8.1. IP Network

The heart of an IP telephone system is an IP network that connects all of the components together. This network represents the transport layer for the system's primary switching fabric.

10.8.2. Network Gateways

The IP telephone system must interconnect with circuit-switched and/or packet-based telephone circuits. This is accomplished through one or more media gateways and, for networks with out-of-band signal-

Figure 10-27. Fully modularized iPBX

ing, associated signaling gateways. These are responsible for translating and conveying media streams and signaling between the IP telephone system and the connected networks. (See section 10.7.)

10.8.3. Stations

The principal endpoints for a telephone system are telephone stations. iPBX stations may be connected directly to the network in the form of IP telephones and computer workstations.

IP telephones are similar to the digital sets used for ISDN and digital PBX telephone stations in that they have built in media processing and digital signaling capabilities. However, unlike these other digital stations, IP telephones connect directly to an ethernet network rather than to a dedicated telephone circuit.

Computer workstations on the IP network can be configured to operate as PC telephones. They simply run the same voice coders and protocols as IP telephones and utilize their speakers and microphones in place of a telephone station handsets.

10.8.4. Station Servers

Companies that prefer to use existing telephone stations, fax machines, etc., may use station servers as gateways between these telephone stations and the IP network. Station servers implementations may include converted PBXs, servers running on CT backplanes with station interface cards, and dedicated components known as *IP telephony terminal adapters*.

An IP telephony *station server* bridges the gap between telephone station equipment based on traditional telephone line circuits and the new world of IP switching fabric technology. A station server appears on the network as a proxy for one or more terminal devices and acts as a media and signaling gateway specifically for those devices. Just as gateways that provide access to external networks do, it interconnects the appropriate packet-based IP telephony media streams to the media stream channels for the appropriate line circuit. Likewise, it interprets and translates the signaling from the line circuit and delivers it to the IP telephone system on behalf of the telephone station while carrying out instructions from the system to generate appropriate signaling to the telephone station.

IP telephony station servers implementations include:

- Converted PBX

  Customers that wish to migrate to an iPBX from a PBX with proprietary digital stations will generally have to dispose of both the PBX and the associated telephone sets they are used to. However, vendors of certain conventional PBXs are helping with the migration by providing hardware and software that converts the PBX into a station server so that the telephone sets and associated wiring remain where they are.

- ECTF H.110 Backplane Cabinet

  General purpose station servers designed to connect standard analog or ISDN station equipment may be implemented using a standard compact PCI chassis and the ECTF H.110 TDM backplane. A processor card in the chassis controls the server and H.110 boards known as line cards provide the line interfaces.

- PCI-based Computer

  A general purpose computer with a PCI bus can be configured as a station server on the IP network. It can utilize line cards plugged into its PCI bus connected together using the ECTF H.100 TDM bus.

- Terminal Adapter

  An internet telephony *terminal adapter* is a dedicated station server product. It may be in the form of a small box designed to sit on or under a desk or may be in the form a rack-mountable box. Terminal adapters are similar in appearance to network routers and small PBXs. In addition to a power connector, they have an ethernet connector to connect to the IP network and one or more jacks to which a telephone station can be attached. Like routers, they are typically administered over the IP network, but some may also have a local serial port or front panel for administration.

- Residential Gateway

    A *residential gateway* is a terminal adapter with only one or two analog telephone ports designed for residential applications.

10.8.5. Media Servers

Media servers represent the third type of media stream endpoint that may be found on an IP telephone system. They may be present if media processing is desired over and above that already embedded in the media gateways and station servers in the system.

10.8.6. Switching Control Server

The *switching control server*, known as a gatekeeper in some implementations, is responsible for managing all of the IP switching fabric components distributed across the IP network.

10.8.7. Call Processing Server

At the heart of the whole iPBX is the *call processing server*. The call processing server implements call control for the system. It directs the switching control server, handles all requests for call control services, manages call timers, reports call details through appropriate reporting and billing interfaces, and supports CTI applications through a CTI interface. The system may have multiple call control and switching control servers to function as back-ups in the event that one fails.

10.8.8. CTI Clients

CTI clients distributed across the IP network augments the functionality of the call processing server. CTI software may be installed on user workstations to provide a software-based interface for handling telephone calls and it may be installed on servers to automate some aspect of call processing.

10.8.9. Administration Software

Administration software is used to start and stop system components, configure call processing and other components, to collect performance data, and to monitor for component failures.

10.9. Telephone Service Providers

As described earlier (in Chapter 3, section 3.2.5), a LEC, or local exchange carrier, generally provides the connection(s) between your telephone system and the PSTN. In other words, the LEC owns the switch or switches that extend lines or trunks to your telephone equipment, for local calling and typically for access to IXCs (interexchange or long-distance carriers). You may, however, connect directly to one or more IXCs through what is referred to as carrier bypass. (Note that the notion of bypass applies only in the context of competing carriers. In countries where a single telephone company provides all of these services, this notion does not apply.)

A switch owned by the LEC to which your lines or trunks connect is known as a *class 5 central office switch*, (*CO switch* for short), or a *public exchange*. The class 5 CO switch is the workhorse of the carrier industry as it is the component in the PSTN that provides the on-ramps and off-ramps for virtually all telephony traffic. A typical class 5 switch must therefore be equipped to support the thousands of individual homes and businesses served by a given central office. This means it must support a wide variety of line interfaces and be equipped with batteries and hardware to drive the analog lines that make up the majority of subscriber lines.

The combination of line interfaces and telephony features and services available for subscription on a given CO switch represents the carrier's *service offerings*.

10.9.1. Alternate Wireline Providers

All of the services described in this section are provided by traditional telephony common carriers. These are companies and organizations that manage the switches making up the PSTN (the worldwide public telephone network). In some locations, however, these services or equivalent services also can be accessed through alternate wireline providers. These are organizations that provide an alternative communications path for delivering telephony services. Cable TV companies are among the first to begin providing telephony services; power and other utility companies may do likewise. ISPs and those providing telephony services through the Internet represent yet another form of alternate telephone company.

10.9.2. Alternate Non-wireline Providers

A very popular alternative to the traditional delivery of telephone services through copper or fiber cable is wireless telephone service. Cellular carriers are the most prevalent form of wireless providers. however, other forms of wireless subscriber loops involving delivery of Internet services in addition to telephony services are also popular.

10.9.3. Individual Subscriber Lines

One way to obtain telephone service from a carrier is simply to order one or more *individual subscriber lines*. These are effectively extensions associated with the carrier's CO switch, and each has a unique telephone number. At a minimum, these lines generally offer the basic POTS functionality of placing, answering, and dropping calls.

Your telephone company also may offer various supplementary telephony features and services that can be activated for some fee. The set of available services, the names under which they are marketed, and their cost vary widely between countries, states, telephone companies, and between different brands of CO switches used by a carrier. A well-defined group of functions in North America are referred to as *CLASS*, or *Custom Local Area Signaling Services*. These services include callerID and features that take advantage of other call associated information.

10.9.4. Centrex Services

Centrex services are a variation on individual subscriber lines, wherein an entire group of lines is purchased at a time and the lines all behave as if they were connected to a private (CPE) switch. Centrex is generally positioned as an alternative to buying your own telephone system; a portion of the CO switch is dedicated to your telephones and simulates the operation of a self-contained switch.

The chief advantage and disadvantage of Centrex are opposite sides of the same coin. The advantage is that you are able to rely on the telephone company to manage and administer your telephone system for you. The disadvantage is that you are dependent on them and have little or no local control over how and when administration is carried out. Organizations most likely to use Centrex are those that have multiple physical locations in the same geographical area, all of which should be connected to the same telephone system. If all of the extensions in all of the sites would have to be connected via a CO switch in any case, Centrex generally is a more efficient approach than having a CPE switch. Theoretically the overall capacity of the CO switch is likely to mean fewer limitations to the expansion of your telephone system than a smaller CPE switch, but this is not always the case.

10.9.5. Combination Trunks

A *combination trunk* is a trunk line that allows *two-way* (inbound and outbound) operation. In other words, it is basically just an individual subscriber line that corresponds to the network interface device of your switch, rather than the line associated with a piece of telephone station equipment.

Combination trunks can be formed into an *incoming service group*, or *ISG*, which is a hunt group within the CO switch that allows an incoming call to use the next available trunk (network interface device) on the receiving switch. Your business' main number is assigned to the hunt group so that your callers will never get a busy tone, assuming you have enough trunks in the ISG to handle your peak number of simultaneous incoming calls.

10.9.6. DID Trunks

DID trunks are trunks that support only incoming calls and that support the DID feature described in Chapter 5, section 5.6.2. DID allows an incoming call to be directed automatically to the appropriate extension by having the calling switch indicate the desired extension to the answering switch; it does so by dialing the last two, three, or four digits of the number originally dialed. DID trunks are always provided as a set of lines associated with an ISG (incoming service group or hunt group) that directs each new incoming DID call to the next available DID trunk.

10.9.7. Tie Lines and Private Networks

One type of service you may obtain from a carrier is a *tie line* that connects two CPE switches in different locations, thereby creating an extended private network. This type of point-to-point line is treated as a trunk by both of the CPE switches concerned, and it is up to you to set up the dial plan rules and network interface groups for each switch to determine how the trunks will be used.

 For example, if a switch in Toronto and a switch in Ottawa are linked with a tie line, extensions on the switch in Ottawa can place calls to extensions on the switch in Toronto as if they were themselves in Toronto. This is illustrated in Figure 10-28.

A tie line is referred to as a *dedicated circuit* or *leased line* if it is a facility that is for the exclusive use of the subscriber. It is guaranteed to be available at all times even though it is passing through one or more

Figure 10-28. Tie line example

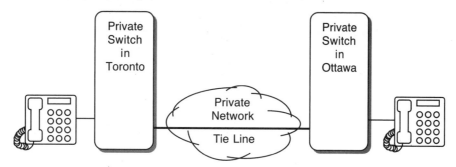

carriers' networks. A virtual private network (see section 10.9.11) provides tie lines without physically dedicating a particular set of facilities. Instead, the facilities are allocated on an as-needed basis.

10.9.8. Foreign Exchange (FX) Lines

A *foreign exchange line* is like a tie line except that it connects your telephone system directly to a CO switch (public exchange) in a distant location. A foreign exchange line allows your telephone system to receive and place calls locally in two or more different geographical locations. This is typically implemented for trunks, but it could be applied to a subscriber line.

For example, if a switch in San Jose has a foreign exchange line to Sacramento, callers in Sacramento can place incoming calls to the switch using the local number corresponding to the switch. Likewise, extensions on the switch can place calls to Sacramento local numbers without incurring toll charges. This is illustrated in Figure 10-29.

10.9.9. Off-Premises Extensions (OPX)

An *off-premises extension* is a dedicated circuit that connects an extension on a CPE switch to a subscriber line associated with a particular CO switch. This is illustrated in Figure 10-30. A telephone station that is connected in this fashion behaves in every way as though it were directly connected to the switch on which it is an extension.

Figure 10-29. FX line example

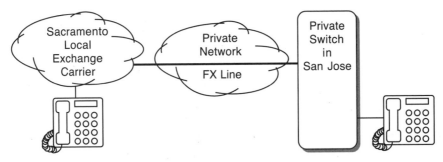

Figure 10-30. OPX line example

 One example of OPX use would be the arrangement used by a doctor associated with a small clinic who also manages a practice at home. In this example, one extension on the clinic's switch has its telephone station at the doctor's home practice.

 Another popular use of the OPX feature involves telecommuting. A car rental company could allow its reservations agents to work from home by placing extensions corresponding to devices in the call center's ACD group at their homes. The reservation agents then could log onto the ACD group, handle customer calls for reservations requests, and log off at the end of their shifts—all from home.

10.9.10. Toll-Free Numbers

Toll-free numbers allow a caller anywhere within a defined geographical area to dial the number without paying long-distance charges. The resulting call is then routed to any designated line, or ISG and is billed to the called number.

In North America, toll-free numbers, originally marketed as *IN-WATS* (INward Wide Area Telephone Service) numbers, are dialed using 1-800 or 1-888 followed by the seven unique digits of the toll-free number.

These numbers can be configured to deliver calls to lines dedicated to handling toll-free numbers, or they can be directed to any general-purpose line that supports incoming calls. In addition, subscribers to these services have the option to obtain the DNIS (dialed number identification service) and/or ANI (automatic number identification) information with each toll-free call placed, independent of any callerID blocking in place on the caller's logical device. The mechanisms used for delivering this call associated information vary, depending on how call delivery is configured and which telephone company is providing the service. (See Chapter 5, sections 5.5.1 and 5.5.2 for more information on ANI and DNIS.)

In many ways, the reliable delivery of ANI and DNIS information is the most valuable part of toll-free service. The expense of these lines may not be justifiable purely on the basis of customer/caller convenience. However, the cost savings that result from installing a CT system that takes advantage of this information can often justify the cost of obtaining an 800 or 888 number.

10.9.11. VPN

A *virtual private network*, or *VPN*, is a service provided by a carrier or other service provider which creates a wide area network that is accessible only to a given organization or set of users. In telephony, the term has traditionally referred to a service in which multiple PBXs were interconnected without the expense of dedicated tie lines. In

IP networking, the term refers to creating a secured IP tunnel through a public or semi-public network. Next generation carriers offering both telephony and Internet services typically provide their services by creating a VPN for a customer and then hosting services for that customer within the VPN.

10.9.12. Softswitches and Internet Telephony Gateways

A *softswitch* is the name given to an IP telephony product that provides the functionality of a class 5 CO switch at the edge of the PSTN. A softswitch is to a LEC what an iPBX is to an enterprise. Just as an enterprise uses a PBX to interconnect trunks from the LEC to telephone stations, a LEC uses a CO switch to interconnect the rest of the PSTN and its subscribers.

Competitive local exchange carriers, or CLECs, must connect to the public network with a switch that appears to the PSTN as a Class 5 CO switch. However, if all of their subscribers are connected using VoIP trunking, all they require is a softswitch that includes the appropriate media and signaling gateways along with switching and call control. A CLEC can then benefit from the CTI, media services, and administrative interfaces supported by the softswitch by providing a unique and differentiated service to their subscribers.

These telephony service providers may also host an *Internet telephony gateway*. This is a VoIP gateway that provides their subscribers with access to VoIP networks on the public Internet.

10.9.13. Hosted IP Telephony

The software components of an iPBX can be installed on any appropriate server(s) having IP access to all of the other components of the IP telephone system. As businesses of all sizes now know, it is often more cost effective to host mission-critical applications in off-site server farms where they can be maintained and managed by a service provider with a centralized data center.

A *hosted iPBX* is an iPBX deployment where the mission critical, centralized iPBX software is installed on a server that is hosted in a service provider's data center rather than on the customer premise. While a traditional, monolithic PBX must be located at the customer site, the iPBX is modular so only the telephone stations and station servers of the iPBX need be on the customer premise. This greatly reduces the complexity of the equipment and maintenance required at the customer site and lower's the initial costs. It also simplifies the businesses networking architecture and eliminates the need to perform capacity planning for separate voice trunks.

LEC Hosting

A LEC can provide iPBX hosting much the same way that it can provide centrex. However rather than having to provide dedicated circuits for each subscriber extension, a single IP network connection is used to connect the service provider and the customer site. The equipment at the customer-site is all IP-based and can be managed by the customer's network administrator, and the LEC needs only a softswitch (rather than a traditional CO switch) for PSTN access.

Figure 10-31. LEC Hosting

ISP Hosting

As virtually every business moves to deploy an ecommerce presence, each must come to terms with their need for both Internet and telephone network connectivity. As these companies discover the need to incorporate CTI functionality into their ecommerce solutions, they will begin to demand IP telephony services directly from their network service providers. Where LECs are responsible for providing voice network access, ISPs provide internet access. However, as the distinction between IP networking and voice networking disappears with the deployment of IP telephone systems, the lines between LECs and ISPs blur. ISPs may therefore choose to host their customer's iPBX as a means of providing enhanced services.

Figure 10-32. ISP Hosting

ASP Hosting

Business that don't want any of the capital costs and management costs associated with setting up their own CTI solution may turn to an *application service provider*, or *ASP*, offering not only Internet and IP telephony services, but a complete turnkey solution for CRM or some other CTI application. ASPs are service providers that offer turnkey application hosting. They effectively combine all the services provided by software developers, LECs, ISPs, and system integrators. They deploy and manage software applications and systems on behalf of their customers.

The ASP deploys the iPBX software along with CTI software, web server software and the other elements of their software offering at a central hosting center. A single IP network pipe connects the ASP's hosting center to the customer. This pipe delivers both voice and internet services to the business. All the customer needs are IP telephone stations and workstations running appropriate client software. The customer uploads their web content, customer databases, etc., to the ASP's servers and configures all of the components to further customize it. The customer is not required to buy any servers or even to pay for traditional telephone service. The ASP simply receives a single payment covering all of the different applications being hosted and all of the services delivered.

Figure 10-33. ASP Hosting

Service Provider Benefits

The iPBX hosting model offers service providers with a number of significant benefits. These include:

- Reduced customer turnover (known as churn)
- Simplified bandwidth planning
- Lowered barrier to the voice market

Service providers who offer hosted iPBX services are more likely to retain customers in an increasingly competitive market for telephone service. By hosting a customer's most mission critical application, their telephone system, that customer is much less likely to switch to a different service provider based on price. Furthermore, service providers that offer exclusive access to special CTI software can easily differentiate their iPBX offering in order to reduce subscriber defection.

iPBX hosting also allows service providers to deliver telephony services in a fashion which is independent of geography. If a subscriber changes location there is no need to change service providers and telephone numbers. Instead, the IP connection is simply relocated and business continues. In this way, iPBX service providers can retain fast growing businesses throughout their life cycle.

Because both data and voice are concentrated over a single high bandwidth IP network link, neither the customer nor the service provider must perform the traditional trunk capacity planning exercises to determine the exact number of simultaneous voice calls that will be supported. If additional capacity is required, the immediate impact will be a minimal degradation in Internet access. This can be inexpensively remedied with a more generic IP bandwidth rather than dedicated voice circuits.

Since iPBX hosting does not require a CO switch, LECs, ISPs, and ASPs are able to effectively compete with incumbent carriers without huge investments in switching equipment.

10.10. Review

In this chapter we have explored the services, products, and technologies that can be combined to build a telephone system. In particular, we have seen how the tangible services and products offered by telephone companies and equipment vendors are all easily abstracted using the concepts covered in Chapters 3, 4, and 5.

The telephony features and services available to a given individual are a function of the features, services, and channels provided by one or more telephone companies, combined with the features and services provided by the individual's *CPE* or *customer premises equipment*. The CPE might range from a single telephone station to a network of CPE switches, each with many associated telephone stations.

A *telephone switch* is a set of telephony resources connected to one or more other switches with trunk lines that correspond to network interface devices and extensions, or subscriber lines, that correspond to its station devices. CPE switches include *front-end switches, key system units (KSUs), private branch exchanges (PBXs), IP-based PBXs (iPBXs), application-specific switches*. PBXs generally implement a complete set of telephony features and services, while traditional KSUs and front-end switches simply connect and disconnect telephone stations from trunks and take advantage of telephone features and services supplied by the CO switch to which it is connected.

Telephone stations and *telephone station peripherals* are implementations of telephony resources that connect to lines from switches. Telephone stations generally rely on their switch for telephony feature and service implementation. The most common line interface types used by telephone station equipment are analog, ISDN BRI, proprietary digital, and wireless. Telephone stations include a physical device element that provides at least one auditory apparatus for accessing the media stream of a call. The use of lamps, buttons, ringers, and a display in the design of a telephone set varies dramatically, as a result both of the device configuration(s) and of the aesthetic, cost, and usability trade-offs made by product designers.

Telephony Equipment and Services

A *telephone service provider* or *telephone company* provides connectivity from a *class 5 central office switch* (or *CO switch*). In addition to providing media stream channels, the CO switch provides certain telephony features and services that are *subscribed* to from the telephone company. Conventional telephone companies including *inter-exchange carriers* (IXCs) and *local exchange carriers* (LECs) and PTTs are beginning to face competition from competitive local exchange carriers (CLECs), internet service providers (ISPs), and application service providers (ASPs).

11.
CT System Configurations

This chapter describes how CT systems are assembled from CT hardware components, the communication links that connect them, and the CT protocols that flow between them. We'll explore the broad range of CT system configurations that can be assembled with increasingly interoperable CT system components.

Many different configurations for CT solutions are presented here, but they are all just examples, or a subset, of the unlimited range of CT system configurations that are possible. The goal is to provide a collection of building blocks and "serving suggestions" that you can use to plan and implement your own CT systems.

A CT system consists of components and communication links between them. Components may be hardware (switches, CTI servers, telephone stations, PDAs, personal computers, and hybrids) or software (processes running on a hardware component). Messages are passed over communication links between hardware components (LANs, dial-up, cable, infrared, etc.) or programmatic interfaces between software components.

Graphical Notation for CT System Configurations

This book uses a standardized graphical notation for describing CT system configurations. The next two sections describe the general classes of hardware components and the classes of communication links that can be established between them. Accompanying the description of each component and communication link are the icons and graphics that are used to represent them in CT system configuration diagrams. These icons and graphics are also summarized on the inside of the back cover.

In general, the icons representing hardware components include any hardware or software necessary for a given hardware component to do its job—to expose a CTI or media services interface and to establish the communication link shown in a particular CT system configuration. For example, if a component is shown with a modem-based connection to another component, the modems are not shown explicitly. The modem, modem software, serial ports, cables, etc., at each end of the connection are considered part of the corresponding hardware components.

11.1. Hardware Components

In describing CT system configurations, this chapter deals with discrete hardware components—the individual CT products—that make up tangible CT systems. It does not deal with the abstract logical devices that form the basis for the telephony abstraction or the configuration of the software components that are actually installed on the hardware components within a system. (Refer to Chapter 12 for information on the configuration of CT software within a CT system.)

This section defines the basic forms that CT hardware components can take, explains their role in a CT system, and defines the icons used to represent them. The basic types described can also be hybridized, or combined, to form specialized hardware components. These are described at the end of this section.

11.1.1. Personal Computer

Personal computers are the most pervasive type of computing product today. They are relatively inexpensive computers that range widely in form factor and performance. Several features distinguish personal computers from the other categories of computing products used in CT systems: They are primarily intended as tools for individual users; they are based on general-purpose, mainstream operating systems; and they support the installation of third-party (add-on) application software and operating system extensions.

For the purposes of a CTI system, a personal computer is considered a *client computer* if it has a software component, referred to as a *CTI client implementation,* that allows application programs to participate as components in the CTI system. (This is described in Chapter 12.)

11.1.2. Multi-user Computer

Multi-user computers are like personal computers in most ways except that they are shared by more than one user simultaneously. Each user interacts with the multi-user computer using a terminal or application-specific client-server LAN protocol. The icons above represent multi-user computers and their associated terminals.

Like the personal computer, a multi-user computer which includes CTI client implementation software (see Chapter 12) is considered a client computer for purposes of a CTI system. In general, all client computers are interchangeable in CTI system configurations, so most configuration examples use personal computers.

553

11.1.3. PDA

A *personal digital assistant* (*PDA*) is another form of computing product. While it is like a personal computer in the sense that it is intended as a personal tool for an individual, it is quite different in most every other way. PDAs are special-purpose products that typically have only enough computing power and storage capacity to carry out the specific tasks for which they were designed. As a result, PDAs generally have special-purpose operating systems or no operating system (in the traditional sense) at all. PDAs may or may not allow the installation of third-party software.

PDAs are usually in the form of highly portable products such as electronic organizers, electronic clipboards, intelligent wristwatches, and smart remote control units. However PDAs can also be embedded in automobile dashboards, refrigerator doors, and hospital beds.

11.1.4. Telephone Stations

Telephone stations are the devices people use to directly access media streams from telephone networks. The range and variety of telephone stations is even greater than that of personal computers. They include home phones, desk phones, attendant consoles, cellular phones, pay phones and even fax phones and video phones. Refer to Chapter 10 for more information about telephone station products.

If the telephone station provides a CTI interface, the switching domain that can be accessed (depending on the implementation) may encompass the entire switch to which it is attached, or it may treat the switch as an external network and provide control over just the lines (logical devices) that are directly connected. The CTI interface may or may not provide access to the associated physical device element.

Typically a telephone station will expose a CTI interface in one of two ways:

- Built-in CTI interface

  Some telephone stations, so-called *smart phones*, have the necessary built-in intelligence and communications hardware to allow a CTI client to connect directly to the physical telephone set itself.

- Add-on CTI interface

  Some telephone stations do not have a built-in CTI interface but were designed with a CTI interface as an option. If available, these modules allow a separate CTI interface component to be added via a snap-on module, upgrade card, or some other mechanism. This add-on exposes an interface for use with a standard communication link (as described in section 11.2).

These CTI interface variations often provide media services of some type, most notably asynchronous data and possibly isochronous digital data or audio. They also may have built-in data or fax modem functionality.

11.1.5. Telephone Station Peripheral

A *telephone station peripheral* is a computer peripheral that can be interfaced to one or more telephone lines in place of, or in addition to, a telephone station.

If the type of line involved is analog, ISDN, or another nonproprietary type, the telephone station peripheral may be from any vendor and the product is likely to be in the form of a common modem or ISDN terminal adapter. If the type of line is proprietary, however, the switch vendor generally is the exclusive source of any telephone station peripheral that can be connected. These peripherals may be attached to a computer system through an external connector, or may be in the form of an add-in card.

Telephone station peripherals typically provide support for media services. Often these peripheral products either include or provide access to data and fax modem resources, and may also provide access to resources for audio streams and raw isochronous data.

If the telephone station peripheral provides a CTI interface, then, depending on its implementation, the switching domain that can be accessed may encompass the entire switch (or network) to which it is attached, or may be restricted to just the lines (logical devices) that are directly connected. The peripheral may or may not be able to control and observe the physical device element(s) associated with telephone station(s) on the same line(s).

11.1.6. CTI Server

In a CTI system configuration, a *CTI server* is a computer of some sort that is dedicated (from the perspective of the CTI system) to playing the role of a logical CTI server to a set of logical clients in the form of client computers and PDAs. The CTI server hardware may be any type of computing platform running any type of operating system. The only essential characteristics are that it physically has the ability to connect to its logical server and logical clients and that it supports a software component known as a CTI server implementation which carries out its role as a CTI server. (CTI server implementations are described in section 11.4 and in Chapter 12, section 12.3.)

A CTI server generally plays one or more of the following roles in a CTI system:

- A CTI server is usually a fan-out component that channels CTI control and status messages between its logical clients and logical server.

- A CTI server also may act as a secure gatekeeper for its logical server's CTI interface. A switch that does not authenticate the clients that connect to its CTI interface can be front-ended by a CTI server that provides this capability. The security added may be limited to restricting access to authorized clients, or it may include customizing the view of the switching domain that each client sees.

11.1.7. Media Server

A *media server* is a computer of some sort that is dedicated (from the perspective of the CT system) to providing media services to a set of logical clients in the form of client computers, PDAs, and CTI servers. A media server may work in conjunction with an associated logical CTI server. The media server may act as a CTI client in order to provide basic CTI services for media clients. The CTI server may use the media server to provide basic media services and to support media binding (described in Chapter 6) for requesting access to a particular media service instance.

The CTI component that is providing the media services binding function for a particular media server makes the media server appear as part of the switching domain to its logical CTI clients.

11.1.8. Telephony Gateways

Telephony gateways, which include both *media stream gateways* and *signaling gateways* are system components that contain switching fabric resources for interconnecting media streams and conveying associated signaling information. The switching control function within telephony gateways may be controlled centrally by a switching control server, or may be autonomous and work in cooperation with other switching resources in the system based on signaling requests. Telephony gateways include station servers that provide a gateway between a given switching fabric implementation and conventional telephone station equipment.

11.1.9. Call Processing and Switching Control Server

Call processing and switching control servers are servers that implement the core of a softswitch, an iPBX, or other distributed PBX implementation. This component is similar to a conventional switch except that it does not directly interact with media streams. However, the combination of one or more call processing and switching control servers with one or more telephony gateways is the functional equivalent of a conventional switch and can be treated as a set of components or as a single entity.

11.1.10. CO Switch

For purposes of this chapter, a *CO (central office) switch* is any switch operated by a common carrier (i.e., telephone company). It is typically located in one of the carrier's central offices and may provide wireline service, wireless service, or both. It may be implemented as a conventional monolithic switch or as a softswitch utilizing telephony gateways for interconnecting media streams and translating signaling information. (See Chapter 10 for more information.)

11.1.11. CPE Switch

A *CPE (customer premise equipment) switch* may be a front-end switch, a KSU,[11-1] a PBX, a complete iPBX, or an application-specific switch; and it may support telephone stations through wired (analog, ISDN, ethernet, proprietary) or through wireless connections. Although iPBXs are made up of many components distributed across a network, as are other more conventional forms of distributed PBX implementations, their components can be treated as a single whole relative to the other components of a CT system configuration.(Refer to Chapter 10 for more information on the forms of CPE switch products.)

11-1. Hybrid switch — In some cases a CPE switch may be a so-called hybrid switch that is a combination PBX/KSU, but these are generally treated as PBXs.

559

11.1.12. Hybrids

Hybrid components combine the features of two or more basic hardware components. Described below are a few of the most compelling hybrids that may be formed by blending hardware components described previously. Not all the hybrids described here are explicitly featured in the CT system configurations in this book, however they can be constructed by simply making appropriate substitutions.

- Client computer + CTI server

  The hybrid product equivalent to a client computer and a CTI server is a single computer that is able to both run CTI applications and provide fan-out of CTI services to other logical clients in a CT system. The dual role of this computer is transparent to CTI software components running on it. From their perspective, this hybrid is equivalent to a client computer. Configurations featuring this type of hybrid can be constructed by substituting it for a CTI server. (An example using this type of hybrid is presented in Chapter 13.)

- Media server + CTI server

  The hybrid product equivalent to a media server and CTI server is a single hardware component in which both types of servers are simultaneously active. A component of this type is effectively a CTI server that is able to provide media services to its clients more easily by incorporating the media services rather than by binding to them in another, independent server. (Examples of this type of hybrid are discussed in section 11.7.3 and in Chapter 13.)

- Switch + CTI server

  A hybrid product with the functionality of a switch and a CTI server is a switch of some sort that provides the connectivity associated with a CTI server and also includes a CTI server implementation software component. This is a compelling hybrid because it provides the scalability associated with a CTI server, while simplifying the configuration by substituting for a switch-CTI server combination in any given system.

- Switch + media server

    The hybrid product with the functionality of a switch and a media server is a switch of some sort that provides media access services to its clients more easily by incorporating the media services resources rather than by binding to them using an external media server. This type of hybrid is compelling because it eliminates the complexity associated with connecting an independent media server to a switch. CT configuration using hybrids of this type can be constructed by substituting for any switch-media server combination.

- Telephony gateway + media server

    The hybrid product that is equivalent to a telephony gateway and media server is a computer that is very similar to the switch-media server combination described above. The same physical hardware is required to build a media server and a telephony gateway; the only difference is in the software that they run. This hybrid is therefore a natural and likely combination because it involves simply running both types of software on the same machine. A hybrid of this sort can be substituted for any telephony gateway.

- PDA + telephone station

    Hybrid products with the functionality of a PDA and a telephone station are sometimes referred to as "intelligent telephones" or even "intelligent communicators" and can be found in both desk set and wireless formats. They effectively embed some level of personalized computing functionality in a hardware component that is otherwise a telephone station. From the perspective of CT system configurations, they are functionally equivalent to a telephone station and can be substituted accordingly. From the perspective of any software components running inside the PDA portion of the product, there is effectively a PDA inside the telephone station in a direct-connect relationship with the telephone station portion.

11.2. CT Communication Links

CT communication links, or *CT links*, are the data connections that two hardware components use to establish an inter-component boundary with one another in a CT system.

A CT communication link consists of both the physical connectivity (cabling, hardware transceivers, etc.) and the session/transport protocol stack[11-2] used above it.

11.2.1. CTI Session and Media Services Session

The relationship between two hardware components on each side of a particular communication link is illustrated in Figure 11-1.

The communication link between two hardware components is used to carry CT messages (service requests, events, etc.) between the logical CT client and the logical CT server that exist on each side of the service boundary that the communication link represents.

The sequence of CTI messages travelling between a logical CTI client and a logical CTI server is referred to as a *CTI session*. Likewise the sequence of media services messages between a media services client and a media server instance are referred to as a *media services session*.

In the scenario illustrated in Figure 11-1 the logical server hardware is hosting both a CTI server implementation as well as a media server and the switching domain implementation includes support for binding to one or more media services. After the logical CTI client uses media binding services to bind to one or more of them, the resulting *media services sessions* may also be carried by the same communication link, if appropriate.

11-2. Session/transport protocol stack — In ISO terminology the session/transport protocol stack consists of data link, network, transport, and (optionally) session layers (2 to 5). A communication link represents ISO layers 1 to 5.

Figure 11-1. Anatomy of a communication link

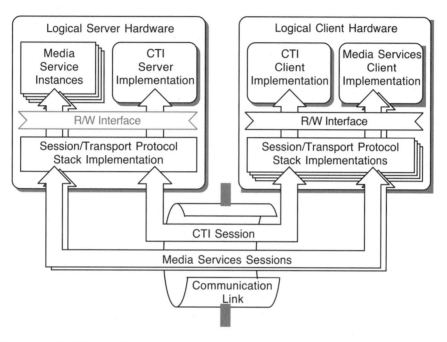

Note that in Figure 11-1 the computing domain implementation is shown as having multiple session/transport stack implementations, while the switching domain is shown as having only one. This generally is the case, but there may be some exceptions. Most switching domain implementations have proprietary internal structures, so the notion of a R/W interface[11-3] really is not applicable. By the same token, they tend to implement support for just one type of communication link (preferably one supported by all computing domain implementations). On the other hand, computing domain implementations typically are based on operating systems that provide R/W interfaces and multiple protocol stacks.

11-3. R/W interface — R/W interfaces are programmatic interfaces that provide access to session/transport protocol stack implementations. They are described in Chapter 12, section 12.4.1.

11.2.2. LAN

A *LAN* (*local area network*) communication link uses a LAN protocol stack and multi-point communication infrastructure to establish links between CT components.

The physical network used for LAN communication links in a given configuration may be wired or wireless, may have any topology (structure), and may support thousands of components or as few as two. Components may be directly connected to the LAN or may be indirectly connected using a dial-up bridge. The key feature characterizing a LAN's physical layer is that it represents a shared medium over which any number of connected components can setup, tear down, and simultaneously use, communication links with any number of other components.

LAN communication links may be established using any session/ transport protocol stack that provides reliability and virtual circuits (sessions) of some sort. The stack optionally may support authentication and encryption. Examples of the popular LAN protocols satisfying these requirements include:

- TCP/IP (Internet protocol family)

- IPX/SPX (Novell's protocol family)

- ADSP (Apple's AppleTalk protocol family)

The graphical notation for a LAN communication link is shown above (symbolically it appears as a daisy chain although it might be any topology). Figure 11-2 shows a LAN communication link between a personal computer and a CTI server.

Figure 11-2. LAN connection

11.2.3. Serial Cable and Serial Bus

Serial cable and *serial bus* communication links are used to establish sessions between two CT components in close proximity. At the physical level, these are inexpensive cables that carry serial data over short distances. One serial cable may constitute the whole physical layer or a single segment in a serial bus.

Serial options include:

- RS-232 (or V.24)
    The universally supported point-to-point serial interface.[11-4]

- USB (Universal Serial Bus)
    Multi-point/multi-channel communication

- GeoPort
    Multi-channel isochronous communication

- FireWire (IEEE 1394)
    Multi-point/multi-channel isochronous communication

11-4. Multiplexed serial — The Versit CTI Encyclopedia specifies a number of mechanisms that can be used to reliably multiplex CTI and media services sessions over an RS-232 or equivalent connection in a standard fashion. This includes a specification for implementing standard CTI protocols in products that also operate using the Hayes AT command set.

The graphical notation for a serial cable/bus communication link is shown above (symbolically it appears as a single span although it might be part of a multi-point serial bus). Figure 11-3 shows a serial cable/bus communication link between a multi-user computer and a media server.

Figure 11-3. Serial connection

11.2.4. Infrared (IR)

>>>>>>>>>>>

Infrared (IR) communication links are used to establish sessions between two CT components without requiring cables. At the physical level, the connection is established by infrared light transmitters and receivers that are built into each component. Infrared signaling, long used for remote controls on consumer electronics products, is now a proven technology for data exchange, particularly between mobile products such as notebook computers and PDAs.

The Infrared Data Association (IrDA) has defined standard IR protocol stacks for interoperability between the products of different vendors.

The graphical notation for an infrared communication link is shown above and Figure 11-4 shows its use in a configuration involving a PDA and a pay phone.

Figure 11-4. Infrared connection

11.2.5. Bluetooth

>>>>>>>>>>

Bluetooth wireless technology is another technology for establishing communication links between two CT system components without requiring cables. It creates a *personal area network* (*PAN*) around an individual carrying bluetooth-capable devices. At the physical level, connections are established using low power, low range radio-frequency transceivers. This technology is specifically designed to allow communication among mobile products such as notebook computers, PDAs and wireless telephones.

The graphical notation for an infrared communication link is shown above and Figure 11-5 shows its use in a configuration involving a PDA and a pay phone.

Figure 11-5. Bluetooth connection

One key difference between IR and bluetooth is that IR is a line-of-site technology and bluetooth is an omni-directional technology. For example, the user of IR-based PDA can identify a particular pay phone to connect with by pointing the PDA at the pay phone and getting close enough that no other pay phones are in the IR beam. With bluetooth technology all the bluetooth pay phones in the area would be accessible and the user would have to pick the desired one from a list presented by the PDA. One advantage of bluetooth is that it was specifically designed to carry three voice media stream channels in addition to data. This allows media resources to exist in the client device. (See Media Service Instances in section 11.5.)

11.2.6. Dial-up

A *dial-up* communication link is one that takes advantage of a wide area telecommunications network (*WAN*). (See the sidebar "Telephony and Telecommunications" on page 93 for more information.) As with the other types of communications links, once a dial-up communication link has been established between two points using one of these well-defined standards, the communication link operates as if the two components were directly connected using a serial cable.

Typical dial-up communication links include:

- Modem connections (modulated data)

- SVD modem connections (supporting simultaneous voice and data)[11-5]

- Digital communication links (such as ISDN)

- Packet data virtual circuits (such as X.25)

The particular telecommunications technology being used determines how many dial-up communication links a single physical connection to the network (e.g., telephone line and modem, ISDN line and TA, X.25 PAD, etc.) can support.

The graphical notation for a dial-up communication link is shown above. Figure 11-6 shows its use in a configuration involving a notebook computer and a CTI server.

11-5. Simultaneous voice and data — Simultaneous voice and data (SVD) involves using modulation and compression technologies to allow both voice and other media data to travel simultaneously over a single voice media stream channel.

Figure 11-6. Dial-up connection

11.3. Proprietary CT Protocols

The presence of a proprietary CT protocol traveling over a communication link in a CT system configuration diagram, is indicated by shading the appropriate communication link symbol in the fashion illustrated in Figure 11-7.

Figure 11-7. Proprietary CTI protocols traveling over communication links

As described in Chapter 1, section 1.10, CT products based on proprietary CT protocols take two forms, corresponding to the first two phases of CT evolution:

- Custom CT systems

- API-centric products

In addition, CT products using proprietary CT protocols may be made to interoperate with CT components that are based on standard protocols by using *mappers*.

Custom CT Systems

CT components relying on proprietary protocols may be formed into a closed, custom CT system that can be viewed only as a single, turnkey CT solution. All the CT components making up the solution must be from the same vendor or be developed to that vendor's specifications. This is illustrated in Figure 11-8.

Figure 11-8. Custom CT solution using proprietary CT protocol

All the software and hardware making up such a CT system is interdependent, so the solution can be used in only the configuration or configurations for which it was developed. The vendor of such a system may base its configuration on any of the system configurations described in this chapter, but system integrators and customers are not in a position to reconfigure, expand, or add to the system.

API-centric CT Products

Another approach for building CT solutions involving proprietary CT protocols is through the use of special pieces of *API-centric adapter software*.[11-6] These are able to establish appropriate communication links, interact with the proprietary protocol, and make their functionality accessible through a programmatic interface (typically a procedural API). These special pieces of code must be written by or for the vendor of the proprietary protocol, and must be developed and tested for each operating system platform and system model to which the CT component is to be connected. CT hardware components for which the vendor chooses not to develop this special software, and

---

11-6. API-centric adapter software — An implementation of API-centric adapter software is a type of CT client implementation in the CT software framework that is presented in Chapter 12.

those that do not have standard APIs (or operating systems), have no access at all to CT components that use proprietary protocols. Depending on the platform, this special piece of adapter software is referred to as a *driver, service provider,* or *telephone tool.* This is illustrated in Figure 11-9.

Figure 11-9. Proprietary CT protocol with adapter software

The CT component with the proprietary CT protocol and its associated adapter software are interdependent, so the CT component in question can be used only in the configuration or configurations for which it was developed. The vendor of such a system may base its configuration on any of the system configurations described in this chapter, but system integrators and customers are not in a position to reconfigure, add new platforms, or upgrade these CT systems without first obtaining the appropriate adapter software (if available).

The significant improvement that this represents over custom CT systems is that off-the-shelf applications that are compatible with the particular API chosen (and the behavior of the proprietary protocol under that API) may be used instead of custom-developed application software.

11.3.1. Protocol Mappers

CT component implementations designed for CTI Plug & Play interoperability generally will support only standard CT protocols across the R/W interface (refer to Figure 11-11). However, a CT component that is not CT Plug & Play because it implements only proprietary protocols can still achieve interoperability by providing a *protocol mapper*, or *mapper*, for short.

Mappers are software or hardware components that transform a proprietary CT protocol into a standard CT protocol so that it can be used by a standards-based component. They do this in a fashion that is entirely invisible to the CT Plug & Play component.

The principal advantage of mappers over the other approaches to supporting proprietary protocols is that they allow for virtually any arbitrary CT system configuration (as exemplified in this chapter) to be assembled.

11.3.2. Protocol Mapper Hardware

Hardware mappers are physical adapters that sit between the two components in question and are transparent to both. The icon for a hardware mapper is shown above and an example of a CT configuration involving a protocol mapper is illustrated in Figure 11-10.

Figure 11-10. Hardware mapper example

Despite the incremental cost associated with an extra piece of hardware in a given configuration, hardware mappers generally are a more attractive option than software mappers because only one version needs to be developed. This single implementation will work with any CT Plug & Play component—in contrast to the software mappers, described below, which must be developed for each and every CT component platform. (For some types of components, such as an IR-based remote control, this is impossible.) It is also an attractive approach for many vendors because it allows them to incorporate the hardware mapper into the CT product in its next generation.

11.3.3. Protocol Mapper Code

CT components that are not CT Plug & Play and do not support hardware mappers need to provide *mapper code*, or software, that runs on each of the CT components with which they need to interoperate.[11-7] Software mappers are invisible to CT software components because they are implemented as session/transport protocol stack modules. So, like hardware mappers, they appear to be within the communication link itself. (Software mappers are discussed further in Chapter 12.)

Although conceptually a piece of protocol mapper software is very similar to the adapter software used in API-centric models, it has two key advantages. The first is the ability to flow the resulting standard protocol to any number of downstream CT Plug & Play components using a fan-out component. This means that a CT product with proprietary protocols can be used in a much broader range of configurations. An equally significant advantage of mappers over API-centric adapter software is that mapping to a standard protocol causes CT behavior to become normalized, which in turn eliminates restrictions on the range of applications that work reliably using that particular API-CT component combination. One disadvantage

11-7. Mapper economics — The economics of developing mapper code might not seem evident at first glance, as it requires that mapper code be written and tested for many different platforms, operating system versions, and hardware configurations. Some vendors have determined it to be a faster migration path, however.

software mappers share with API-centric adapter software is the fact that neither can support logical CT clients that lack traditional operating systems (such as PDAs and consumer electronics products).

In the CT system configuration diagrams in this book, the presence of mapper code installed on a given CT component to map from a proprietary CT protocol is indicated with the symbol shown above.

11.4. CTI Sessions and CTI Protocols

CTI protocols define the structure, contents, use and flow of PDUs carrying CTI messages for a given CTI session. CTI protocols are high-level protocols[11-8] like the protocols used to send electronic mail, print to a printer, retrieve files from a file server, or browse the World Wide Web. Like these other protocols, they are designed for transmission over a reliable, connection-oriented (i.e., guaranteed-delivery) session/transport protocol stack referred to as a *communication link*. (Refer to the previous section for more detail on communication links.) CTI protocols may travel between any two CTI components (hardware or software) across any appropriate communication links. While CTI protocols represent the only way to convey CTI messages between two hardware components they are equally applicable to two software components within a single hardware component.

A CTI component capable of interoperating with other components using CTI protocols must include a module known as a *CTI protocol encoder/decoder*. As shown in Figure 11-11, a CTI protocol encoder/decoder is responsible for managing CTI sessions (setting up communication links, sending and receiving PDUs, etc.) and for encoding and decoding the PDUs based on knowledge of the CTI protocol in use. When implemented as a software module running on an open computer system, CTI protocol encoder/decoders utilize a

---

11-8. ISO layering — In ISO terminology, CTI protocols are application layer protocols and their definitions include the presentation layer encoding. CTI protocol definitions correspond to ISO layers 6 and 7.

system's R/W interface in order to access the session/transport protocol stack used to deliver the CTI session. (Refer to Chapter 12 for more on software configurations.)

Figure 11-11. CTI component using CTI protocols

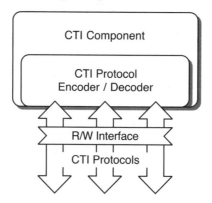

A given interoperable CTI component may be either a logical server, a logical client, or both a logical server and a logical client simultaneously. (These concepts are illustrated in Chapter 6, section 6.3.) A hardware component that acts as a logical client with respect to one component and as a logical server to one or more logical clients contains a software component known as a *CTI server implementation*. As illustrated in Figure 11-12, a CTI server implementation interprets and keeps track of all the CTI service request, event, and acknowledgment messages from both sides and, as appropriate, conveys them to their correct recipient.

For example, if a particular logical client sent a CTI service request message, the CTI client implementation would pass the service request to the logical server, receive the acknowledgment, and forward the acknowledgment to the logical client that originally made the request. In this fashion a CTI server implementation acts as a proxy[11-9] for its logical clients.

11-9. Proxy server — The operation of a CTI server implementation can be compared to the operation of a proxy-based Internet firewall implementation. Both are servers that appear to be providing a certain service to their clients; in reality they access the actual service on behalf of those clients.

Figure 11-12. CTI server implementation component

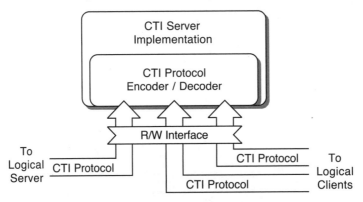

If the CTI protocol encoder/decoder in a given CTI server implementation supports multiple CTI protocols, it is able to effectively act as a translator between logical clients using one protocol and a logical server using another. (Refer to Chapter 12, section 12.3. for more information on the features and responsibilities of CTI server implementations.)

Standard CTI Protocols

CTI Plug & Play interoperability is made possible through the use of standard CTI protocols. CTI Plug & Play means that no software specific to a particular logical server needs to be installed on a logical client. This type of interoperability between any two CTI components is achieved when each CTI component implements both:

- Standardized CTI protocol(s)

- Standard communication link(s)

A CTI component that does not satisfy both of these requirements requires special software to be installed and thus is not CTI Plug & Play. (For more information on proprietary implementations and software configurations, see the next section and Chapter 12 respectively.)

Three standard CTI protocols have been defined in the Versit CTI Encyclopedia.[11-10] They are:

(1) **"Versit CTI Protocol 1"**
CTI Protocol 1 is optimized for use between a switch and a CTI server

(2) **"Versit CTI Protocol 2"**
CTI Protocol 2 is optimized for use between client computers and CTI servers

(3) **"Versit CTI Protocol 3"**
CTI Protocol 3 is optimized for use between telephone stations, PDAs, and client computers

The Versit-defined protocols all represent the same application layer protocol, but they are optimized for different implementations through the use of different encodings, or presentation layer protocols.[11-11] Each Versit CTI protocol may be used across any service boundary, and each is capable of carrying the same information, though they have been optimized for use with specific types of CTI system components. For example, CTI Protocol 3 is optimized to use buffers as small as 80 bytes, which is necessary for implementing CTI components such as CTI-enabled pay phones, cellular phones, desk phones, and PDAs. In contrast, CTI Protocol 1, which is intended for a dedicated link between the CTI interface on a switch and a CTI server, uses ISO's ASN.1 encoding and requires buffers that may be as large as 2000 bytes.

A fourth standard protocol has been defined by ECMA[11-12]:

(1) **"ECMA CSTA III Protocol"**
The ECMA CSTA phase III protocol is very similar to Versit CTI Protocol 1 and thus is optimized for use between a switch and a CTI server.

11-10. Versit CTI Encyclopedia — Versit Computer Telephony Integration (CTI) Encyclopedia (Versit, 1996).

11-11. Application and presentation layer protocols — In the ISO model, the presentation layer (ISO layer 6) is concerned with the syntax and semantics used to encode application layer data. The application layer (ISO layer 7) is concerned with conveying parametrized information reflecting the abstraction of some service or capability.

ECMA's CSTA phase III specification is based on the Versit CTI Encyclopedia but reflects ECMA's desire to maintain consistency with their older CSTA phase II specifications rather than incorporate all of the improvements found in the Versit specifications. Differences between the two are minimal although only the Versit protocols support true CTI plug-and-play operation in all environments.

Graphical Notation for Standard CTI Protocols

Communication links indicated in the CTI system configuration diagrams in this book carry CTI sessions using standard CTI protocols unless indicated otherwise (as described in section 11.3). The flow of a standard CTI protocol between various components using different communication links is indicated as shown in Figure 11-13.

Figure 11-13. Standard CTI protocols traveling over communication links

When indicating the use of a specific standard protocol for a particular communication link in a CTI system configuration, the protocol's number is used as illustrated in Figure 11-14. In this example a LAN communication link is shown carrying a CTI session using CTI Protocol 2.

Figure 11-14. Example CTI protocol

11-12. ECMA — ECMA, formerly known as the European Computer Manufacturers Association, is a recognized European standards body. ECMA Task Group 11, a group of PBX manufacturers that develop specifications for PBX CTI interfaces, has published ECMA-269, Services for Computer Supported Telecommunications Applications (CSTA) Phase III 3rd edition (ECMA 1998) and ECMA-285 Protocol for Computer Supported Telecommunications Applications (CSTA) Phase III (ECMA 1998).

11.5. Media Services Sessions

Media services sessions carry media services messages over a communication link between a component containing a media services instance and a corresponding media services client (Figure 11-1). In the graphical notation for CT system configuration diagrams, the presence of a media services session on a particular communication link (PDUs encoded according to the corresponding media services protocol) is indicated with the symbol shown above.

Standard Media Services Protocol

The ECTF S.200 specification defines a standard protocol for accessing media services. S.200 is an operating system and transport independent application layer protocol that complements the ECTF S.100 specification for media services.

Media Service Instances

Media service instances may be implemented on any type of hardware component described in section 11.1, not just on dedicated media servers. The presence of a media service instance in a given CT component is indicated with the symbol above. One or more logical CTI servers may support binding to this media service instance and their logical CTI clients will see a switching domain that includes this media service instance. In Figure 11-15 a personal computer accesses the media service instance in a telephone station.

Figure 11-15. Media service session traveling over a communication link

Media Service Mappers

If a given media service instance is not available for binding by any available logical CTI server, or if the CTI session and media services sessions are not delivered over the same communication link, media access binding may be implemented using *media service mapper code*. The presence of a media service mapper is indicated with the symbol shown above.

By simultaneously establishing communication links both to a media server using an appropriate media services session and to a logical CTI server using standard CTI protocols, this type of mapper creates the illusion of a switching domain that includes media binding capability. As a mapper, it tracks the messages traveling in either direction on both sessions and presents them to its own logical clients in an integrated fashion.

11.6. Direct-Connect Configurations

Direct-connect configurations, as the name implies, involve a direct connection between a user's client computer or PDA and a telephone station or telephone station peripheral.

All of the configurations in this section are presented using the standard graphical notation. Refer to the inside of the back cover for a summary of the symbols.

11.6.1. Basic Direct-Connect Configuration

In a *direct-connect configuration*, a client computer or PDA is connected directly to a telephone station or telephone station peripheral that supports a CTI protocol for access to CTI services. The logical client is connected using a serial cable (RS-232), serial bus (FireWire, GeoPort, or USB), infrared communication link, or a product-specific communication link such as an add-in card for a particular computer architecture. Figure 11-16 shows an example of a CT Plug & Play direct-connect configuration.

Figure 11-16. Direct-connect configuration example

Telephony features and services available through the resulting CTI session may be restricted to just the logical and/or physical device elements corresponding to the connected telephone station. This is referred to as *first-party call control* and involves limiting the scope of the switching domain to just a single device or device configuration. This is illustrated in Figure 11-17. If the switching domain contains additional telephony resources, it is referred to as *third-party call control*. The switching domain in this case may consist of station devices and/or additional telephony resources within the switch in addition to the connected telephone station, as shown in Figure 11-18. In either case, the CTI session from the switch is delivered through a communication link with a telephone station or telephone station peripheral.

Figure 11-17. Direct-connect first-party call control

Figure 11-18. Direct-connect third-party call control

Other station devices   All telephony resources

11.6.2. Direct-Connect Mapper Configurations

If the telephone station or telephone station peripheral does not implement a standard CTI protocol, it may have appropriate mapper hardware or mapper code available; otherwise, the telephone station must be used in conjunction with API-specific adapter code or custom application software.

Direct-Connect Mapper Hardware

Figure 11-19 presents an example of a CTI system configuration involving a piece of direct-connect *mapper hardware*. This special hardware component (labeled "P") sits between a telephone station that employs a proprietary CTI protocol and a client computer that supports standard CTI protocols. In this particular example, the mapper hardware supports CTI Protocol 3 and the telephone station's proprietary protocol and translates between the two.

Figure 11-19. Protocol mapper hardware configuration

As shown here, a direct-connect hardware mapper is a miniature CTI server of sorts that can be installed between two otherwise incompatible CTI components. The combination of the telephone station and the protocol mapper hardware is functionally equivalent to a single CT Plug & Play telephone station (except to those in the CT value chain that must obtain, install, and manage two components instead of one). In this example the hardware mapper is connected using a serial cable or bus to connect to each side, however both communication links need not be the same. For example, the mapper hardware could be designed to connect to the telephone station using a serial link and to the client computer or PDA using infrared.

Direct-Connect Mapper Code

Figure 11-20 presents examples of proprietary *mapper code* being used in direct-connect configurations. The telephone stations in these configurations implement a proprietary CTI protocol. In order to establish a CTI session with them, each of the different client computers and PDAs must use a piece of mapper code (labeled "P") that corresponds to the proprietary protocol for the telephone stations being connected.

Figure 11-20. Direct-connect protocol mapper code configuration

The disadvantage associated with mapper code is that it must be developed for every popular version of every popular operating system (e.g., DOS, Mac OS, Newton OS, Pen Windows, Unix, Windows, Windows NT, etc.). It also assumes that a given client product will actually have an operating system and a CTI client implementation[11-13] that supports the installation of mappers.

11-13. CTI client implementation — CTI client implementations are CTI software components in the CTI software framework that is presented in Chapter 12.

11.6.3. Direct-Connect Media Access Configurations

Media access of some sort is quite common in direct-connect configurations because the communication link used for delivering the CTI session often can be shared easily for delivering a media services session as well.

CT Plug & Play Media Access

Figure 11-21 shows a CTI configuration in which a telephone station provides access to a media service instance through its CTI interface. The logical client in this example, a personal computer, is using standard CTI protocols for call control, etc. After requesting media services binding, a media services session was formed over the same serial communication link used by the CTI session.

Figure 11-21. Direct-connect CT Plug & Play media access configuration

Mapper Hardware Media Access

In the system configurations presented in Figures 11-22 and 11-23 media access functionality has been added to the basic mapper hardware configuration previously presented in Figure 11-19.

In the example shown in Figure 11-22, the media service instance is in the telephone station and the telephone station's proprietary protocol mapper supports media access services using proprietary messages which are translated by the mapper. In contrast, Figure 11-23 shows a case where the media service instance is actually in the mapper hardware itself. In the latter case, mapper hardware has access to the isochronous media streams associated with the telephone station and is therefore able to provide the media services itself.

Figure 11-22. Mapper hardware media access configuration

Figure 11-23. Mapper and media access hardware configuration

Mapper Code Media Access

Figures 11-24 and 11-25 correspond to the mapper software versions of the mapper hardware examples shown in Figures 11-22 and 11-23 respectively. In Figure 11-24 the telephone station provides the media service instance so the mapper code simply translates protocols appropriately.

Figure 11-24. Mapper code media access configuration

Figure 11-25. Mapper and media access code configuration

In the second case, the media service instance is actually implemented within the mapper code running on the client computer. The communication link to the client computer is FireWire, GeoPort, or an equivalent technology that permits direct access to the isochronous media streams associated with the telephone station. The mapper code is then able to terminate the media services session itself, in addition to providing CTI protocol mapping.

11.6.4. Smart Phone Serial Cable/Bus Configurations

A *smart phone* is a telephone station that has built-in or add-on CTI interface (i.e., it is "smart"). A client computer or PDA connects to a smart phone using a serial cable (RS-232), GeoPort, FireWire, or USB serial bus, as shown in Figure 11-26. In this configuration, standard CTI protocols are supported by the telephone stations so no proprietary mapper code is present.

Figure 11-26. Smart phone serial cable/bus configuration

This is a likely configuration for office and home situations. It also is a likely configuration for instances where a notebook computer is docked for the night with a hotel phone or is connected to a cellular phone or the seat-side phone in an airplane or train.

Mappers may be used as described in section 11.6.3, but are practical only in home or office situations. Mappers are not practical for mobile users who are likely to encounter many different telephone products from many different vendors during the course of each day and who cannot acquire and install special software for even a fraction of them.

11.6.5. Smart Phone Infrared and Bluetooth Configuration

Infrared (IR) is a significant and important alternative to serial cables for connecting CTI products. It allows mobile products to connect and disconnect without requiring physical contact. This is critical for CTI solutions in any kind of public setting, such as an airport, hotel lobby, or vehicle. IR is also a very attractive communication link in home environments, where CTI-enabled remote control units can interact with the CTI-enabled functionality of telephone stations built into consumer electronics products such as televisions. Finally, IR used in office environments allows for instant docking of notebook computers and PDAs with telephone stations in conference rooms, lobbies, and work areas.

Two examples of IR-based CTI system configurations are shown in Figure 11-27. Here an IR communication link is used to connect a PDA with a pay phone and a laptop computer is connected to a desk set telephone station.

Figure 11-27. Smart phone infrared configuration

Telephone stations supporting IR tend to be in high traffic areas and in contexts where people will establish and tear down communication links with these products frequently. In most cases, mappers which must be specially installed and configured are not practical for IR-based products.

A bluetooth RF link will provide the same benefits as IR but has the additional benefit that the two devices involved need not be aimed at one another. The user of the bluetooth device simply chooses from the logical CTI servers in range and the logical CTI client initiates the link.

11.6.6. Serial-Based Telephone Station Peripheral Configuration

In this direct-connect, serial-based CTI configuration, a telephone station peripheral providing a CTI interface connects to one or more lines from a switch. There may or may not be one or more telephone stations on the same line and, depending on the implementation, the CTI interface may or may not be able to observe or control their physical elements. As shown in Figure 11-28, the peripheral may be attached by a serial cable or serial bus using a standard raw serial stream. The configuration is CT Plug & Play and requires no special software on the client computer or PDA.

Figure 11-28. Serial telephone station peripheral configuration

Inexpensive, so-called *dumb peripherals* deliver raw media streams from the telephone line across the serial interface. They require installation of appropriate mapper code on the client computer or PDA to translate these raw media streams into independent CTI sessions and media services sessions. This configuration is shown in Figure 11-29.

Figure 11-29. Mapper code serial telephone station peripheral configuration

Another variation involves having a telephone station in tandem, so that the telephone station peripheral is between the switch and the telephone station (Figure 11-30). This arrangement gives the peripheral greater control over the telephone station. Depending on the implementation, the peripheral can take over the role of the switch controlling and interpreting commands from the telephone station.

Figure 11-30. Tandem serial telephone station peripheral configuration

11.6.7. Add-In Board Configuration

The *add-in board telephone station peripheral* configuration is very similar to the serial-based telephone station peripheral configuration. The only difference is that, rather than being connected to the client computer through a serial cable or serial bus, the peripheral is in the form of an EISA, ISA, VESA, MCA, NuBus, PCI, or PCMCIA add-in card.

Because these boards generally do not have an associated session/transport protocol stack, they usually require a mapper if standard CTI protocols are to be used. In most cases, however, the mapper is a simple virtual serial port implementation that allows the CTI interface on the board to be accessed as if it were the type of serial-based peripheral described in section 11.6.6. This mapper-based configuration is shown in Figure 11-31.

Figure 11-31. Mapper code add-in board configuration

One case where a mapper is not required is the case of a PCMCIA implementation that uses the same mechanism as a modem card to expose a serial interface. In this case the configuration is CT Plug & Play, as shown in Figure 11-32.

Figure 11-32. CT Plug & Play add-in board configuration

As with the serial-based peripheral, the add-in board also may be arranged in tandem with a telephone station. This is most notably the case where the add-in card attaches to a digital line and provides an analog line interface. This is illustrated in Figure 11-33.

Figure 11-33. Tandem mapper code add-in board configuration

Add-in board configurations are applicable primarily in scenarios where the line interface from the switch is nonproprietary (as with analog and ISDN lines).

11.6.8. Other Implementation-Specific Ports

A variation of the proprietary add-in board configuration involves the use of a parallel port or some other implementation-specific computer port to attach a CT component. Like the add-in board, this type of CT component must be accompanied by a mapper of some sort, which at a minimum will allow access to the CT session provided by the component.

11.7. Client-Server Configurations

Client-server configurations, as the name implies, involve an indirect communication link between a user's client computer or PDA and a telephone station or other telephony resources.

All of the configurations in this section are presented using the standard graphical notation. Refer to the inside of the back cover for a summary of the symbols.

11.7.1. Basic Client-Server Configuration

In the simplest of *client-server configurations*, a client computer or PDA establishes a communication link with a CTI server which acts as a proxy in obtaining CTI functionality from a switch.

Figure 11-34 shows the *logical integration* of a client computer and a telephone station in a client-server configuration involving a single client computer. From the perspective of a user controlling the functionality of their telephone station, the indirect flow of CTI messages through the CTI server is functionally equivalent to the direct flow of messages found in direct-connect configurations (assuming that the CTI interfaces themselves have the same functionality).

In the configuration shown, both the switch and the CTI server provide standard CTI protocols, so no proprietary software is required on either the CTI server or the client computer.

Figure 11-34. Client-server configuration example

Logical
Integration

If the telephony features and services available through the resulting CTI session reflect a switching domain that is limited in scope to a single device or device configuration, then the configuration involves first-party call control. This is illustrated in Figure 11-35. If the switching domain contains additional telephony resources, then it supports third-party call control. The switching domain in this case may consist of other station devices and/or additional telephony resources within the switch (Figure 11-36). In either case, CTI messages from the switch are appropriately delivered through a LAN-based communication link between the CTI server and the client computer or PDA.

Figure 11-35. Client-server first-party call control

One of the most significant benefits of client-server CTI system configurations involves taking advantage of the fan-out capability of a given CTI server. By making a given CTI server available on a LAN, a

Figure 11-36. Client-server third-party call control

Other station devices   All telephony resources

client-server configuration can be easily scaled to any size and can include any type of PDA, personal computer, multi-user computer, or other logical client. This is illustrated in Figure 11-37. The CTI server in

Figure 11-37. Client-server LAN configuration

the configuration shown supports standard CTI protocols so any CTI Plug & Play hardware component can establish a communication link with it (after using the appropriate user ID and password if necessary).

11.7.2. Client-Server Mapper Configurations

Three different kinds of mappers may be found in client-server CTI system configurations because they encompass both switch-to-CTI server links and CTI server-to-client links:

- Switch-server mapper hardware

- Switch-server mapper code

• Server-client mapper code

Switch-Server Mapper Hardware

The role of a switch-server mapper hardware component is illustrated in Figure 11-38. In this example, the mapper hardware (labeled "P") allows the switch, which supports only a proprietary CTI protocol, to communicate with the CT Plug & Play CTI server, which supports CTI Protocol 1. The combination of the switch-server mapper hardware and the switch is functionally equivalent to a CT Plug & Play switch.

Figure 11-38. Mapper hardware configuration

Switch-Server Mapper Code

Figure 11-39 shows a client-server configuration in which the proprietary mapper code, written and provided by the switch vendor, is installed on the CTI server so that it can encode and decode the proprietary CTI protocol provided by the switch.

Figure 11-39. Switch-server mapper code configuration

Server-Client Mapper Code

In the system configuration depicted in Figure 11-40, the CTI server uses a proprietary protocol across the LAN communication link and thus does not support integration with CT Plug & Play components unless proprietary mapper software is installed on each client computer, PDA, etc., on the LAN. Figure 11-41 illustrates the magnitude of the challenge associated with supporting server-client mapper code for a large network of diverse products. If a CTI server does not support a standard CTI protocol, and is therefore not CT Plug & Play, mapper software may or may not be available for every type of product on a given network.

Figure 11-40. Server-client mapper code configuration

Figure 11-41. Server-client mapper code LAN configuration

11.7.3. Client-Server Media Services Binding Configurations

When media access service requests are used to bind a media service instance to a call, a media access device (which may be a station device as far as the switch is concerned) that is associated with the appropriate media server is connected to the call in question. The media access device may be added (resulting in a multi-point call) or it may take the place of a device previously participating in the call.

CT Plug & Play Media Binding

Figures 11-42, 11-43, and 11-44 present three different CT system configurations in which media binding services are supported using standard CTI protocols and media services protocols. The CTI sessions and media services sessions flow between the CTI servers and their logical clients without requiring proprietary software, making these configurations CT Plug & Play.

The system configuration shown in Figure 11-42 features a hybrid product consisting of a CTI server and a media server. The media service instance and the CTI server implementation are integrated on a single component so all binding of media session identifiers is simplified.

Figure 11-42. CTI server with media access resources

The system configurations shown in Figures 11-43, and 11-44 involve distinct CTI and media servers that cooperate to present the client computer shown with a media-capable switching domain. In each case one server front-ends the other to provide the media binding function. (Note that the communication links between the servers are shown above one another to better illustrate their relationship. Typically both servers would be connected to the same LAN.)

Server Mapper Media Binding

Figure 11-45 depicts the use of media server mapper code (as defined in section 11.5). In this system configuration, there is no direct interaction between the CTI server and the media server. It is the responsibility of the special media server mapper code installed on the client computer (labeled "S") to provide the media binding functionality. The fact that the CTI server in this configuration

Figure 11-43. CTI server front-ending media server

Figure 11-44. Media server front-ending CTI server

supports CT Plug & Play means that, in this case, mappers are only required to integrate the media services with the standard CTI protocol provided by the CTI server. Media access services can then be used by other software components running on the client computer.

Figure 11-45. Media service mapper code

In Figure 11-46, a variation on the previous system configuration is shown. In this case, the CTI protocol does not support CT Plug & Play and requires that its logical clients use mapper code to work with its proprietary protocol. In this case, the same media service mapper (labeled "S") from the previous example continues to work but it is actually layered over the mapper code for the CTI server (labeled "P"). The CTI server's mapper code delivers a standard CTI protocol which can, in turn, be used by the media server mapper.

Figure 11-46. Layered media service mapper code and CTI mapper code

11.7.4. LAN Dial-Up Bridge Configuration

Dial-up bridges allow hardware components to remotely connect to a LAN. Figure 11-47 illustrates a variation on the basic client-server system configuration depicted in Figure 11-34 in which the personal computer is located in a remote location from the telephony resources being observed and controlled.

Figure 11-47. LAN remote access configuration

This configuration is useful for remotely monitoring a particular telephone station or telephony resources such as ACD groups. For example, an employee working at home could monitor calls being routed to her office phone and selectively transfer them to her home.

11.7.5. LAN Dial-Up Bridge Configuration/OPX

This configuration is a variation on the basic dial-up LAN remote-access configuration described above. The extension being monitored in this case is an off-premises extension (OPX) that is co-located with the remote CTI client. This is illustrated in Figure 11-48.

Figure 11-48. LAN remote access configuration with OPX

This configuration is a standard approach for work-at-home call center agents. With this arrangement, any user can have all the telephony functionality he or she would have in the workplace.

11.7.6. LAN Dial-Up Bridge Configuration/SVD

This configuration is yet another variation on the LAN remote-access configuration. The dial-up bridge in this case utilizes simultaneous voice and data (SVD) technology to provide access to the voice channel associated with the given user's phone line on the PBX, despite the remote location. This functionality, shown in Figure 11-49, is equivalent to the OPX line approach but requires only a single media stream channel (or analog phone line).

Figure 11-49. LAN remote access configuration with SVD

11.7.7. CO-Server Dial-Up

The ability to provide CTI and media services capabilities to remote users is not restricted to users of CPE switches. LECs may also make CTI access available for their CO switches. Figure 11-50 shows a CTI configuration in which client computers use a dial-up communication link to reach a LAN on which the CTI server is located. This LAN could be a private LAN managed by the carrier or it could be a segment of the public Internet. Figure 11-51 shows a simplification of this approach where the CTI server itself supports dial-up communication links and no LAN is required. Both these configurations utilize dual media stream channels (e.g., two analog lines, two ISDN channels) at the user's location. For this reason CO-server dial-up is optimal for multi-channel lines such as ISDN BRI or cable TV.

Figure 11-50. CO-server remote access configuration using dial-up bridge

Figure 11-51. CO-server remote access configuration using dial-up server

LEC-provided CTI services are already available for certain Centrex customers, but they represent a very attractive offering for home businesses and telecommuters. LECs acting as ISPs can create combined service packages that feature CTI along with the usual offers for e-mail and web service.

11.7.8. CO-Server Remote Access/SVD

SVD technology can be used to overcome the need for two phone lines in CO-switch CTI configurations. SVD technology may be incorporated into a dial-in bridge (Figure 11-52) or into a specialized CTI-server that provides support for SVD dial-up (Figure 11-53).

Figure 11-52. CO-server remote access configuration using SVD dial-up bridge

Figure 11-53. CO-server remote access configuration using SVD dial-up server

11.8. Client-Client Configurations

A common scenario in many office environments involves using a pair of products, such as a personal computer and a laptop, or a personal computer and a PDA, only one of which can be directly integrated into a supported CTI configuration. *Client-client configurations* involve using a special software component installed in one of the products, perhaps the personal computer, to let the other product act as a secondary client.

Figures 11-54 and 11-55 depict two typical client-client system configurations. In both, a PDA uses an infrared communication link to send CTI messages to a personal computer. The personal computer acts as a proxy in what are otherwise basic direct-connect and client-server configurations. Any of the system configurations presented in this chapter can be extended in this fashion.

Figure 11-54. Client-client configuration, direct-connect case

Figure 11-55. Client-client configuration, client-server case

11.9. Review

In this chapter we have seen that a limitless variety of CT system configurations can be assembled from a set of basic *CT hardware components*, given standard *communication links* and *CT protocols*.

CT hardware components include *client computers* (*personal* and *multi-user*), *personal digital assistants* (*PDAs*), *telephone stations*, *telephone station peripherals*, *CTI servers*, *media servers*, *switches* (*CPE* and *CO*), *telephony gateways*, and *hybrids* of all of these basic types.

Communication links, which include both the physical link layer and the complete session/transport protocol stack, are broadly categorized as *local area network (LAN)*, *serial cable* and *serial bus*, *infrared*, and *dial-up*.

The communication links between adjacent CT hardware components in a CT system are used to send a *CTI session* using a *CTI* protocol and/or one or more *media services sessions* for any media services that are accessed. CT protocols are either *standard* or *proprietary*. Standard CT protocols allow for CT Plug & Play operation.

Protocol mappers are components that translate between a proprietary CT protocol and a standard CT protocol in order to let a hardware component that relies on proprietary protocols interoperate with CT Plug & Play hardware components. They may be implemented as *mapper hardware* that can connect within a communication link and will work with any hardware components, or they can be implemented as *mapper code* that must be installed and run on every hardware component configured as a logical client, and that must be developed for each platform individually.

CT hardware components that implement proprietary protocols may provide *API-specific adapter software* for specific operating systems that offer programmatic CT interfaces (APIs). Names for adapter software vary depending on the operating system, but include *driver*, *service provider*, and *telephone tool*. Adapter software may be developed to support CT system configurations described in this chapter, but these can be used only in the configuration(s) and on the operating system(s) for which they were implemented.

CT hardware components that implement proprietary protocols also may be found within turnkey *custom CT solutions*. These solutions may be developed to encompass one or more CT system configurations described in this chapter, but once again can be used only in the configuration(s) and on the operating system(s) for which they were implemented.

12.
CT Software Components

In this chapter we explore the many different types of software components that may play a role in a CTI or media services system. These range from protocol stack implementations, to various APIs, to different categories of application software. This chapter covers the layering and relationships among the various types of CT software components and describes the role of each type of component in a CT solution.

While the overall framework for CT software components is independent of the operating system involved, the implementations of certain components represent significant de facto standards in their own right. Sections of this chapter specifically describe certain portions of the CT software framework as they are implemented for the two leading personal computing operating systems: Apple Computer's Mac™ OS and Microsoft's Windows™ operating system. Apple's framework for CT is called the *Macintosh Telephony Architecture*, or *MTA*. Microsoft's set of specifications is called *Windows Telephony*. The ECTF framework and associated specifications provide complementary platform-neutral interfaces for 'C' and Java on both of these platforms as well as all others, including Linux and variations of Unix.

12.1. CT Software Component Hierarchy

Most CT hardware components are based on computer technology and thus are managed by software that is responsible for sending, receiving, interpreting, generating, and handling CT messages. In assembling CT systems and building individual CT components, however, we are concerned only with software environments designed for the installation or addition of software[12-1] developed by third parties. Within such an environment, CT software components are arranged in functional hierarchies—a *CTI software framework* and a *media services software framework*, that reflects the CT value chain.

12.1.1. CT Value Chain

The CT value chain (introduced in Chapter 2) represents the various specialized groups that contribute components to any given CT system. At this point in our exploration of the technologies that support this value chain, we have looked primarily at the lower levels of the pyramid, where the telephony resources are located. In Chapter 11, we explored how these are embodied and physically connected to form systems of hardware components. In this chapter, we turn to the upper layers of the value chain, where software vendors and system integrators add their value to solutions.

In these tiers of the value chain we are concerned primarily with:

- Operating system and/or network operating system software components that support CT

- Products from telephony software developers

12-1. Closed CTI hardware components — CTI hardware components that are not open to the addition of software components range from switches to PDAs. Internally, any of these components may be built using off-the-shelf computer hardware, operating systems, and standards-based telephony resource components (such as S.100 components). The internal architecture of these devices (and related standards) are beyond the scope of this book because they are not of direct consequence to a CTI system.

- Mainstream applications

- Scripts, utilities, and other software developed by systems integrators, customers, and individuals customizing their systems

Historically, CT integration has taken place exclusively in the realm of software, without consideration for the plug-and-play interoperability of hardware components. For this reason, the generalized software framework also allows for software components that are provided by individual telephone equipment vendors.

12.1.2. Modularity

A CT system consists of a collection of individual CT components in which each communicates with its neighbor(s) using CT messages that travel across an inter-component boundary (as described in Chapter 6).

Modularity is just as important for software components as it is for hardware components. System integrators, customers, and individuals want to be able to plug and play with their software, just as they can plug and play with their hardware. This is fundamental to the concept of the CT value chain.

For CT hardware components, the inter-component boundaries are in the form of CT protocols that travel over communication links between hardware components (as illustrated in Chapter 11). For software components they are in the form of programmatic interfaces through which CT messages are communicated.

12.1.3. Programmatic Interfaces

Just as hardware components use transmitters, receivers, cables, and connectors to support the communication links that carry CT sessions, two distinct software components running on the same hardware component can link with one another and communicate using a *programmatic interface*.

Figure 12-1. Programmatic interfaces

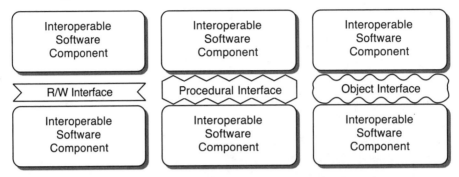

Programmatic interfaces are software-based inter-component boundaries as shown in Figure 12-1. Software components interoperating through a programmatic interface boundary are typically arranged vertically in diagrams to reinforce the fact that both software components exchanging messages through a particular programmatic interface are on the same computer.

Three distinct categories of programmatic interfaces are of concern to CT systems:

- Read/write interfaces

> R/W Interface

  Read/write interfaces, or just *R/W interfaces* for short, are simple programmatic interfaces that allow a software component to obtain access to a *protocol stack implementation* (or a communications driver) in order to open or close a communication link to another software component (on the same or a different hardware component), and to read or write a stream of data across the communication link. If the stream is for carrying CT messages, it is referred to as a *CT session* and the messages are delivered in the form of encoded CT *protocol data units*s, or *PDUs*. R/W interfaces and protocol stack implementations are described in further detail in the next section and in section 12.4.1.

- Procedural Interfaces

Procedural Interface

Procedural interfaces are frequently referred to as Application Programming Interfaces or APIs.[12-2] Procedural interfaces allow two software components to communicate through a set of well-defined function calls. Unlike a R/W interface, a procedural interface often uses many different function calls for exchanging messages. Message parameters are placed in structures or simple variables and passed across the interface through references in function parameters.

- Object Interfaces

Object Interface

Object interfaces allow a software component to access telephony functionality by manipulating software objects. Unlike procedural interfaces, formal object classes are used to define the programmatic interface and the way in which information is exchanged.

12.2. CTI Software Framework

The CTI software framework that forms the basis for integrating CTI software components on a given hardware component is shown in Figure 12-2.

12-2. API — In general usage the term API (which technically stands for Application Programming Interface) refers to any programmatic interface that allows two independent software components to interact. It is not limited to application software programming. To avoid any confusion, however, this book uses the term programmatic interface to refer to interfaces between software components. The term SPI (for Service Provider Interface) and other acronyms are used by some vendors to refer to programmatic interfaces that are not intended for application developers.

Figure 12-2. CTI software component framework

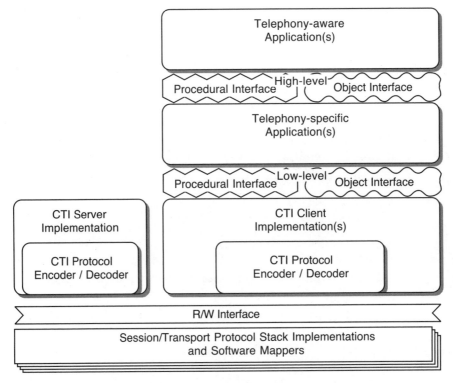

Software component types are listed below, starting from the bottom and working up.

- Session/transport protocol stack implementations

  Session/transport protocol stack implementations and associated communications drivers are usually provided either by operating system vendors or network operating system vendors. In cases where the desired CTI session must be accessed through a piece of proprietary hardware, or a proprietary protocol stack, one or more of these components must be provided by the vendor of the logical CTI server.

- Software mappers

  Mappers (introduced in Chapter 11, section 11.3) translate between proprietary CTI protocols and standard CTI protocols. They are implemented as session/transport protocol modules so that they are transparent to all of the other software components. These are, of course, provided by the vendor of the logical CTI server to be used in a given instance.

- R/W interface

  The *R/W interface* on a particular platform typically is provided by the operating system vendor.

- CTI server implementations

  CTI server implementations typically are provided by telephony software developers and network operating system vendors. Client-client configurations (as described in Chapter 11, section 11.8) are a special case in which simple CTI server implementations may be provided by operating system vendors.

- CTI client implementations

  CTI client implementations are software components that manage the communication link to a particular logical CTI server; they translate the CTI protocol that flows through the resulting CTI session for access through a low-level procedural CTI interface, a low-level object CTI interface, or both. There are two types of these software components. *CT Plug & Play client implementations* support standard CTI protocols, which flow across communication links either from connected logical servers or from software mappers, installed below the R/W interface, that translate from a proprietary protocol. They are provided by operating system vendors or telephony software developers. On the other hand, *API-specific adapters* are provided by an individual telephony equipment vendor for use with the proprietary protocol or driver associated with their product. They are called API-specific because each is written to support a specific programmatic interface (typically called an API) on a specific operating system.

CT Software Components

- Low-level procedural and low-level object interfaces

  - *Low-level procedural CTI interfaces* and their object-based counterparts, *low-level object CTI interfaces*, are programmatic interfaces that represent the service boundary used by telephony-specific applications to access the switching domain for a particular CTI system. The three most popular low-level CTI interfaces are TAPI, Telephone Manager, and TSAPI.[12-3]

- Telephony-specific applications

  Telephony-specific applications are software components with observation (and optionally control) of telephony resources as their primary mission. Telephony-specific applications can be categorized as either being either *screen-based telephone* applications or *programmed telephony applications*.

- High-level procedural and high-level object interfaces

  High-level procedural CTI interfaces and their object-based counterparts, *high-level object CTI interfaces*, are programmatic interfaces designed to support telephony-aware applications. They are optimized to allow mainstream application developers, system integrators, and end users to easily add telephony support to their products, solutions, and work environments. While the functions associated with high-level CTI interfaces generally are implemented and defined by operating system vendors, these interfaces are actually used to access functionality in a designated telephony-specific application. For this reason they are layered above telephony-

---

12-3. Popular CTI interfaces — TAPI stands for Telephony Application Programming Interface. It is provided by Microsoft as part of the Windows operating system. Telephone Manager is provided by Apple Computer as part of the Mac OS. TSAPI stands for Telephony Services Application Programming Interface. The definition of TSAPI is in the public domain, so any operating system vendor, network operating system vendor, or telephony software developer is free to implement TSAPI. See section 12.5 for details on each of these procedural interfaces.

specific applications, and they represent the service boundary between telephony-aware applications and telephony-specific applications.

- Telephony-aware applications

  Telephony-aware applications (also called *telephony-enabled applications*) are, by definition, any application or other piece of software (such as utility or script) taking advantage of one of the high-level CTI interfaces. Despite the fact that they may be very much removed from the core of telephony functionality and telephony resources, they represent the primary vehicle through which system integrators, customers, and individuals see much of the immediate benefit of telephony solutions. As a result of the ease with which telephony-aware applications can be written (in marked contrast with every other layer in the software framework), the number of available telephony-aware applications will be many orders of magnitude greater than telephony-specific applications. However, each plays an important role in a complete CTI solution.

An instance of each of these CTI software component types may or may not exist in any given CTI system implementation. For any particular software component to actually work, however, all of the components it depends on must be in place below it.

It is likely for a complete CTI solution to involve many different applications, and for there to be only one (or fewer) of each of the other types of software components. However, it is possible to have any number of each of these component types running on the same hardware platform.

12.3. CTI Server Implementations

 CTI server implementations are the actual software components that correspond to the CTI server hardware components described in Chapter 11. A CTI server implementation may play one or more of the following roles in a CTI system:

- It acts as a fan-out component (described in Chapter 6, section 6.3.3)

- It acts as a security firewall between logical clients and an unsecured CTI interface (such as one found on a switch).

- It translates among different protocols, allowing logical clients using one protocol to communicate with a logical server using a different protocol.

- It augments the functionality of its logical server to provide a richer set of telephony features and services to its logical client.

A generic CTI server implementation operates as follows:

- It is configured to connect to one or more logical servers (often directly to the CTI interface of a switch), and it maintains these communication links over time.

- It is configured to accept communication link requests from one or more (typically many) logical clients using one or more session/transport protocol stacks. If the CTI server implementation supports security features, the protocol stacks used must support authentication. Authentication may be as simple as layering an initial user ID and password exchange mechanism on top of a non-authenticating protocol, or it may involve a more sophisticated mechanism such as the exchange of encrypted keys as implemented in such protocols as Secure Sockets Layer, orSSL.[12-4]

12-4. Secure Sockets Layer — Secure sockets layer is a protocol for secure communication over the Internet.

The authenticated protocol may or may not also employ encryption for complete protection of all information traveling across the CTI session.

- If the CTI server supports security, it is also likely to support user privilege restrictions. This involves linking an individual's authentication identity to a database of privileges. Depending on the implementation, privileged information for each user may include:

  - What devices in the switching domain are visible. For example, a secretary may be permitted to see only her phone and her manager's phone.

  - What services may be requested for each device that is visible. The secretary above might have access to all services for her own phone, but only the *start monitor* and *stop monitor* services for her manager's phone.

- When a logical client (normally a CTI client implementation, but possibly another downstream CTI server implementation) attempts to establish a communication link to the CTI server implementation, the attempt is first authenticated if security is being used. The pair of communicating components then exchange protocol and version negotiation packets to determine whether or not they can interoperate and, if so, which protocol to use. Assuming they agree on a protocol, and if the server indicates to the client that it is in an operational state, the logical client completes the start-up process by requesting capabilities information from the CTI server implementation. The CTI server implementation provides the appropriate capabilities information based on:

  - The capabilities of its logical server;

  - Any capabilities that it adds; and

  - Any privilege restrictions that apply to the logical client.

- At this point, the CTI session between the CTI server implementation and its new logical client continues until the logical client shuts down the communication link or the communication link fails for some reason. The CTI server is responsible for acting as the full proxy for the logical client. This includes, for example:

    - Ensuring that for every service request the logical client issues, it receives the appropriate positive or negative acknowledgment;

    - Ensuring that every event the CTI server implementation receives (from its logical server) that applies to one of its logical client's active monitors is delivered to the logical client with the correct monitor cross-reference identifier.

CTI server implementations support CT Plug & Play, or use proprietary interfaces, or may be a combination of the two.

12.3.1. CT Plug & Play Servers

A *CT Plug & Play server implementation* is one that is based on standard CTI protocols. It uses the appropriate R/W interfaces for the platform on which it is running, and connects with both logical servers and logical clients using standard protocols over these communication links. This is illustrated in Figure 12-3.

A logical server that delivers a proprietary protocol across the communication link may provide a software mapper in order to interoperate with CT Plug & Play server implementations. This is illustrated in Figure 12-4.

A CT Plug & Play server implementation generally supports two or all three of the standard CTI protocols for communicating with both logical servers and logical clients. At a minimum, it will support CTI Protocol 1 for the CTI session with its logical server (typically a switch) and CTI Protocol 2 for the CTI session with the logical clients. It then translates these protocols (a process which involves only minimal changes to the encoding) between one boundary and the other.

Figure 12-3. CT Plug & Play server implementation

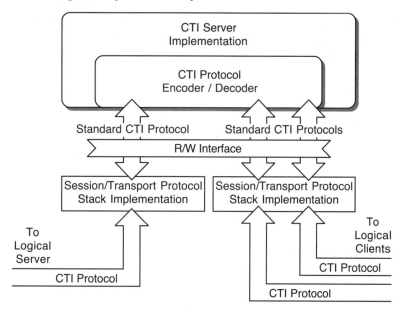

Figure 12-4. CT Plug & Play server implementation with mapper

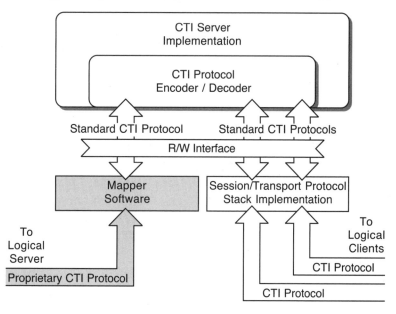

12.3.2. Proprietary Interface Servers

A CTI server implementation with proprietary interfaces assumes that its logical server is a switch and its logical clients are client implementations; it uses proprietary interfaces to connect with each. This is illustrated in Figure 12-5.

Figure 12-5. CTI server implementation with proprietary interfaces

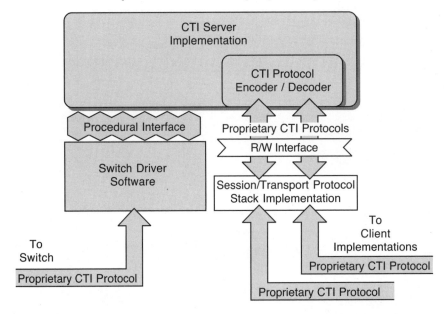

The CTI server implementation vendor defines a procedural API for connecting to the switch using special switch driver software, a form of API-specific adapter software for server implementations. The switch driver software may be written by the switch vendor or by the vendor of the CTI server implementation. Proprietary protocols are then fanned out to corresponding CTI client implementations.

12.4. CTI Client Implementation

The *CTI client implementation* is the software component in a CTI system that is responsible for ensuring that a particular logical CTI server can be accessed through one or more programmatic interfaces.

12.4.1. R/W Interfaces

The programmatic interface used to access session/transport protocol stacks (across which CTI protocols are passed) is called a *R/W programmatic interface*, or *R/W interface*. A R/W interface allows software components to open and close communication links and to read and write the corresponding streams of data using session/transport protocol stack implementations.

Features of a R/W interface include:

- Activating (opening) and deactivating (closing) session/transport protocol stack implementations;

- Layering session/transport protocol stacks on top of one another;

- Supporting multiple streams and multiple protocol stacks simultaneously;

- Sending (writing) and receiving (reading) to and from streams; and

- Translating a reference for a desired communication link destination into the correct protocol stack and destination address.

CTI messages travel through a R/W interface in the form of buffers of encoded (and thus self-contained) PDUs (Figure 12-6). The R/W interface lies between a CTI protocol encoder/decoder and a session/transport protocol stack implementation instance managing an active CTI session.

Figure 12-6. R/W interface

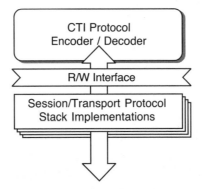

Software components, typically CTI client implementations and CTI server implementations, use R/W interfaces to establish communication links among hardware components and to exchange CTI messages as encoded PDU information. Unlike procedural and object CTI programmatic interfaces, R/W interfaces are general purpose mechanisms designed to allow manipulation of session/transport protocol stacks. In contrast to the other programmatic interface types, using the R/W interface itself is quite simple and any complexity comes only in the process of encoding and decoding the PDUs.

Common session/transport protocol stacks that may be accessed through R/W interfaces (as applicable to a given computer system) include the following, as shown in Figure 12-7:

- ADSP

- Infrared

- IPX/SPX

- RS-232 and virtual serial ports

- TCP/IP

Figure 12-7. Session/transport protocol stack implementations

A very important feature of R/W interfaces is the ability to layer one protocol stack implementation on top of another. Figure 12-8 shows an example where a TCP/IP protocol stack implementation has used PPP to establish a remote IP session over a modem protocol stack which is communicating with modem hardware using an RS-232 driver.

Mainstream operating systems support standard R/W interfaces for managing and using session/transport protocol stack implementations and communication drivers. Examples include:

- Macintosh: Open Transport[12-5]

- Windows: Windows Sockets[12-5.]

- Unix: Streams

12-5. Standard R/W interfaces — Secondary R/W interfaces for Mac OS include the Connection Manager and the serial port interface. Secondary R/W interfaces for Windows include the file and communications functions. CTI client implementations (and CTI server implementations) may support as many R/W interfaces as they wish, but they must at least support the primary R/W interfaces listed in order to support software mappers.

Figure 12-8. Example of layering with R/W interfaces

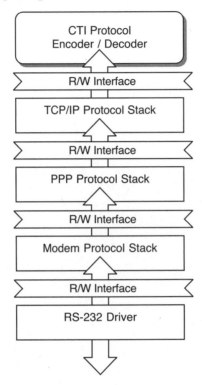

12.4.2. Software Mappers

Software mappers are software components that are constructed as session/transport protocol stack implementations for the R/W interface on every platform. The role of a software mapper is to translate between a proprietary CTI protocol and a standard CTI protocol in order to transparently support CTI client implementations (and CTI server implementations) that are CT Plug & Play.

Software mappers appear to the R/W interface just like any other session/transport protocol stack. They are therefore used just like any other protocol stack implementation. When a CTI client implementation wants to connect to a logical server that uses a mapper, it simply opens a communication link as it normally would, but it specifies the mapper as the protocol stack to use. Figure 12-9 provides two examples of a software mapper in operation. In both cases the Phones-R-Us logical server is accessed using the Phones-R-Us software mapper. In the case on the left, the Phones-R-Us logical server happens to use the TCP/IP protocol stack to transport its proprietary protocol, and the Phones-R-Us mapper layers itself on top of this protocol stack and generates a standard CTI protocol. In the case on the right, the CTI logical server is actually an add-in board that is accessed through a driver.

Figure 12-9. Software mapper example

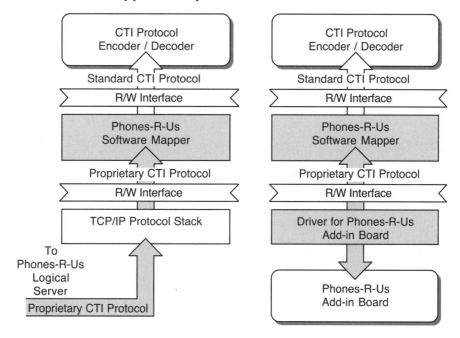

12.4.3. CT Plug & Play Client Implementations

A CTI software component is a *CT Plug & Play client implementation* if it both supports one or more low-level programmatic CTI interface and contains a CTI protocol encoder/decoder in order to support one or more standard CTI protocols as shown in Figure 12-10.

Figure 12-10. CT Plug & Play client implementation

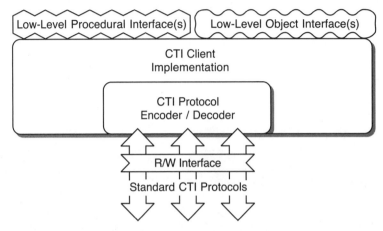

CT Plug & Play client implementations are developed specifically for a given operating system platform. They support the programmatic interfaces (both CTI and media services) appropriate for that operating system.

Support for low-level procedural interfaces involves mapping the CTI messages traveling across the R/W interface in the form of PDUs into a collection of procedural function calls. Support for an object interface involves doing a similar mapping operation using formal object classes that correspond to the CTI abstraction. In either case, this also involves making sure that multiple applications using any of the exposed interfaces are able to simultaneously access the same switching domain without having to establish multiple communication links. In other words, CTI client implementations are software fan-out components.

In addition to supporting low-level CTI interfaces by appropriately mapping the contents of CTI messages, the CT Plug & Play client implementation is also responsible for translating references to media services to the appropriate programmatic interface for a given media type on the operating system in question.

As with every other CTI component in a CTI system, the CT Plug & Play client implementation software can incrementally add value to the switching domains that it is representing to logical clients across the programmatic interfaces. For example, if a given CTI client implementation determines that a particular switching domain does not have support for canonical phone number format, and the CTI client implementation itself is able to perform the necessary translation, it can report to its logical clients that their switching domain does support canonical numbers.

12.4.4. API-Specific Adapter Software

API-specific adapters are CTI client implementations that support a single language/platform-specific programmatic interface and interact directly with the CTI resources they represent, typically through a proprietary CTI protocol of some sort. Because these pieces of software are API-specific, they need to be described in the context of the APIs with which they are associated.

Windows Telephony (TAPI): Telephony Service Providers

Windows Telephony defines the components used to provide access to a particular source of CTI functionality to clients of the TAPI interface as *telephony service providers*. They are software components that are implemented as Windows *dynamic link libraries*, or *DLLs*. In this model, a complete CTI client implementation typically consists of the telephony service provider and some associated drivers that interact with the CTI logical server, and may in fact be made up of several different modules, including media drivers and proprietary hardware drivers.

Microsoft's specifications for Windows Telephony define a pr ocedural interface called the *telephony service provider interface*, or *TSPI*, that represents the boundary between the TAPI implementation and the telephony service provider.[12-6] This is illustrated in Figure 12-11.

Figure 12-11. Windows telephony service provider

Macintosh Telephony Architecture (Telephone Manager): Telephone Tools

The low-level portion of the Macintosh Telephony Architecture is based on the Telephone Manager. The Telephone Manager defines the vendor-specific software components it uses to access CTI functionality as *telephone tools*. In the Macintosh Telephony Architecture, telephone tools are coupled with a *device handler* and one or more media drivers. This is illustrated in Figure 12-12.

The device handler portion is responsible for handling the media streams associated with a given CTI service, and for ensuring that media access is shared and appropriately reflected by both the media drivers and the telephone tool. The structure for the device handler is not defined by Apple.[12-7]

12-6. Telephony Service Provider Interface — The Telephony Service Provider Interface, or TSPI, is documented in the book Telephony Service Provider Programmer's Reference available from Microsoft through the Microsoft Developer Network Library.

Java Telephony (JTAPI): JTAPI Providers

Figure 12-12. Mac OS telephone tools

Java Telephony (JTAPI): JTAPI Providers

The Java language is object oriented so the Java telephony API, JTAPI, is defined in terms of objects and the methods that can be used to operate on them. However, this is not the only difference between JTAPI and the procedural interfaces discussed in this chapter. JTAPI's objects are defined[12-8] in terms of interfaces that must be implemented by a set of objects that constitute the CTI client software. TAPI and the Telephone Manager are pieces of operating system software that actually act as the CTI client implementation. They provide interfaces to adapter software (TSPs and tools) which is different from the interface that they expose to client applications. In contrast, JTAPI represents a zero-thickness interface because the client application interfaces directly with the client implementation using the methods defined by JTAPI.

12-7. Telephone Tool specifications — Documentation for developing telephone tools is contained in the book Telephone Manager Developer's Guide that is part of the Mac OS Software Developer's Kit. This product is available from Apple through APDA. (Telephone Manager Developer's Guide, Apple Computer, Inc., 1991.)

12-8. JTAPI specifications — documentation for developing JTAPI peers is contained in the JTAPI documentation which is available from the Java web site at: java.sun.com/products/jtapi/index.html

627

A JTAPI client implementation is encapsulated by an object known as a *JtapiPeer*. When a JTAPI client application desires access to a particular logical CTI server, it instantiates the corresponding JtapiPeer object (using a *JTAPIPeerFactory* object). The JtapiPeer object is responsible for establishing communication with the logical CTI server and for instantiating a JtapiProvider object that, in turn, provides an application objects for interacting with calls, connection, devices and other resources in the switching domain.

Java software is designed to be run on any computer that provides a Java virtual machine. Given the nature of Java, JTAPI was designed to be layered above popular procedural CTI APIs in addition to CTI protocols and local CTI server implementations as shown in Figure 12-13.

Figure 12-13. JTAPI peer object

12.5. CTI Low-Level Application Programming Interfaces (APIs)

Programmatic interfaces for CTI fall into three categories:

- Implementation-specific

    Implementation-specific CTI interfaces are those in which a proprietary interface is defined and implemented as part of a particular CTI client implementation.

- Operating-system-specific

    Operating-system-specific CTI interfaces are defined and implemented as part of an operating system implementation. The two most popular personal computing operating systems, Mac OS and Windows, each have their own low-level APIs for CTI. These are the Telephone Manager and TAPI respectively.

- Platform-independent

    Platform-independent CTI interfaces are those that are designed to work with any platform. JTAPI[12-9] and TSAPI[12-10] are the only platform-independent low-level CTI interfaces.

TSAPI is characterized as a very thin API layer that provides direct access to standard CTI protocols. TSAPI's primary goal is ensuring that every piece of available information and every accessible capability of a given switching domain implementation is made available to client software, independent of operating system platform. The specification of the TSAPI interface is focused strictly on accessing and using the CTI functionality represented by the standard CTI protocols. It provides a layer above the raw protocol that simplifies software access by providing functionality for encoding and

12-9. JTAPI — All references to JTAPI are to the versions and packages of JTAPI identified in the ECTF C.100 specification.

12-10. TSAPI — All references to TSAPI are to the definition of Versit TSAPI as specified in the Versit Computer Telephony Integration (CTI) Encyclopedia. Earlier versions of TSAPI, including Novell's TSAPI implementation for the NetWare environment, are effectively subsets of Versit TSAPI.

decoding and managing the queues of incoming and outgoing messages, without providing any interpretation of the protocol that might lead to loss of information or accuracy.

JTAPI is the Java language's standard API for CTI (and also for telephony media services as described in section 12.6.1). Like TSAPI, it is based on the standard CTI protocols and is designed to support the full range of CTI functionality on any operating system where a Java virtual machine is running. However, it is also similar to the OS-specific interfaces in that its design is optimized for a particular operating environment; the Java virtual machine in this case, and for use by programmers familiar with the conventions of standard Java APIs. JTAPI was also designed so that it could be used with proprietary protocols by taking advantage of available adapter software written for a given OS-specific interface that it could be layered upon.

On the other hand, OS-specific interfaces were designed and developed with slightly different goals. The principal focus of these APIs is to simplify the programming task for the majority of CTI application developers who are using the facilities of the operating system in question. This has resulted in simplifications that may help most application developers, but also has reduced the information available for certain types of applications. In other words, operating system interfaces attempt to provide an additional layer of functionality above the raw messages from the switching domain. These extras include:

- Providing a simplified first-party call control model;

- Presenting distinct interface subsets for the three different aspects of CTI (call control, telephone control, media access);

- Tracking switching domain capabilities;

- Tracking call identifiers on behalf of applications;

- Mapping switching domain identifiers into handles and local identifiers;

- Tracking state, status, and setting information on behalf of applications;

- Tracking what services are and are not applicable to a connection at a given instant;

- Tracking service completion on behalf of applications;

- Abstracting differences between digital data and voice calls;

- Supplying mechanisms for arbitration of call ownership between applications; and

- Integrating tightly with OS-specific media service APIs.

Of all these features, the first is the most significant distinction. Existing operating-system-specific interfaces (TAPI and the Telephone Manager) derive much of the simplification they offer to application developers by abstracting all switching domains as first-party. (Refer to Chapter 6, section 6.5.2 for the definition of first-party call control.) This simplification reflects the fact that most CTI applications are concerned only with a single device. Depending on the service provider or telephone tool being used, support for logical CTI servers offering third-party call control is accomplished either by representing every device as a distinct switching domain, or by representing it as many devices that are part of a large device configuration.

In practice, either type of low-level API can be used for most applications, assuming that any needed functionality factored out by the OS-specific API still can be accessed through an escape mechanism in that API. When developing a telephony-specific application, the following trade-offs must be considered in choosing between using a platform-independent interface (TSAPI) or an OS-specific interface (TAPI and the Telephone Manager):

- First-party call control

  If an application only requires the ability to observe a single device or device configuration, either type of API may be used. If an application needs the ability to track calls that travel between multiple devices, it probably will be easier to develop using TSAPI.

- Application short cuts

  TAPI and the Telephone Manager provide interfaces in which much of the work associated with tracking information from the switching domain is handled automatically. When the information is needed by an application, it is immediately available. On the other hand, applications using TSAPI must themselves track all of the pertinent pieces of information that arrive in messages and might be needed later. This includes tracking information to determine what service requests can be applied to a call given its state, and the presence of dynamic feature presentation information if provided.

- Veracity

  TSAPI provides the most complete view of what is taking place in the switching domain. While the functionality of the other APIs is quite good, they do involve many more simplifications (and hence mappings) than TSAPI, and ultimately they represent a subset of the full feature set that TSAPI has access to through standard CTI protocols.

- Portability

  While implementations of TSAPI on different platforms must differ slightly to adapt to the way each operating system works, the main benefit of a platform-independent interface is to allow greater portability between platforms. Telephony-specific software that is not otherwise using OS-specific functionality (graphical user interface support, native file management, etc.) can be ported between platforms with little or no effort. Realistically speaking, however, most applications do take advantage of OS-specific capabilities regardless of what CTI interface they happen to use, so portability benefits are limited only to the CTI portion of an application.

12.5.1. ECTF C.100: JTAPI

JTAPI, the Java telephony API provides an object oriented interface for CTI that is platform independent and provides access to the majority of CTI functionality. The JTAPI specification is the responsibility of the ECTF. The ECTF has published *ECTF C.100* which identifies the approved portions of JTAPI that relate to CTI.

The JTAPI interfaces support:

- Instantiating and disposing of JTAPI client implementations;

- Enumerating available switching domains and connecting to one after providing an optional login name and password;

- Determining the capabilities of the switching domain (which is accessed through a *provider* object in JTAPI);

- Instantiating objects to monitor for events;

- Obtaining a snapshot of all physical and logical device elements (known as *terminal* and *address* objects in JTAPI) and determining the associated device configuration;

- Obtaining a snapshot of the active calls in the switching domain;

- Obtaining dynamic capabilities information about existing objects in the switching domain using capabilities objects;

- Making CTI service requests directed towards specific objects; and

- Accessing the proprietary vendor-specific features of a particular implementation.

As an object oriented interface, all services are carried out by invoking methods associated with software objects that correspond to the objects in the switching domain. JTAPI defines interfaces for call, connection, address, terminal, and terminal connection objects. The associated object model is illustrated in Figure 12-14.

Figure 12-14. JTAPI object model

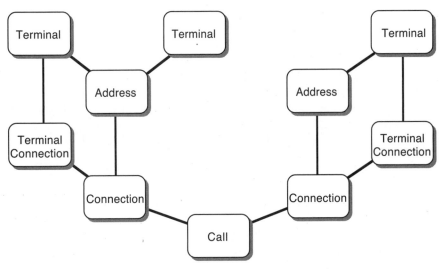

In addition to these CTI objects, JTAPI defines capabilities objects which may be used to determine the dynamic capabilities; that is, the precise set of applicable and non-applicable methods at a given point in time.

JTAPI is really several different interfaces all in one definition. In an effort to simplify use of the interface, the JTAPI specification partitions CTI functionality into a series of packages and defines different versions of the object interfaces for each package, including different views of the event sequences.

Figure 12-15. JTAPI packages

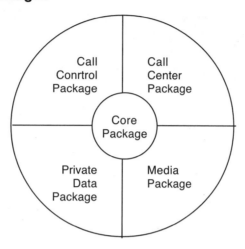

JTAPI defines the following packages:

- Core Package

  The core package supports the basic call control operations of making a call, answering a call, and disconnecting a call. It also supports only a limited set of connection states.

- Call Control Package

  The call control package supports the supplementary call control services.

- Call Center package

  Adds support for routing devices and certain specialized services most often associated with call center applications.

- Private Data package

  Provides access to vendor-specific extensions.

Media Services

  The media services package provides media services binding and access to media services. (This is discussed in Figure 12.7.2.)

JTAPI is becoming one of the most popular CTI interfaces because it is platform independent, supports the richest set of CTI functionality, and is universally accessible.

12.5.2. Windows Telephony: TAPI

At the center of the Windows Telephony specification (part of the Windows Open Services Architecture, or WOSA) is *TAPI*: the *Telephony Application Programming Interface*.[12-11] Microsoft includes implementations of TAPI with its various releases and versions of the Windows operating system.

TAPI allows telephony-specific applications[12-12] to access CTI functionality at a low level. TAPI defines functions for:

- Initializing and shutting down instances of the TAPI implementation. This includes the ability to register message handlers that are used by TAPI to deliver event messages to applications;

- Negotiating the appropriate version of each portion of the API that will be used by an application;

- Identifying, opening, and closing a connection to a particular switching domain (referred to as a *line device* in TAPI 1.0 through 2.1 and as a *TAPI object* in TAPI 3.0);

- Performing capabilities exchange, or *Caps()* functions;

- Allowing snapshots of current call or device information to be obtained;

12-11. TAPI documentation — TAPI is documented in the book Telephony Application Programming Interface Programmer's Reference, available from Microsoft through the Microsoft Developer Network Library.

12-12. TAPI support for telephony-aware applications — In Windows Telephony, telephony-aware applications are supported by a special portion of TAPI known as Assisted Telephony. This part of TAPI is a high-level CTI procedural interface, and is described in section 12.10.

- Establishing monitoring and specifying which events are to be filtered (and also keeping track of what monitoring has been requested on the application's behalf);

- Issuing requests for basic and supplementary telephony features and services;

- Detecting and generating telephony tones and DTMF tones;

- Observing and controlling the components of a physical device element (referred to as a *terminal* or *phone device* in TAPI); and

- Supporting vendor specific functionality or extended services, including versioning and exchanging private information.

Additional functionality provided by TAPI addresses issues related to running multiple applications under Windows, making use of Windows media access interfaces, and performing common application tasks. These include functions for:

- Performing housekeeping tasks such as allocating/deallocating data structures for tracking calls, and assigning particular objects to numeric identifiers;

- Allowing the CTI client implementation (telephony service providers) associated with a particular switching domain (line device) to display a configuration dialog box. This allows CTI client implementations that are not autoconfiguring the ability to be administered from within a TAPI application;

- Performing dial plan management and address translation; and

- Supporting media handling and the ownership of a call by a particular TAPI client application. This allows incoming calls to be directed to specific applications that can determine the type of media stream they are carrying and perform a *handoff* to another application if appropriate. TAPI keeps track of the media type that is identified, and has extensive functionality for identifying the correct application to work with the call and the appropriate media service API to use.

TAPI 1.0 through 2.1

In versions 1.0 through 2.1, TAPI's function calls and data structures are designed for use with switching domains that contain only a single logical device or device configuration, that is, ones that support only first-party call control. This significant simplification (relative to the full abstraction of telephony) allows for application implementations to be streamlined somewhat. When observing and controlling a first-party switching domain, calls, call appearances, and connections are all effectively interchangeable. With this insight, TAPI defines an *address* as the logical device in the switching domain (TAPI line device) that can be observed and controlled. A device configuration consisting of multiple logical devices is therefore represented by multiple addresses.

Applications are given visibility only to the calls associated with an address, but not to the associated appearances and connections. While this generally simplifies the programming model for most simple applications, it requires that all of the connection states associated with a particular call be reflected as a single *call state*. As an aid to programmers, call state definitions also incorporate context information. Call states can be translated to connection states as shown in Table 12-1.

Table 12-1. TAPI call states

TAPI Call State	Local Connection State	Connection state of device outside switching domain
Accepted	Alerting (ringing mode)	N/A
Busy	Connected	Fail (w/ busy cause)
Conferenced	Connected (after conferenced)	N/A
Connected	Connected	Connected
Dialing	Initiated (after digits dialed)	N/A
Dialtone	Initiated	N/A
Disconnected	Connected	Null (after connection cleared)
Idle	Null	N/A
Offering	Alerting (offered mode)	N/A

Table 12-1. TAPI call states

TAPI Call State	Local Connection State	Connection state of device outside switching domain
Onhold	Hold	N/A
On hold pending conference	Hold (consult purpose of conference)	N/A
On hold pending transfer	Hold (consult purpose of transfer)	N/A
Proceeding	Connected	Unknown [connected] (after network reached)
Ringback	Connected	Alerting (ringing mode)
Special info	Connected	Fail (w/other cause)

TAPI 3.0

TAPI version 3.0 adds an object oriented interface to TAPI using Microsoft's COM technology and provides a different model for media services binding. (Media services interfaces are discussed in section 12.7.4.) The TAPI 3.0 architecture is illustrated in Figure 12-16.

Figure 12-16. TAPI 3.0

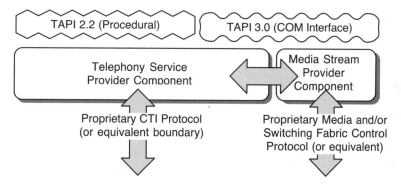

The most significant improvement in TAPI 3.0 is that it provides access to an object that corresponds to a call. Microsoft refers to this object as a *call hub*. The call hub object allows an application to enumerate the other connections associated with the call and examine their states independently. Though the labels are different, the TAPI 3.0 object model, illustrated in Figure 12-17, reflects the standard CTI object model.

Figure 12-17. TAPI 3.0 call control objects

12.5.3. Macintosh Telephony Architecture: Telephone Manager

The *Telephone Manager* is the procedural interface designated by the Macintosh Telephony Architecture for writing screen-based telephony applications and most programmed telephony applications. The Telephone Manager[12-13] is distributed as part of Mac OS.

The Telephone Manager provides CTI applications with functions for:

- Identifying the appropriate version of the Telephone Manager API to use.

- Identifying, initializing, opening, and closing a connection with a particular switching domain (known as a *terminal* in the Telephone Manager API).

- Establishing, clearing, setting filters for, and tracking the status of message handlers. The Telephone Manager interface delivers messages to separate handlers according to what they reference. There are message handlers for messages applying to the whole

---

12-13. Telephone Manager specifications — Documentation for developing applications using the Telephone Manager is found in the Telephone Manager reference documentation that is part of the Mac OS Software Developer's Kit.

switching domain (terminal), for logical devices (referred to as a *directory number* in the Telephone Manager), and for call appearances.

- Obtaining the capabilities of a switching domain (terminal), a physical device (associated with the terminal), or a logical device (directory number).

- Obtaining a snapshot of all the calls associated with a device or all the information about a call. In both cases this information includes dynamic feature presentation data, so the application knows what service requests apply to the device.

- Issuing requests for telephony features and services.

- Detecting and generating telephony tones and DTMF tones.

- Observing and controlling the components of a physical device element.

- Accessing vendor specific functionality (referred to as *other features* and *other functions*).

The Telephone Manager also provides applications with a number of value-added functions that simplify application development. They include:

- Functions for initializing the Telephone Manager.

- A suite of functions that allow the CTI client implementation (telephone tool) associated with a particular switching domain (terminal) to display a configuration dialog box. This gives non-autoconfiguring CTI client implementations the ability to be administered from within a Telephone Manager application. A CT Plug & Play client implementation uses this mechanism to determine what session/transport protocol stack implementation and address to use to connect to the desired CTI logical server. In addition, through the use of additional functions CTI client implementations are able to display windows and menus that coexist with the application's.

- Functions that perform housekeeping functions such as locating, allocating, and deallocating data structures for tracking calls and devices.

The Telephone Manager allows telephony-specific applications to access CTI functionality through a first-party call control model. This means that the switching domains (terminals) that applications may observe and control through this API are represented by a single device configuration, where each logical device in the device configuration is referred to as a directory number. The first-party nature of this interface also allows for a simplification in which call appearances, connections, and calls are treated as equivalent.

The relationship between the Telephone Manager's call appearance states and connection states is presented in Table 12-2. These call appearance states reflect the seven different states that the local connection can take and, for the ease of programmers, provide variations on the *alerting, connected, hold,* and *initiated* connection states to incorporate recent event information into the state itself. In addition, there are five call appearance states that correspond to a local connection state of *connected* in combination with the connection state and context of a called device

Table 12-2. Telephone Manager call states

Call Appearance State	Local Connection State	Connection state of called device
Active	Connected	Connected or Hold
Alerting	Alerting (ringing mode)	N/A
Busy	Connected	Fail (w/ busy cause)
Conferenced	Connected (after conferenced)	N/A
Conferenced held	Hold (consult purpose of conference)	N/A
Dialing	Initiated (after digits dialed)	N/A
Dialtone	Initiated	N/A
Held	Hold	N/A
Idle	Null	N/A
In use	Fail (w/blocking cause)	N/A

Table 12-2. Telephone Manager call states

Call Appearance State	Local Connection State	Connection state of called device
Offering	Alerting (offered mode)	N/A
Queued	Queued	N/A
Reorder	Connected	Fail (w/reorder cause)
Ringing	Connected	Alerting (ringing mode)
Waiting	Connected	Unknown [connected] (network reached)

12.5.4. TSAPI

TSAPI is the name of the procedural interface that is based on the standard CTI protocols. The latest version of the TSAPI specification is defined as part of the *Versit CTI Encyclopedia* (Versit, 1996) and is referred to as Versit TSAPI. Other TSAPI versions that predate Versit TSAPI are effectively subsets of it that do not support CT Plug & Play. This book concentrates on the latest TSAPI specification, which is in the public domain and can be implemented by any vendor.

The functions making up the TSAPI interface are quite simple and easy to use because TSAPI is defined to operate on any operating system platform, and because its focus is exclusively on providing a consistent API for access to the standard CTI protocols. TSAPI does not offer extra value-added services that might mask the accurate flow of CTI control and status information. The application, or other client software, is itself responsible for interpreting capabilities exchange information, events regarding the merging of calls, and generally for keeping track of the information that is reported in the messages provided by the switching domain.

TSAPI is divided into two functional parts: ACS and CTI.

Application Control Service (ACS) functionality

ACS functionality provides applications with the ability to establish
and manage a CTI session to a desired source of CTI functionality (a
logical CTI server). It includes functions to identify the appropriate
API version and to locate switching domains that may be accessed.
The balance of the ACS functions obtain the information necessary to
communicate with a switching domain, open and close a CTI session,
and retrieve messages sent across the CTI session.

CTI functionality

CTI functionality corresponds to the service requests,
acknowledgments, and events defined for the standard CTI protocols.
TSAPI provides access to CTI messages through TSAPI function calls
and TSAPI application events.

TSAPI function calls correspond to messages that are to be sent across
the CTI session. Each TSAPI function is defined as generating either a
specific service request or a specific acknowledgment to a service
request sent by the switching domain.

TSAPI application events correspond to messages received across the
CTI session. Each TSAPI application event is defined as corresponding
to a CTI event message, a positive acknowledgment, a negative
acknowledgment, or a service request sent by the switching domain.

 TSAPI application events must not be confused with the CTI event
messages that are sent by the switching domain. CTI event messages
are one type of CTI message, whereas TSAPI application events
represent any type of message sent from the switching domain.

12.6. Media Services Software Frameworks

The software framework that forms the basis for integrating media services client software components on a given hardware component is shown in Figure 12-18.

Figure 12-18. Media services client software framework

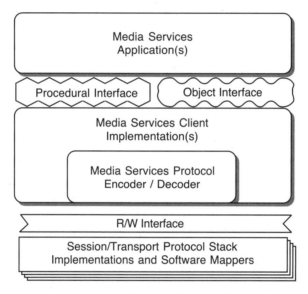

Software component types included in the media services client software framework are listed below, starting from the bottom and working up.

- Session/transport protocol stack implementations
  Session/transport protocol stack implementations and associated communications drivers are usually provided either by operating system vendors or network operating system vendors.

- Software mappers
  Mappers (introduced in Chapter 11, section 11.3) translate between proprietary media services protocols and a standard media services protocol. They are implemented as session/transport protocol modules so that they are transparent to all

645

of the other software components. These are, of course, provided by the vendor of the logical media server in question.

- R/W interface

  The R/W interface on a particular platform typically is provided by the operating system vendor.

- Media services client implementations

  Media services client implementations are software components that manage the communication link to a particular logical media server; they translate the media services protocol that flows through the resulting media services session for access through a procedural media services interface and/or an object-based media services interface. There are two types of these software components. CT Plug & Play client implementations support a standard media services protocol. Server-specific client implementations use a proprietary protocol or communicate directly with a local media server implementation.

- Procedural and object media services interfaces

  Procedural media service interfaces and their object-based counterparts, *object media service interfaces*, are programmatic interfaces that represent the service boundary used by media services applications. These interfaces provide applications with access to media services such as data communications, audio recording and playback, speech, and video. This set of interfaces includes the general purpose media service clients and media services interfaces defined by the native platform that can be bound to telephony media streams.

- Media services applications

  Media services applications are software components that are concerned with observation and control of media resources and their interaction with media streams. *Media–only applications* are only concerned with using media resources to carry out a particular interaction for each call that is presented to them. *Integrated media applications* use both CTI and media services interfaces and are concerned with both controlling and with manipulating both the media streams of a given call and the call itself.

An instance of each of these software component types may or may not exist in any given system implementation. For any particular software component to actually work, however, all of the components it depends on must be in place.

It is likely for a complete CT solution to involve many different applications, and for there to be only one (or fewer) of each of the other types of software components. However, it is possible to have any number of each of these component types running on the same hardware platform.

12.6.1. Media Services Available for Media Binding

The low-level programmatic CTI interfaces in the CTI software framework are complemented by all applicable media service interfaces available for use with telephony on a given platform. Regardless of the CTI interface used, applications that want to access media information associated with a particular call must use the CTI interface to identify what media service instances are available for use with that call, and then use them to bind an instance of the desired media service. Once this is done, the application uses the appropriate media services interface for the operating system being used to access the selected media service instance. The application is provided with a media stream identifier in the form of a device identifier, session reference, driver specification, object, or some other identifier that is specific to the media services interface. This is provided by the CTI interface to specify to the media services interface what media stream is desired.

Media binding is implemented in CT Plug & Play environments using well-defined media services type values. Each media services type value corresponds to a:

- Specification for how the media stream identifier that is delivered in the CTI session is to be interpreted in order to establish a media service stream for access to the desired media service instance.

- Specific protocol (or interpretation of data) that is to be used with the resulting media service stream.

Table 12-3. Standard media service types

Media Service	Media Stream Identifier Represents
Live Sound Capture - Analog	Analog jack used
Live Sound Transmit - Analog	Analog jack used
Live Sound Capture - FireWire	FireWire channel used
Live Sound Transmit - FireWire	FireWire channel used
Live Sound Capture and Transmit - GeoPort	GeoPort stream used
Live Sound Capture and Transmit - ATM	ATM virtual channel used
Live Sound Capture and Transmit - ISDN	ISDN bearer channel used
Sound Capture and Transmit - Rockwell ADPCM	Address to be used on same link used for CTI session
Sound Capture and Transmit - API	Identifies mapper-supplied sound drivers to be used with OS API
Data/Fax Modem	Address to be used on same link used for CTI session
Digital Data - Isochronous - FireWire	FireWire channel used
Digital Data - Isochronous - GeoPort	GeoPort stream used
Digital Data - Isochronous - ATM	ATM virtual channel used
Digital Data - Isochronous - ISDN	ISDN bearer channel used
Digital Data - API	Identifies mapper-supplied drivers to be used with OS API
ECTF S.100 Media Services	Call channel resource identifier or S.100 session identifier

The media service types listed in Table 12-3 are part of the CTI standard protocols. They allow basic media services to be accessed by a logical CTI client in a CT Plug & Play fashion, that is, without requiring any implementation-specific media access drivers.

In CTI solutions involving mappers or API-specific adapters, these software components include (or can be tied to) specific media binding drivers that are invoked when media binding takes place.

12.7. Media Services Application Programming Interfaces (APIs)

Media services APIs are used to allow applications to communicate with media service instances in order to perform media processing. In the case of media-only applications that do not utilize any CTI functionality, these APIs are the only ones required. Integrated media services applications require these APIs in addition to a CTI API.

12.7.1. ECTF S.100

The ECTF S.100 API is at the heart of the media services portion of the ECTF's framework. S.100 is a "C" language programming interface that provides applications with access to media services. It is designed to be platform independent and the specifications are freely available to any developer.

S.100 supports integrated applications that use a CTI interface and media binding services as well as media-only applications.

The S.100 specification defines a procedural API that corresponds directly to the media services model presented in Chapter 7. The model abstracts implementation details of call processing hardware, switching fabric implementations, and system configuration to enable media applications to be written for maximum interoperability, flexibility, and modularity. The API's functions and events allow an application to configure a group of vendor-independent media processing resources and perform operations with them.

The S.100 API includes functions for:

- Allocating, assembling, storing, and managing KVSet (Key Value sets) data structures. These data structures are the primary data type used throughout the S.100 API. A KVSet consists of a list of key-value pairs. The key portion of the structure contains a symbol which identifies the parameter stored in the value field. KVSets may be nested by as a value field may itself contain a KVSet.

- Establishing and concluding a session between the application and the media service instance.

- Registering event handlers for asynchronously delivered events, polling for events, and requeueing events.

- Registering to receive messages from, and sending messages to, other media clients.

- Creating and destroying media groups.

- Configuring and reconfiguring, allocating and deallocating, the media resources within a media group.

- Connecting and disconnecting the inter-group switching between two media groups.

- Creating, deleting, moving, copying, and accessing parameters for data objects and containers in the media service instance's local data storage.

- Moving data to and from the media service instance's local data storage.

- Performing basic call control functions.

- Performing media processing operations using the functionality defined for each of the resources that have been configured into a given media group. This includes: T.30 and T.611 Fax, Automatic Speech Recognition (ASR), Player (including Text To Speech), Recorder, Signal Detector, and Signal Generator resources.

• Specifying run time control (RTC) parameters and arbitration of intra-group media streams.

S.100 refers to acknowledgment messages as *completion events* and to event messages as *unsolicited events*. S.100 completion events must not be confused with the unsolicited event messages that are sent by the media service instance to indicate changes in the state or status of resources. Media services event messages are one type of message, whereas S.100 events represent any type of message sent from the media service instance.

The implementation of an S.100 API consists of an Application Interface Adapter which corresponds to the media services client implementation on a given machine. The AIA is responsible for implementing all of the local services such as the management of KVSet data structures, establishing and managing the client side of the media services session, and managing the queue of received events. All other functions correspond to messages that the AIA sends to the media service instance using a standard protocol (ECTF S.200), a proprietary protocol, or through interprocess communication. A typical AIA implementation will typically support only one or two of these communication options.

12.7.2. ECTF S.410: JTAPI Media

ECTF S.410 refers to the JTAPI media package(1.3 or later)[12-14], or JTAPI Media for short. JTAPI is the set of telephony APIs for the object oriented Java language. JTAPI media represents a platform independent and portable API that allows both the media services client implementation and the applications that use it to run on any Java virtual machine.

12-14. JTAPI 1.3 Specification — Specifications for ECTF S.410, the JTAPI media package are available both from the ECTF web site at http://www.ectf.org and from the Java web site at http://java.sun.com/products/jtapi/index.html

Like S.100, S.410 is based on the media services model presented in Chapter 7. S.100 and S.410 are complementary interfaces that effectively provide the same functionality but S.100 does so in a fashion which is appropriate for 'C' language procedural programming and S.410/JTAPI Media does so in a fashion which is appropriate for object oriented programming in Java.

The set of interfaces defined by JTAPI Media is implemented by a JTAPIPeer object just as with the CTI related JTAPI packages (described in Chapter 12, section 12.5.1). In fact, a typical implementation would involve doing both sets of interfaces in the same JTAPIPeer.

In addition to the basic services of the JTAPI peer, JTAPI Media includes functions for:

- Creating and destroying media groups and binding these to calls;

- Configuring and reconfiguring the media resources within a media group;

- Performing media processing operations using the functionality defined for each of the resources that have been configured into a given media group; and

- Specifying run time control (RTC) parameters.

JTAPI Media refers to acknowledgment messages as *transaction completion events* and to event messages as *non-transactional events*. JTAPI Media transaction completion events are not to be confused with the event messages that are sent by the media service instance to indicate changes in the state or status of resources.

As an added convenience to programmers, JTAPI Media defines a BasicMediaService class which packages all of the interfaces in a single class so that it can be inherited and used as the basis for a JTAPI Media application.

12.7.3. Mac OS Media Access Interfaces

The Macintosh Telephony Architecture specifies a number of media services and corresponding media service interfaces to be used in conjunction with telephony. These are listed in Table 12-4.

Table 12-4. Mac OS media service interfaces

Media service	Media service interface
Stream-based data	Open Transport
Fax/data modem	Connection Manager
Sound Capture	Sound Input Manager
Sound Playback	Sound Manager
Text to Speech	PlainTalk TTS
Multi-track media	QuickTime

12.7.4. Windows Media Access Interfaces

Windows Telephony specifies a number of device classes and corresponding media service interfaces that can be used to access the media streams associated with active calls. See Table 12-5 for the list of these interfaces.

Table 12-5. Windows telephony device classes and media service interfaces

Device Class	Media service interface
comm	File and communications functions
comm/datamodem	File and communications functions
wave/in	Wave functions
wave/out	Wave functions
midi/in	MIDI functions
midi/out	MIDI functions
ndis	Network driver media access control functions

Windows Telephony: TAPI 3.0

TAPI 3.0 expands TAPI to include an architecture for media stream providers as shown in Figure 12-16. The new COM-based media services interface for TAPI 3.0 applications work in a fashion similar to the mechanism for earlier versions of TAPI described above but allow greater flexibility and permit new media stream types and allow processing of media streams.

12.8. Screen-Based Telephone Applications

Screen-based telephone applications (*SBTs* for short) are responsible for providing the on-screen user interface that allows an individual with a computer to manage telephone calls. SBTs represent the most visible category of telephony-specific application.

Individuals use their screen-based telephone both to do the things that they might previously have done with their physical telephone, and also to take advantage of telephony features and services that previously were never easy or possible to use. In order for the transition to using an SBT to be successful, however, the SBT must do everything that the telephone set can do and be:

- Easier to use,

- Faster to use,

- Just as reliable, if not more.

Screen-based telephone applications that meet these basic requirements quickly become the most important and most relied-upon application on people's computers. It generally is the one application that is set to launch whenever the computer starts up, and is the last application to be shut down if the computer is turned off.

The basic features of a screen-based telephone application include:

- Click-and-drag dialing

    Click-and-drag dialing involves selecting a name or phone number on the screen and dragging it to the screen-based telephone (or equivalently selecting a special dial menu item or function key) to call that number or person. Typically the SBT will support recognition and correct interpretation of number fragments. Dialing names involves having the SBT look up the numbers for people (or simply pass the names to the switching domain if it supports dial-by-name).

- Custom speed-dialing lists

    Most telephone users have a small number of people whom they call with great frequency. While many telephone sets have the ability to program speed-dial (or rep-dial) buttons, few people take advantage of this feature because of the user interface barrier. SBTs make setting up speed-dial lists easy. A good speed-dial list implementation not only lets you specify the people you want in your speed-dial list, but optionally keeps track of additional *most frequently called* and *most recently called* lists. Another feature to look for in a speed-dial implementation is the ability to keep track of all the numbers for a given person; if the person isn't in one location, you can try another without resorting to a different directory to look up the alternate number.

- Presenting callerID information

    A good screen-based telephone application will present callerID information in some fashion if it is available. It also should present the name of the person normally associated with the number, if known, rather than just the number itself.

- Customized call announcement

    A popular SBT feature is the ability of the application to provide an alternative notification mechanism for incoming calls. The SBT does this by setting the ringer volume to zero on the physical element in use (if any) using the *set ringer status* service, and then playing a customized sound or speaking a customized announcement to indicate the phone is ringing. For example, Bob's screen-based telephone application could be set to say, "Bob, there's a call for you!" every time the phone rings. By using a text-to-speech engine, the SBT can speak the name of the caller out loud using callerID information.

- Support for Set Microphone Mute

  The ability to support the *set microphone mute* service is a critical feature for a screen-based telephone. In contrast to the user of a traditional handset, who can place a hand over the microphone for privacy, the SBT user relies on the application's user interface to get immediate privacy if interrupted. If the switching domain being used does not support the *set microphone mute* service, the *hold call* service is a good fallback.

- Support for Transfer Call

  Of all of the common telephony features, transfer is the one with which people most frequently have problems. Effortless transferring of calls is something from which everyone in a business environment benefits.

Screen-based telephone applications have three basic responsibilities in the context of a CTI solution. They provide:

1. A telephony user interface to an individual

2. Full access to desired telephony functionality

3. Support for telephony-aware applications

Screen-based telephone applications should, in general, be as focused and modular as possible. Too many bells and whistles detract from the usability and universality of a product. If needed by a particular individual, these features can be added as customization using telephony-aware applications (described in section 12.10). Keeping an SBT design "focused" means that individuals will have the ability to use the screen-based telephone application that best meets their needs, without having to trade off against features that might be implemented best in a separate, complementary application. Depending on the customer, however, it might be quite appropriate to have many features in a single product. For example, a screen-based telephone to be used in a home environment might integrate screen-based telephony with voice mail (which is a programmed telephony function).

12.8.1. User Interface

User interface is the most important part of a screen-based telephone application. A computer user's screen-based telephone application in effect *becomes* their telephone, and that means it is the focal point of communications activities. The user interface design for SBTs is the place where much of the value is added.

It is very important to note that there is no such thing as the "optimal user interface design" for screen-based telephone applications. User interface studies have shown that an SBT product with a given feature set cannot have its user interface optimized for the general population (as is the case for many categories of software). This is the case because the expectation, work patterns, mental models for telephony, and user preferences are simply too diverse. A trip to the telephone department of any consumer electronics store underscores the fact that people have very diverse preferences when it comes to the look and feel of their telephones. This insight has a number of implications for people at different layers in the CTI value chain.

- For CTI application developers it means that there is great opportunity in developing screen-based telephone applications, because it is unlikely that any single design will dominate this field.

- Mainstream application developers adding telephony support to their applications should not assume that their product will be used in conjunction with any particular screen-based telephone application.

CT Software Components

- Telephone network providers, telephone equipment vendors, hardware vendors, and operating system vendors considering bundling or promoting a screen-based telephone application with their products should consider structuring such deals as sampler promotions, where customers get a chance to try out a number of different SBTs. Choosing just a single SBT to work with means that a significant portion of customers will, at best, be ambivalent, and, at worst, will actually be turned off.

- System integrators and organizations implementing CTI solutions should assume that people using the CTI system will want the ability to choose their own screen-based telephone application. Solutions should be built such that individuals are given a range of SBT options from which to choose.

- Individuals should be choosy. If the screen-based telephone application you're using doesn't feel right, then try another one.

Another important insight into screen-based telephone applications is that users have (and need) only one. Though this might seem to be an obvious statement, some application designers overlook it. Just as people rarely have multiple telephone sets on their physical desks, they need or can effectively use, only one SBT on their virtual, computer desktops. The implication for application developers reinforces the fact that SBT functionality should be narrowly focused. Some mainstream developers, in trying to add telephony support to their applications, actually turned them into screen-based telephone applications rather than telephony-aware applications. This approach really isn't practical because the user would end up with multiple SBTs if other mainstream applications started doing this.

When considering options for screen-based telephones (or when designing them), keep in mind that an SBT needs to be lightweight in its use of memory and other computing resources. In order for users to have a positive experience, they need to be able to rely on their application being available at all times; that means it has to be in memory all the time.

In this section we'll look at a number of different user interface paradigms for screen-based telephone application design. The SBTs that you evaluate (or build) may fall cleanly into these categories; more likely, however, they will contain elements from multiple schools of design. Again, it is important to keep in mind that there is no right answer when it comes to SBT design. There is only the need for sufficient diversity to satisfy the preferences of a diverse population.

Figure 12-19. Phone-under-glass design approach

Phone Under Glass

One approach taken in the design of SBTs is to make the on-screen view of the virtual telephone look and feel much like the physical telephone that it replaces. This is referred to as putting the *phone under glass* (Figure 12-19).

The thinking behind this approach is that users get all the benefits of screen-based telephony described above, but they still continue to work with an interface with which they are familiar. The traditional rough edges of a phone's design can be polished through online help dialogs and help balloons that explain what all the buttons do.

Designers who don't like this approach point out that it preserves the limited user interfaces that aren't very effective anyway, and it results in the consumption of much more screen real estate than is really necessary. Another criticism is that users don't want to click on-screen buttons representing a dial pad when they can just type numbers very quickly on the keyboard. As with all of these schools of thought, there is no single right answer.

Figure 12-20. Button panel design approach

Button Panel

The *button panel* approach to SBT design involves putting all of the features and services that are accessible using the SBT in a window or menu of some kind as an array of buttons, along with a simple text area that provides status information. When a user wants to place a call, for example, he or she just presses the "Dial..." button and is prompted for the number to dial. A number then can be typed or dragged from somewhere (see Figure 12-20).

In this type of design, the window or menu may be hidden from view (collapsed or actually absent from the screen) until needed. Upon the arrival of an incoming call, it might appear automatically with the "Answer" button highlighted as the default and the display area providing the callerID. For outgoing calls, a function key or "hot area" for the mouse on the computer display could be used to pop the SBT to the foreground.

Proponents of this design style maintain that this is the best approach because every option is plainly labeled and it is very efficient in terms of screen real estate. Dissenters say that it is good only for linear thinkers because of its reliance on text labels and a text status display. Again, there are no right answers.

Figure 12-21. Minimalist design approach

Minimalist

Yet another design approach is the graphical minimalist approach. In this approach, the emphasis is on minimizing screen real estate but using a graphical rather than text view. A postage-stamp-sized window contains an icon that indicates the status of any telephony activity. The window floats in front of all other windows so that it is always accessible, but it is tiny so it doesn't get in the way. If there is an active call, the icon reflects that status and the name(s) and phone number(s) of the participants appear. Calls can be placed (or added to) at any time by dragging telephone numbers or names to the window. Services and features can be invoked using simple mouse gestures and pop-up menus. This design approach is illustrated in Figure 12-21.

Supporters of this approach say that it strikes the best balance of screen real estate and easy-to-understand graphics. Detractors find fault in the fact that the capabilities of the application are not readily apparent because the menu must be popped up and the gestures must be learned.

Figure 12-22. Direct manipulation/visualization design approach

Direct Manipulation or Visualization

The direct manipulation/visualization approach involves changing the user paradigm for telephony from the traditional, feature-oriented view to a view that reflects what is really taking place in the switch.

The user interface is based on a graphical representation of the devices, with calls being used much as they are presented in Chapter 5. The users of this type of SBT are able to see the devices they control, the connected calls, and the other devices in these calls. They can use the mouse to manipulate connections directly, without having to think in terms of actual feature or service names. For example, to add a person to a two-party call, the user just drags the name of the person to the icon representing the call (rather than pressing a button to invoke the *single step conference call* service). Figure 12-22 shows an example of this design approach.

When this approach was user-tested, a surprisingly large group was very enthusiastic. An equally large group disliked it. As with all of the design approaches, each is good for a certain part of the population.

Speech Interface

The last example we'll look at here is one that switches from the visual interface to the audio interface. In this case the SBT's primary user interface is through speech and other audio input/output. This type of screen-based telephone has a very small visual indicator in the corner of the screen, just to indicate status and the fact that it is running. The primary interface for controlling it is speech recognition, and the primary means for feedback is text-to-speech and prerecorded sound clips. When an incoming call arrives, for example, the caller's name is announced. To answer it, the user simply says, "Computer, answer the phone." To place a call, the user just says, "Computer, call Frederic." The computer looks up Frederic in its preconfigured speed-dial list and initiates the call.

As with all of the other examples, some people like this approach and others dislike it.

12.8.2. Functionality and Feedback

A second area of responsibility for screen-based telephone applications is functionality. The range of functionality supported by different SBTs will vary, depending upon the type of user their designers had in mind. It is important to note that not every user wants access to the full range of telephony features. Nonetheless, a good screen-based telephone application doesn't make arbitrary decisions about what subset of telephony features and services it chooses to support. A screen-based telephone designed for mainstream use should support the full range of telephony features and services, as well as supporting control of physical device elements. When the application connects to the switching domain, it should then scale back its user interface to reflect the actual features and services available. A design that provides for users who prefer a reduced feature set might additionally allow users to deactivate other features and further scale back the functionality that appears in the user interface.

As we have already seen, observing is at least as important as controlling. In the context of screen-based telephone applications, this translates into putting at least as much emphasis into the implementation of feedback mechanisms as into control. Regardless of the user interface design employed, an SBT must squeeze out every ounce of information about the status of relevant items in the switching domain, and it must provide this information as feedback in an immediate fashion.

12.8.3. Support for Telephony-Aware Applications

The third area of responsibility for screen-based telephone applications is supporting telephony-aware applications. From the perspective of the CTI value chain, the screen-based telephone application is near the top. From the perspective of an individual user, however, the screen-based telephone application is the foundation and everything below it is transparent. Individuals see the screen-based telephone application as the starting point for the customized, integrated telephony workspace they will form using telephony-aware applications and scripts. A good screen-based telephone application is designed to support this role as an anchor point in a larger CTI solution. (Telephony-aware applications and the high-level programmatic CTI interfaces that support them are described in section 12.10.)

12.9. Programmed Telephony Applications

Programmed telephony applications are the complement to screen-based telephone applications. Where screen-based telephone applications provide an interface to the user who is placing calls using a particular computer, programmed telephony applications present a telephone-based interface to someone who is calling a particular computer. Where screen-based telephone applications are designed to operate under the direct control of a computer's user, programmed telephony applications are intended to run in an autonomous fashion once they have been given their instructions (or have been *programmed*).

Programmed telephony applications are telephony-specific applications that interact with telephone calls in an autonomous fashion based on predetermined rules of any complexity. A programmed telephony application could be as simple as a program that answers calls after 10 rings and tells the caller to try again at a later time. On the other hand, a more complex programmed telephony application might be a call routing product that interacts with callers using speech recognition technology and, after asking a series of questions and interpreting the replies, directs each call to the person most likely to be able to help a given caller. In fact, the possibilities for programmed telephony are limited only by the designer's imagination and the sophistication of available technology.

A programmed telephony application may be *CTI-only*; that is, written to use only a CTI interface, or *media-only* because it uses only a media services interface, or as *integrated* applications that use both CTI and media services together.

CT Software Components

In carrying out their tasks, programmed telephony applications capture and use one or more of the following pieces of information relative to a call:

- Call associated information (e.g., ANI, callerID, DNIS, correlator information, and user data)

- Date, day, and time of day

- Resource availability (e.g., whether an agent is ready or not)

- DTMF tone or dial pulse information

- Recorded audio

- Speech recognition

Information captured from the call, along with information from databases and the rules that govern the operation of the programmed telephony application, are combined to determine what actions the application is to take.

The first three items on the list above do not require any interaction with the caller. The latter three involve prompting the caller in some fashion to obtain the desired response. *Voice processing* applications are the subset of programmed telephony applications that involve this type of interaction with callers. Information and prompts provided to a caller might use:

- Prerecorded (digitized) sound

- Text-to-speech

- Concatenated speech

Information also can be returned to callers using methods that include fax, either within the call itself or on a separate call; data transfer using simultaneous voice and data technology (assuming the caller has the appropriate software running); or electronic mail.

12.9.1. Programmed Telephony Application Categories

Programmed telephony applications can be categorized generally as *call logging* applications, *call routing* applications, *messaging* applications, *information access/capture* applications, and *notification* applications.

Call Logging

The simplest form of unattended programmed telephony application is one that observes only. A *call logging* application does not interact with the call except to extract the information that it will record.

Call Routing

Call routing applications comprise a class of programmed telephony solutions that automate the routing of calls. These applications typically take advantage of the routing services provided by the switching domain (described in Chapter 6, section 6.10). However they also might use the *single step transfer call* service, *redirect call* service, or any other call-control services.

Calls can be routed purely on the basis of call associated information, or through interaction with the caller. For example, a call routing application might route all calls from a particular area code to one group of agents during the day and to another group during the evening. In another example, involving interaction, a customer calls a technical support line and the programmed telephony application asks for the customer's ID number. If the ID number has expired, the customer's call is directed to the sales department. If the number is good, the call is directed to the technical support group.

Other examples of call routing applications include:

- Auto-attendants
- Call screening
- Selective blocking
- ACDs

Messaging

Messaging is the most ubiquitous example of a programmed telephony application. Messaging involves:

- Interacting with the caller to identify the destination of the message, if necessary. (In most cases this is not necessary because it is either implicit or is provided through call associated information.)

- Playing an appropriate message prompt or greeting.

- Capturing the voice information representing the message.

- Forwarding the message to the appropriate mailbox.

Messaging also may apply to the receipt of fax documents.

Information Capture

Information capture applications are those that are designed to capture from callers information beyond a simple message. Examples of information capture systems include:

- Order entry (i.e., a caller is buying something)

- Dial-in questionnaires and surveys

Information Access

Information access applications are programmed telephony applications that allow callers to specify and retrieve information interactively. Examples include:

- Fax-back applications

- Database retrieval

- Document retrieval

- Message retrieval

Notification

Notification applications are programmed telephony applications designed to initiate calls at predetermined times or under certain conditions. For example, a volunteer fire department might set up a notification application to handle notifying all of the volunteer firefighters. After being triggered with the information about a fire, this application calls each volunteer and plays back the message describing the fire and its location. Other examples include:

- Wake-up call system

- Remote network or system monitoring

- Reminder of overdue items

12.9.2. Commercial Programmed Telephony Applications

Commercially available programmed telephony applications fall into three basic categories: single-purpose products, customizable products, and application generators.

Single-purpose Products

Single-purpose programmed telephony applications are those whose programming, or the logic governing their operation, are fixed and not subject to further customization by the computer user. These systems generally do provide some means for specifying particular parameters, however.

For example, an answering machine application generally is a single-purpose programmed telephony application. The product allows customization of parameters such as the greeting that is played, the number of rings to wait before answering, and the maximum length of a message. All of the logic that drives the answering machine application is not subject to customization, however; it is hard-coded by the developer of the application.

Customizable Products

Customizable programmed telephony applications are those that are intended for a specific purpose, but allow the customer to define rules that govern its operation, based on a particular set of options.

For example, an auto-attendant application might provide the complete framework and all of the normal algorithms used by an auto-attendant, things such as prompting for a number, looking up an address, transferring calls, parking calls, and playing messages. Final assembly of the logic, however, is left up to the customer. The customer not only records all of the messages (the parameters for the system), but also decides on the logic or rules that will be used for navigating through the automated attendant system.

CTI Application Generators

A *CTI application generator* is a programmed telephony application that provides a complete range of fundamental programmed telephony building blocks, but the application ships unprogrammed. The product is designed to allow the customer to assemble the software building blocks in any arbitrary way for a given solution. The customer is essentially purchasing an erector set that is then used to assemble the desired solution.

CTI application generators are really software development tools in their own right, and they come in many different forms. The easiest to use are those based on graphical programming languages and use icons and connecting lines to show the logical flow between actions the application will take.

CTI application generators are among the most powerful of the commercially available CTI products because they are effectively anything the user wants them to be. With a little bit of effort, one can build any programmed telephony application using these software components.

12.9.3. User Interface Considerations

User interface considerations are extremely important in the design of a programmed telephony application. In this context, however, the user with whom we are principally concerned is the person at the other end of the call, not a computer user. The person who is interacting with a programmed telephony application has a very limited interface:

- The person hears sound generated by the programmed telephony application.

- The person can dial digits.

- The person can speak.

As a result, designing the other-end interface presented by a programmed telephony application can be much more challenging than designing a visual interface for the CTI user, where it is much easier to provide feedback and many more options can be presented simultaneously. This section presents a number of things to keep in mind when evaluating or building a programmed telephony application.

Consistency

The most important consideration in user interface design, regardless of the context, is *consistency*. A programmed telephony application should behave in a consistent fashion no matter where it might be in its logic flow. This provides callers with a sense of security and reliability that makes them much more comfortable about interacting with an automated system. Examples of consistency rules include:

- Reaching an operator
    A single digit should be reserved for reaching a live operator (if appropriate for the application in question). There should be just one value (typically it is "0" or "*") used throughout the application. If the caller presses the designated digit at any time (except perhaps in the middle of entering a multi-digit number), the call should be transferred to a designated operator. If no operator is available, an appropriate message

671

should be played. (For example, after hours on weeknights the message might say, "There is no one available to take your call at the moment. Please leave a message and we'll return your call tomorrow.")

- Reaching the main menu

    Most programmed telephony applications present a hierarchy of menus and allow callers to navigate by pressing digits that identify a desired choice. The root of this hierarchy, the starting point, is referred to as the *main menu*. A single digit may be reserved for returning to this main menu at any time. If this feature is supported, the same key must be used throughout the program and the option must be available at all times.

- Number entry format

    Programmed telephony applications often prompt callers to enter a number of some sort, such as a credit card number, account number, product number, etc. The application should always use the same methodology for collecting a number from the caller. The best approach generally is to use numbers that have fixed numbers of digits. The prompts in these cases would say something along the lines of, "Please enter the six-digit account number now." The caller would then enter six digits (or a time-out would occur).

    Another methodology allows entry of a variable number of digits by using the "#" key to indicate the end of the sequence. (This has become a de facto standard as a result of its use in paging systems.) In this case the prompt would say something along the lines of, "Please enter the account number followed by the pound sign."

    All spoken prompts for a user to enter a number should be identical so as to reinforce that the same methodology is being used consistently. The application might supplement this prompt with a tone (such as the bong tone).

- Date format

    Another type of data that callers are frequently asked to provide is a date. Dates always should be entered numerically (as opposed to using the letters printed on the dial pad buttons). Whatever sequence of prompts is used should be consistent throughout the application.

- Canceling an entry

    The application should expect that callers frequently will make mistakes when using this primitive interface. While supporting a backspace mechanism is unrealistic, a single digit should be used throughout the system to cancel an entry and start over. A good digit for this is the "*". When the designated digit is pressed, the system should discard the partial number or date and play a variation of the last prompt that incorporates the word "re-enter."

- Fail-safe confirmation

    Any time that a significant entry is to be made that cannot be undone, there should be a confirmation step to ensure that the last request was not entered by accident. The consistent support for this confirmation step represents a safety net that will make users much more confident in how they use the application.

Always Allow Interruption

Once a given caller has become familiar with the operation of a particular program he or she will want to move more quickly, without listening to each prompt in its entirety. It is therefore important that programmed telephony applications be prepared to interrupt the playback of a prompt if the response has already been received.

Support for *type-ahead* means that a user can enter a sequence of responses in anticipation of a prompt, so the prompt may be skipped entirely. Type-ahead should be supported except in the case of confirmations.

Presenting Spoken Information

When presenting information in spoken form, it is very important that the most important, desired, or unique information be spoken as early as possible in each statement. The key is to allow the caller to comprehend the desired information as quickly as possible, without abbreviating the language used. Statements should be complete sentences in order to sound natural and to provide the redundancy that is important in spoken communication.

An example illustrating the application of this principle might be the choice of statements in a system that provides order status. If the enquiry was to determine if a particular order was shipped, a system might say, "Your order, #12345, consisting of 2 items was shipped today at 12:02." This could be improved by saying instead, "At 12:02 today your order, #12345, consisting of 2 items, was shipped." Given the likelihood that the person already knows the order number and the number of items ordered, the fact that it shipped should go first. In this example it is also a good idea to have a very different-sounding phrase to state that the order was not shipped. For example, the statement "Order #12345 has not yet been shipped" allows the listener to identify the difference between the two possible cases after hearing just the first syllables of the response.

This principle is applied to menus by listing items in the order of frequency of use; callers can press the digit for the option desired as soon as they hear it, minimizing the time spent listening to prompts. On a related note, the digit corresponding to a particular menu choice should follow the item, not precede it. This frees the listener from having to remember each number while listening to each description.

Assume a Noisy Line

The design of a programmed telephony application always should assume a noisy phone line and therefore the possibility that a caller might be having difficulty making out what is being said. In particular, callers from cellular phones drop out of a call for a few seconds at a time. Especially in cases where a name of some sort (person's name, company name, street name, etc.) is being read back, it is very important to allow the caller to request that the statement be played again. In fact, it is highly desirable to allow the caller to request something be spelled out letter by letter, if necessary using disambiguators (such as the phonetic "N as in Nancy").

Assume Errors

A good programmed telephony application will devote at least half of its code to handling errors. It is certain that, at every step in the logic, something other than the correct input will be received. A good design will anticipate all of the possibilities. This makes for a robust solution and a positive user experience.

12.10. Telephony-Aware Applications

Telephony-aware applications are mainstream applications or pieces of solution software that have been enhanced to support telephony integration by using high-level programmatic CTI interfaces.

Both the Macintosh Telephony Architecture and Windows Telephony define high-level programmatic CTI interfaces. The Windows Telephony approach involves a special, high-level portion of TAPI known as the Assisted Telephony API. The heart of the Macintosh Telephony Architecture is its high-level CTI object-oriented interface, known as the Telephony Apple Event suite.

The capabilities of the two interfaces differ, so the range of functionality available to telephony-aware applications is not the same on both platforms. However, a generic example of how a telephony-aware application works on either platform is as follows:

- Andrew's computer is running a screen-based telephone application and a number of productivity applications.

- An expense form document appears in Andrew's electronic mailbox for his approval. He opens it to find that the form contains a number of questionable expenses, so he decides to call the person who completed the form.

- The expense form application he is using happens to be telephony-aware, so all Andrew has to do is select the "Call sender" menu item from the application menu bar. Using the information it already has built into the form, plus other databases to which it has access, the application sends a command to place a telephone call to the sender of the form.

- The command to place the phone call is transferred through the high-level CTI interface to the screen-based telephone application.

- The screen-based telephone application places the call as if Andrew had activated it directly, looking up the appropriate number and dialing it using the screen-based telephone application itself.

The telephony-aware application is responsible only for initiating new calls and other related telephony tasks. As always, the screen-based telephone application is responsible for managing calls and providing feedback.

Applications that support built-in scripting languages, if their scripting engine provides access, need not support the high-level CTI interface directly in order to become telephony-aware. The same scenario described above could have been accomplished by having the user trigger a script within the application.

Telephony-aware applications typically represent telephone numbers using canonical phone number format (as described in Chapter 4, section 4.6.4). Canonical phone number format is a location-independent telephone address representation that allows telephone numbers to be correctly translated by a telephony-specific application, or by the switching domain implementation at the time the call is dialed.

12.10.1. Windows Telephony: Assisted Telephony

Windows Telephony defines a set of API functions that support what is referred to as *Assisted Telephony*. For developers of mainstream applications, Assisted Telephony allows simple telephony integration support by way of a single program function (*lineRequestMakeCall*) that sends call-placement requests to a telephony-specific application designated for handling such requests. The Assisted Telephony API also includes a second function (*tapiGetLocationInfo*) for returning default location values that a telephony-aware application can use to convert noncanonical numbers into canonical phone number format.

The remaining function calls associated with the Assisted Telephony API are used by the designated telephony-specific application that is handling requests. These are used to register for receiving requests and to retrieve requests.

12.10.2. Macintosh Telephony Architecture: Telephony Apple Events

At the heart of the Macintosh Telephony Architecture is the specification for a suite of *Telephony Apple Events*. Apple Event technology is a capability of Mac OS that supports the exchange of object-oriented messages using a well-defined protocol (referred to as the *Apple Event Interprocess Messaging Protocol*). Apple Events is an object-based programmatic interface, so on the Macintosh operating system the high-level CTI interface is actually object-based.

The telephony Apple Event suite models the switching domain by means of an object, representing the switching domain, that contains one or more directory number objects. These in turn contain zero or more call appearances (at any given time). Telephony Apple Event messages travel between the telephony-aware application and the screen-based telephone application; they describe changes in, or to be made to, the attributes of these objects. The Apple Events that are defined as part of the Telephony Apple Event suite and that manipulate objects include:

- Answer Call

- Conference

- Dial Digits

- Drop Call

- Forward

- Hold Call

- Make Call

- Park Call

- Redirect

In addition, the following Telephony Apple Events are defined to establish monitoring of the objects. These notification-related events allow the telephony-aware application to specify a set of criteria which, if later satisfied by a change in the screen-based telephone application, trigger a notification event to be sent to the telephony-aware application. This mechanism can be established, for example, to log all calls that are not answered, or to notify the application of external incoming calls.

- Notify Dependency

- Register Dependency

- Release Dependency

AppleScript is the system-wide scripting language on the Mac OS. It is an English-like language that allows sequences of Apple Events to be stored as scripts. AppleScript allows any end user to build a telephony-aware applet using a few lines of AppleScript.

Basic Telephony Apple Event Support

The basic category of Mac OS based telephony-aware applications simply use the make call capability, and are equivalent in functionality to applications using Assisted Telephony in Windows. These applications use the *make call* telephony Apple Event to send a message requesting that the screen-based telephone application call the person indicated by the canonical telephone number provided.

Extended Telephony Apple Event Support

Telephony-aware applications that support additional Telephony Apple Events are able to invoke many supplementary services. A good example of an application using that rich feature set is a calendar application for keeping track of telephone conference calls. At the appointed time for a conference call, it places calls to all the participants and conferences them all together. As with all telephony-aware applications, the screen-based telephone application is responsible for actually managing the resulting call(s).

Notification Telephony Apple Event Support

Telephony-aware applications that support the notification events can perform a number of tasks that augment the work performed by a screen-based telephone application. For example, a database application that supports Telephony Apple Event notifications can register with the screen-based telephone application and await incoming external calls. When such a call is received, the screen-based telephone application informs the database application of the new call and its associated callerID information. The database then is able to use this information to automatically find and display records that pertain to the caller—even before the call has been answered.

User Scripted

Mainstream applications that do not support Telephony Apple Events internally, but that may be scripted using AppleScript, can be made telephony-aware by writing scripts that both drive the application and send Telephony Apple Events to a screen-based telephone. These scripts then can be attached to the application or to specific documents used by the application.

AppleScript Application

Finally, entire applications can be built using AppleScript. A telephony-aware application easily can be built from scratch in this fashion. A typical use of such an approach might be building the custom software for a call center agent's work environment. Each agent's computer runs a screen-based telephone, the AppleScript-based application, and other utility applications that are launched as needed. When a call is directed to the agent, the AppleScript-based application acts as the front end. When it receives the notification Apple Event for the call, it opens the necessary databases and other files the agent needs. The AppleScript application acts as a dashboard for the call center agent.

12.10.3. Proprietary High-Level Interfaces

In addition to the OS-defined high-level programmatic CTI interfaces, various screen-based telephone application developers have also developed their own proprietary mechanisms for integrating with mainstream applications.

12.11. Creating Custom CT Solutions: Using Off-the-Shelf Software

In many ways, the truly exciting payoff from the technologies described in this book is the opportunity created for system integrators, customers, and end users to rapidly build powerful, customized CT solutions. For example, the technologies described for conveying CTI messages between components, and ultimately delivering them from a system's basic call processing resources to various applications on computer systems, simply lays the foundation for those who actually derive the value of this infrastructure.

The maturation of standards for CT Plug & Play connectivity and well-defined programmatic APIs will lead to larger and larger numbers of off-the-shelf telephony-specific and telephony-aware applications. The ease of creating telephony-aware applications on either of the mainstream personal computer platforms will mean an abundance of these products.

For all of those on the upper end of the CT value chain, this means:

1. Extremely powerful solutions can be built using the existing generations of screen-based telephony applications, programmed telephony applications (particularly application generators), and telephony-aware applications. Using scripts and other application integration tools, it is possible to integrate a number of off-the-shelf applications into a single, powerful CT solution.

2. It will become increasingly easy to find off-the-shelf telephony-aware applications that require little or no additional customization.

12.12. Review

In this chapter we have seen the various layers that make up the *CTI software framework* and a corresponding *media services software framework*. These frameworks describes the relationships among the various *types of CT software components* that exist in an open computer system capable of supporting CT solutions. We have also reviewed all of the popular programmatic interfaces used between the software components in these frameworks.

CTI software component types include *R/W interfaces* and *session/ transport protocol stacks, software mappers, CTI server implementations, CTI client implementations, low-level* and *high-level procedural* and *object programmatic CTI interfaces, telephony-specific applications*, and *telephony-aware applications.*

R/W interfaces used to connect to logical CTI servers provide direct control over, and access to, CTI sessions. The session/transport protocol stacks used by R/W interfaces may be actual protocol stacks or software mappers that convert from proprietary protocols in a transparent fashion.

CTI client implementations are responsible for providing access to one or more switching domains through one or more programmatic CTI interfaces. *CT Plug & Play client implementations* accomplish this by connecting to logical CTI servers using standard CTI protocols. The alternative approach is to use *API-specific adapters*, which must be provided by the vendor of a particular switching domain implementation, and which work exclusively with a low-level programmatic CTI interface on a particular operating system. With TAPI on Windows, these are referred to as *telephony service providers*. With the Telephone Manager on Mac OS, they are referred to as *telephone tools*.

Low-level programmatic CTI interfaces are designed to support the needs of CTI software developers in writing telephony-specific applications. *TSAPI* is a simple interface that maps directly to the standard CTI protocols without loss or embellishment. *TAPI* and the

Telephone Manager are operating system vendor-provided low-level CTI interfaces for the Windows and Mac OS operating systems, respectively. *JTAPI* is an object-based interface that provides a standard CTI interface for the Java programming language. They provide simplifications for developers at the loss of some functionality and veracity.

The media services client software framework includes many of the same components found in the CTI software framework. Component types include *R/W interfaces* and *session/transport protocol stacks, software mappers, media services client implementations, procedural* and *object programmatic media services interfaces,* and *media services applications. ECTF S.100* is a platform independent media services interface for the 'C' language. *ECTF S.410,* or *JTAPI Media,* is the corresponding interface for the Java language. Developers writing specifically for the Windows or Macintosh platforms can also use CTI media binding functions in conjunction with native media service interfaces such as *QuickTime,* the Windows *wave functions,* and the media services functions of Windows' *TAPI 3.0.*

Telephony-specific applications are either *screen-based telephone applications (SBTs)* or *programmed telephony applications.* Screen-based telephone applications are products that provide computer users an alternative user interface to the telephone set for managing telephone calls. Programmed telephony applications are products that interact with telephony resources in an autonomous fashion once they are configured or programmed with the rules that define their operation. Programmed telephony applications may be *CTI-only, media-only,* or *integrated* applications that use CTI and media services interfaces together. Commercial programmed telephony applications are categorized as being *single-purpose, customizable,* or *CTI application generators.* The latter are full-fledged development tools that can be used to build any type of programmed telephony application.

High-level programmatic CTI interfaces allow telephony-aware applications to request telephony services from screen-based telephone applications. Telephony-aware applications, which by definition include any application that uses the high-level CTI interfaces, request appropriate telephony services (such as placing calls) from the screen-based telephone application. The SBT then provides the user interface for ongoing management of calls.

The benefit of this multi-layered software framework is that CTI system integrators, customers, and individuals are able to integrate off-the-shelf software products into CTI solutions. While already easy to do at present, this integration actually will become easier as more products become available.

13.
CT Solution Examples

We have now explored all the key concepts governing the implementation of telephone systems, the core telephony features and services, and the specific trade-offs with telephone products. We have also learned about CT concepts, CT configurations, and CT software tools and interfaces. It is therefore time to pull everything together and look at how the scenarios presented in Chapter 2 are actually built. (You may wish to go back and review Chapter 2, sections 2.3 through 2.10 at this point.)

Each of the scenarios described earlier can be implemented in many different ways. Just one of the possible solutions will be presented in this chapter for each scenario, in order to demonstrate how the technologies we have explored in this book can be applied. Each solution involves a description of the corresponding CT system configurations and the software components used. To keep the solutions as simple as possible, standard CT protocols were used in each case. As we have seen, however, proprietary protocols can be accommodated in most situations using a variety of approaches.

13.1. Screen-Based Telephony

Andrew, the PR manager for a large public relations agency, set up a CTI solution for himself. It involved connecting his computer directly to his telephone station and using his off-the-shelf contact manager along with screen-based telephone application software. Andrew's scenario is summarized in Table 13-1. (Refer to Chapter 2, section 2.3 for the complete scenario description.)

Table 13-1. Screen-based telephony solution scenario

Name:	Andrew
Occupation:	PR manager
Location:	Corporate office
Switch:	PBX
Telephone station:	Proprietary digital telephone
Type of computer:	Desktop
Type of CT Interface:	Direct-connect
Applications used:	Screen-based telephone
	Contact Manager
Implemented by:	Andrew
Customization:	N/A
Future Plans:	Implement on PDA for cellular phone and home phone
Benefits:	Greater professionalism and responsiveness
	Realize better utilization of computer and telephony resources
	Improves efficiency and reliability
	Improves employee morale
	Reduces staff training times
	Inexpensive: uses off-the-shelf products

13.1.1. CT System Configuration

The system configuration for Andrew's CTI solution is shown in Figure 13-1. Andrew implemented the solution by himself; he didn't have access to the switch, so a client-server solution was out of the question (and also outside his budget). Fortunately, the vendor of the switch that his company uses manufactures a telephone station peripheral for their proprietary digital lines. This is hooked up between the switch and the telephone station as shown. The telephone station peripheral connects to Andrew's computer using an RS-232 cable.

Figure 13-1. Screen-based telephony solution scenario configuration

13.1.2. CT Software Components

The software components that complete Andrew's CTI solution are shown in Figure 13-2. The telephone station peripheral supports standard CTI protocols, so no proprietary software was needed for Andrew's computer. Andrew chose the screen-based telephone application that best fit his needs and preferences; the application connected to the CTI functionality provided by the telephone station peripheral through the CT Plug & Play client implementation installed on his system. Finally, both his contact manager and corporate directory database applications were already telephony-aware, so they were immediately able to take advantage of his new screen-based telephone application once it was installed.

Figure 13-2. Screen-based telephony solution scenario software

13.2. Mobile CTI

Betty, the sales representative on the go, is the happy user of a sales-force automation solution that makes good use of CTI capabilities. It is centered on a sales-force automation utility application that integrates customer and product databases, along with a screen-based telephone application and fax software. Betty's scenario is summarized in Table 13-2. (Refer to Chapter 2, section 2.4 for the complete scenario description.)

Table 13-2. Mobile CTI solution scenario

Name:	Betty
Occupation:	Sales representative
Location:	On the road - hotel - airport - taxi
Switch:	Various
Telephone station:	Various
Type of computer:	Notebook
Type of CT Interface:	Direct-connect (PC-Card modem)
Applications used:	Sales-force automation application
	Customer and product databases
	Electronic mail software
	Fax software
	Screen-based telephone
Implemented by:	System integrator
Customization:	Preferred a different screen-based telephone
Future Plans:	Infrared connections to pay phones and SVD
Benefits:	Eliminate drudgery
	Fast, easy, and error-free dialing
	Calls placed using least expensive method
	Greater professionalism, consistency, and responsiveness
	Morale improvement

13.2.1. CT System Configuration

The CTI system configuration for Betty's CTI solution is shown in Figure 13-3. It takes advantage of the primitive—and yet functionally sufficient—capabilities of the PCMCIA fax modem that is installed in her notebook computer. The system integrator who built the solution chose this particular modem because it supported the standard CTI protocols in addition to supporting the standard, Hayes-compatible AT commands. Using CT Plug & Play components rather than API-specific code allows for components to be exchanged by users of different platforms, and also allows use of infrared and other devices as they become available.

Figure 13-3. Mobile CTI solution scenario configuration

13.2.2. CT Software Components

The software components that complete Betty's CTI solution are shown in Figure 13-4. From Betty's perspective, the principal interface to the solution is the sales-force automation utility application, written by the system integrator using high-level scripting tools. This interacts with the various database, fax, and electronic mail software that are included in the solution by means of platform-specific scripting techniques. It integrates with the screen-based telephone application using the platform's high-level programmatic CTI interface. The CTI client implementation uses standard CTI protocols to access the modem. It also has a value-added feature: It supports dial-plan management and can translate between canonical phone numbers (provided by the screen-based telephone application) and the appropriate dialing digits for the given location.

Figure 13-4. Mobile CTI solution scenario software

13.3. Power Dialing

Chuck, the individual working in the accounts receivable department, now uses a specially developed screen-based telephone application that includes predictive dialing to track down customers with past-due accounts. Chuck's scenario is summarized in Table 13-3. (Refer to Chapter 2, section 2.5 for the complete scenario description.)

Table 13-3. Power dialing solution scenario

Name:	Chuck
Occupation:	Accounts receivable
Location:	Corporate office
Switch:	PBX
Telephone station:	Proprietary digital telephone
Type of computer:	Multi-user
Type of CT Interface:	Client-server
Applications used:	Accounting software
	Special predictive dialing SBT
Implemented by:	Corporate IS department
Customization:	N/A
Future Plans:	Personal computer-based system
Benefits:	Eliminate drudgery / improve morale
	Speeds work process
	Saves money and time
	Easy to develop and use

13.3.1. CT System Configuration

The system configuration for Chuck's CTI solution is shown in Figure 13-5. The accounting software currently used by Chuck's group runs on a multi-user computer as shown. It connects to a CTI server which, in turn, has access to the CTI interface on the PBX being used.

Figure 13-5. Predictive dialing solution scenario configuration

Figure 13-6. Predictive dialing solution scenario software

13.3.2. CT Software Components

The software components that complete Chuck's CTI solution are shown in Figure 13-6. In this case, one of the standard CTI protocols is being delivered from the CTI server using a TCP/IP connection. The multi-user computer is not running one of the common personal computer operating systems, so there are no high-level CTI APIs or off-the-shelf CTI software. For this solution, the corporate IS department has custom-written a simple screen-based telephone application that works in conjunction with the accounting system and utilizes the low-level programmatic CTI interface exposed by the CTI client implementation. This software is built around use of the switch's *make predictive call* service.

13.4. Personal Telephone System

Debbie, the work-at-home consultant, set up a CT solution for herself that involved installing a personal PBX for her home. By maximizing the utilization of two telephone lines, she saved the expense of having to install a third phone line. The personal PBX also provided a number of other significant benefits. Debbie's scenario is summarized in Table 13-4. (Refer to Chapter 2, section 2.6 for the complete scenario description.)

Table 13-4. Personal telephone system solution scenario

Name:	Debbie
Occupation:	Consultant
Location:	Home office
Switch:	Personal PBX
Telephone station:	Analog POTS telephone
Type of computer:	Home computer
Type of CT Interface:	Direct-connect
Applications used:	Client database
	Time and billing database
	Voice mail software
	Call logging software
	Screen-based telephone
	Fax software
Implemented by:	Debbie
Customization:	N/A
Future Plans:	Digital system

Table 13-4. Personal telephone system solution scenario (Continued)

Benefits:	Makes home business possible
	Saves money
	Saves time
	Easy to develop and use
	Provides privacy and enhances professional image

13.4.1. CT System Configuration

The CT system configuration for Debbie's personal PBX solution is shown in Figure 13-7. Debbie's personal PBX connects to the two analog lines coming into her house. (She had the original line switched from a loop-start to a ground-start line and ordered the new line as a ground-start line.) All the telephones in the house are now wired back to extensions on her personal PBX. Anyone picking up a telephone in the house hears the dial tone provided by the personal PBX and is able to call between rooms, or place external outgoing calls if a trunk is available (and if Debbie configures the switch to allow external calls from the station in question). Incoming calls on the residential line ring on the kitchen phone, but they can be picked up by any extension using *group pickup call*. The other trunk is used for incoming business calls that ring only on her extension. If any call is unanswered, forwarding rules determine that the call is redirected to the voice mail/fax extension. The personal PBX is integrated with her home computer using a serial connection that supports isochronous communication. This allows her computer to both control the PBX and access audio media stream data associated with her extension. She also has her computer's voice/fax modem connected to its own dedicated extension. This extension is used for both voice mail and fax reception. Any time that an incoming fax call is detected on either trunk, it is routed to the voice mail/fax extension.

Figure 13-7. Personal telephone system solution scenario configuration

13.4.2. CT Software Components

The software components that complete Debbie's CT solution are shown in Figure 13-8. Debbie's mission-critical applications are the databases that track her clients and do her time and billing. Her client database is tied to her screen-based telephone application through a high-level CTI interface, which allows it to place calls and also to present client information pertinent to incoming calls using callerID. Two programmed telephony applications run on the computer in the background at all times. One is a call-logging application that observes and logs everything that takes place in the switch. The time and billing application monitors the logging application in order to identify calls, placed from anywhere in the house, that can be billed to clients. The other programmed telephony application is a messaging program that handles incoming faxes and voice mails on the fax line. It operates completely independently of the other CT products, using a different CTI service boundary and switching domain. The last key feature of Debbie's solution is that the particular CTI client implementation she has installed supports a speaker phone feature, configured to be activated in conjunction with her extension. This means it is able to use the computer's speakers and microphone as a speaker phone, extending the capabilities of the actual (POTS) physical device element. Debbie's screen-based telephone application perceives a physical device element that has two auditory apparatuses.

Figure 13-8. Personal telephone system solution scenario software

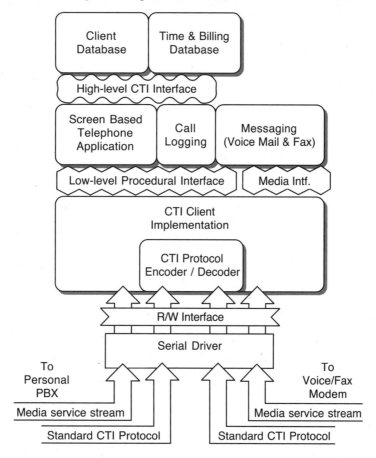

13.5. Personal Telephone Agent

Edmund, a photographer, set up a CT solution for himself that involved customizing a personal telephone agent to handle calls on his behalf. Edmund's scenario is summarized in Table 13-5. (Refer to Chapter 2, section 2.7 for the complete scenario description.)

Table 13-5. Personal telephone agent solution scenario

Name:	Edmund
Occupation:	Photographer
Location:	Small business office
Switch:	Central office
Telephone station:	ISDN with analog terminal adapter
Type of computer:	Desktop computer
Type of CT Interface:	Direct-connect
Applications used:	Personal agent software
	Screen-based telephone
	Fax software
Implemented by:	Third-party developer
Customization:	Prompts and logic customized
Future Plans:	Speech recognition
Benefits:	Edge on the competition
	Establishes more professional image
	Saves money
	Remote access capability

13.5.1. CT System Configuration

The system configuration for Edmund's CT solution is shown in Figure 13-9. It involves an ISDN subscriber line to the central office switch, connected to a telephone station peripheral (a terminal adapter in this case), which in turn is connected to both his desktop computer and his old analog POTS telephone. The ISDN line corresponds to two different logical devices in the central office switch. One corresponds to Edmund's telephone and one corresponds to Edmund's new personal agent. Forwarding rules are established such that if the one associated with the telephone is not answered, or if *do not disturb* is set, calls are forwarded to the agent's logical device. The speaker output from his computer is extended to his studio so that sounds made by his computer can be heard in the studio and darkroom.

Figure 13-9. Personal telephone agent solution scenario configuration

13.5.2. CT Software Components

The software components that complete Edmund's CT solution are shown in Figure 13-10. Edmund has installed a screen-based telephone application that monitors and controls the logical device associated with his telephone number. He uses this application for making and managing calls when he is in his office. He also can use it to invoke the *set do not disturb* service. In addition, he has installed a personal agent application that observes and controls the second logical device associated with the ISDN line. The personal agent has full access to media streams, so it can interact with callers, send faxes, take messages, etc. As described in the scenario in Chapter 2, when the personal agent is taking an important call, it searches for Edmund by announcing the call on the studio speakers and by calling his cellular phone and his home telephone. In each case it uses text-to-speech technology to speak aloud the callerID and other call associated information. It also records the caller's answers to certain questions and plays these recordings back.

Figure 13-10. Personal telephone agent solution scenario software

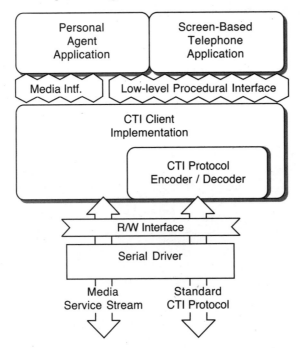

13.6. Interactive Voice Response System

Frances, the school vice-principal, set up a CT solution to provide information to parents using a programmed telephony application for interactive voice response. Frances' scenario is summarized in Table 13-6. (Refer to Chapter 2, section 2.8 for the complete scenario description.)

Table 13-6. Interactive voice response solution scenario

Name:	Frances
Occupation:	Vice-principal
Location:	School
Switch:	Central office
Telephone station:	Analog POTS
Type of computer:	Desktop computer
Type of CT Interface:	Direct-connect
Applications used:	CT application generator
Implemented by:	Frances and school whiz kid
Customization:	Daily updates by teachers
Future Plans:	More phone lines and Internet access
Benefits:	Better community service

13.6.1. CT System Configuration

The CT system configuration for Frances' CT solution is shown in Figure 13-11. It involves an add-in card inside a spare personal computer. The add-in board is connected to a POTS line from the central office switch. There is no telephone set because locally there is no human interaction with calls.

Figure 13-11. Interactive voice response solution scenario configuration

13.6.2. CT Software Components

The software components that complete Frances' CT solution are
shown in Figure 13-12. Like the system configuration for this scenario,
it is quite simple. A CT application generator was used to develop a
special programmed telephony application for the school information
system. Because the hardware component in this case is an add-in
board, a mapper is required in order to access the standard CT
protocols.

Figure 13-12. Interactive voice response solution scenario software

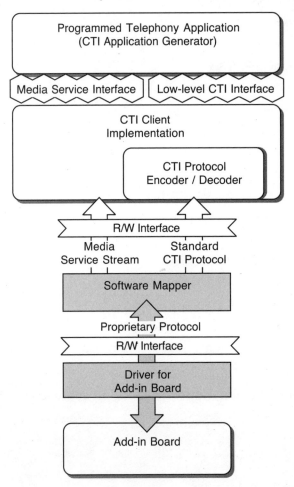

13.7. Help Desk

Gunther, the help desk technician, has constructed an extensive CT solution consisting of a number of different server processes running on a computer that is shared across a local area network. Gunther's scenario is summarized in Table 13-7. (Refer to Chapter 2, section 2.9 for the complete scenario description.)

Table 13-7. Help desk solution scenario

Name:	Gunther
Occupation:	Help desk technician
Location:	University
Switch:	PBX
Telephone station:	Various
Type of computer:	Various
Type of CT Interface:	Client-server
Applications used:	Multi-user call tracking database
	CT application generators (two)
	Web server
	Screen-based telephones
Implemented by:	Gunther
Customization:	Each technician picks own SBT
Future Plans:	Screen sharing
Benefits:	Higher-quality service
	More timely service
	Cost savings

13.7.1. CT System Configuration

The system configuration for Gunther's help desk CT solution is shown in Figure 13-13. Everything in the system is connected together using a LAN, and the university's PBX is used to deliver and queue telephone calls from help desk clients to technicians. A hybrid media/ CTI server is connected to the CTI interface of the PBX using a serial communication path for establishing a CTI stream. This hybrid server provides a fan-out of CTI call control information to authenticated clients, and also allows those clients to access media streams associated with any call by conferencing one of the lines connected to the built-in media server resources into the specified call. The media server that Gunther has installed is a very simple one that is capable only of playing prerecorded messages. The hub of Gunther's solution is the shared computer used to run the call tracking database, a call routing application, a voice processing application, and a Web server.

Figure 13-13. Help desk solution scenario configuration

Help Desk Client — Hybrid Media and CTI Server — Hub Server — Help Desk Technician — Help Desk Technician

13.7.2. CT Software Components

The software components running on the hub machine in Gunther's CT solution are shown in Figure 13-14. These include the following:

- Multi-user call tracking database

  The multi-user database is accessed by all the other applications on this system and by the database clients running on the technician workstations. The database's client-sever protocol (not shown) also travels over the TCP/IP connection.

- Voice Processing

  The voice processing application is a programmed telephony application responsible for interacting with callers to the help desk number. This application uses the CTI interface to

Figure 13-14. Help desk solution scenario hub software

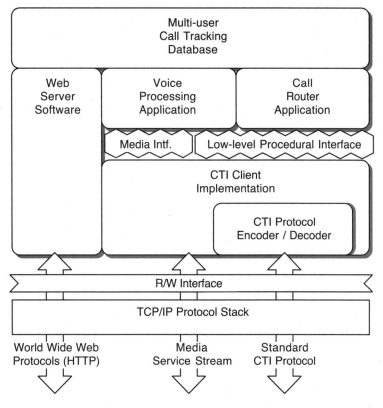

monitor for each new incoming call and, after binding a media service instance to it for playing prerecorded messages, questions each caller about the nature of the problem. The DTMF tones that are detected from the user are translated into the appropriate information to place in fields in the call tracking database. Once the new call has been appropriately classified, it is queued and the routing application takes over.

- Call router

    The call router is a programmed telephony application that uses the CTI interface to observe the calls queued for technicians, and to control routing them to the appropriate technician based on information in the call tracking database.

At the appropriate time, this application also establishes calls to those (like the user of the notebook computer in Figure 13-13) who request a position in the queue via a World Wide Web browser.

- Web server
  The Web server provides an alternate mechanism for joining the queue by directly creating an entry in the call tracking database.

Figure 13-15. Help desk solution scenario client software

The software components running on the technician workstations are shown in Figure 13-15. In contrast to the hub machine, the software configuration of the technician workstations is quite simple. Each has an SBT of the technician's own choice and a database client application that accesses the multi-user call tracking database. (The database's client-sever protocol also travels over the TCP/IP connection, but is not shown here.) When a call is delivered to a particular technician, the appropriate call tracking information is presented automatically.

13.8. Call Center

Henrietta, the manager of a successful virtual travel agency, has built her entire business around CT technology that allows her travel agents to work from their own homes as part of a distributed call center. Henrietta's scenario is summarized in Table 13-8. (Refer to Chapter 2, section 2.10 for the complete scenario description.)

Table 13-8. Call center solution scenario

Name:	Henrietta
Occupation:	Manager of travel agency
Location:	Distributed
Switch:	Central office
Telephone station:	Centrex ISDN, Centrex analog
Type of computer:	Desktop and server
Type of CT Interface:	Client-server
Applications used:	Multi-user client database
	CT application generator
	Screen-based telephones
Implemented by:	Henrietta
Customization:	N/A
Future Plans:	Screen sharing
Benefits:	Makes the business model possible
	Improves customer satisfaction
	Boosts professional image of organization
	Improves employee morale
	Reduces call times and eliminates errors
	Minimizes business overhead

13.8.1. CT System Configuration

The CT system configuration for Henrietta's CT solution is shown in Figure 13-16. Components exist in three distinct locations: the central office, Henrietta's basement, and the travel agent's home.

- The central office
  In addition to the central office switch, the central office houses a CTI server that provides access to a switching domain consisting of all of Henrietta's lines. Henrietta subscribes to Centrex and has both ISDN and analog lines. The analog lines are used for media services (voice processing and sending pager messages). ISDN is used everywhere else.

- Henrietta's basement
  The server that runs the distributed call center is in Henrietta's basement. A single machine is responsible for running everything.[13-1] It is a hybrid media server and CTI server that hosts a local CTI client implementation along with other software needed by Henrietta's business. ISDN is used to establish a dial-up communication path to the CTI server at the central office. Henrietta's basement also contains a dial-in bridge for her LAN which supports dial-in access to the server for all of her agents.

- The travel agent's home
  Each travel agent has an ISDN line and a desktop computer. One ISDN B channel is used to establish a communication path with Henrietta's LAN in order to connect to the server. The other is used for voice and is accessed using an analog phone through a terminal adapter. (Note that the configuration diagram reflects the logical relationships, not the physical connections, so the communication path is

---

13-1. UPS — Yes, Henrietta does have a UPS (Uninteruptable Power Supply) protecting this server.

simply shown between the agent machine and the dial-in bridge, even though it physically passes through the central office switch.)

Figure 13-16. Call center solution scenario configuration

| Central Office | Henrietta's Basement | Travel Agent's Home |

13.8.2. CT Software Components

Most of the software components that complete Henrietta's CT solution are found on her central server as shown in Figure 13-17. These software components include:

- Hybrid media server and CTI server

  The hybrid media server and CTI server, is a single software component. It is a CT Plug & Play CTI server implementation that has added value in the form of full-function media services to several analog lines through H.100-based boards installed in the server machine. The media resources supported include recorder, player, and modem.

- Auto-attendant application

  An auto-attendant application greets callers, identifies an appropriate agent, redirects calls, takes messages, and sends pager messages as necessary. The auto-attendant is a client of the CTI server running on the same machine.

- Multi-user client database

  A multi-user client database tracks all information about the travel agency's clients and their travel plans. In the event that a given travel agent cannot be reached, another can use this database to support a given customer. The database's client-server protocol (not shown) also travels over the TCP/IP connection.

Figure 13-17. Call center solution scenario server software

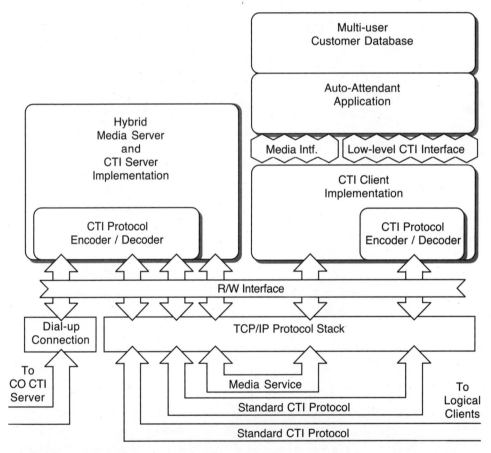

The software components running on the individual travel agents' systems are shown in Figure 13-18. These include:

- Screen-based telephone
    The SBTs used on the travel agent's machines have a unique feature that allows them to automatically navigate through the IVR menus associated with frequently called numbers.

- Client database application
    This application is linked to the SBT through the high-level interface. These allow incoming calls to trigger lookups and outgoing calls to be placed from within the database. The database's client-server protocol (not shown) also travels over the TCP/IP connection.

Figure 13-18. Call center solution scenario client software

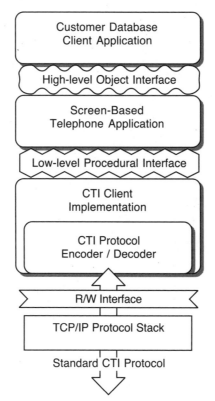

13.9. IP Telephone System

Ian's small but fast growing start-up company is using the services of a telephony application service provider (ASP) in order to efficiently and competitively deploy telephony functionality despite the fast growth of the company. Ian's scenario is summarized in Table 13-9. (Refer to Chapter 2, section 2.11 for the complete scenario description.)

Table 13-9. Internet telephone system solution scenario

Name:	Ian
Occupation:	CEO/Founder of start-up company
Location:	Distributed
Switch:	Hosted iPBX
Telephone station:	Ethernet IP telephone sets and analog sets through station servers
Type of computer:	Desktops and server
Type of CT Interface:	Client-server
Applications used:	Screen-based telephones
	Application generator for media services applications
	Web browser for administration
Implemented by:	Telephony Application Service Provider (ASP)
Customization:	Customized iPBX configuration
	Custom media services applications
Future Plans:	Multiple locations
Benefits:	Allows for fast growth
	Boosts professional image of organization
	Gives employee leading-edge tools
	Minimizes business overhead

13.9.1. CT System Configuration

The system configuration for Ian's internet telephone system is shown in Figure 13-19. Components exist in two distinct locations: at Ian's telephony ASP and at Ian's office location(s).

- The telephony ASP

  The telephony ASP acts as a CLEC to provide Ian's company with an alternative to conventional telephone network access. The ASP has deployed a softswitch as a gateway to the PSTN and hosts an iPBX for Ian's company. In addition, the ASP hosts a number of additional telephony applications and servers that provide media services functionality. The ASP provides the company with access to both the Internet and to the telephone network using DSL circuits that provide high speed IP connections to each of Ian's office locations.

- Ian's offices

  At each of Ian's office locations employees are equipped with desktop computers and ethernet-based IP telephone sets. Fax machines and other analog telephone equipment are connected to the telephone system using terminal servers. Every employee is equipped with screen-based telephony software to manage their telephone and integrate their productivity software with the telephone system. A network administrator in each location uses administrative software to instantly add, delete, or re-assign new extensions and update class of service parameters. The operations group uses a media services application generator package to create and update caller interaction software that makes use of media resources hosted in the ASP's media servers.

Figure 13-19. IP telephone system solution scenario configuration

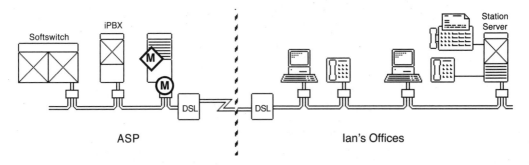

13.9.2. CT Software Components

The software components that constitute Ian's telephone system solution are also widely distributed but the most important elements are the components running on the service provider's iPBX server as shown in Figure 13-20. These software components running on this server include:

- Call processing
    The call processing component is the heart of the iPBX. It uses all of the information in the iPBX configuration database (which is maintained by the administration server software) to react to all requests presented by the CTI interface as well as those presented through signaling delivered by the switching control component.

- Switching control
    The switching control component is responsible for controlling the IP-based switching fabric including the station servers, IP-based telephone stations, and the PSTN gateway provided within the softswitch.

- CTI server
    The CTI server is responsible for implementing the iPBX's CTI interface.

- Media services client

  A media services client component provides call processing with access to media services for performing tone detection and tone generation.

- Administration server

  The administration server is responsible for managing the configuration database.

Figure 13-20. IP telephone system solution iPBX software components

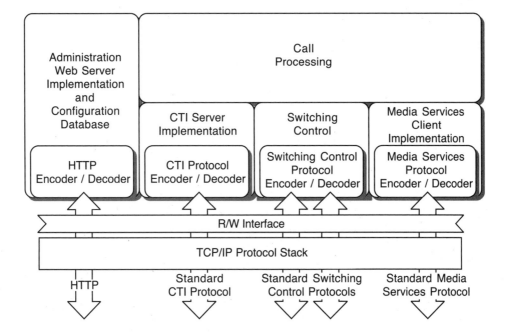

The software components running on the individual employee's systems are shown in Figure 13-21. These are the same types of components found in most non-iPBX scenarios.

Figure 13-21. IP telephone system solution scenario client software

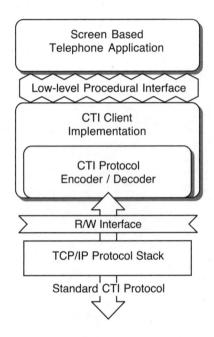

Media services application software runs on selected desktop computers in Ian's offices. Applications, such as auto attendant applications which are created using off-the-shelf application generators, automate the interaction with callers. The client and server components are shown in Figure 13-22.

Finally, all administration is performed using a standard web browser to access to the administration server software on the iPBX server.

Figure 13-22. IP telephone system solution scenario media services software

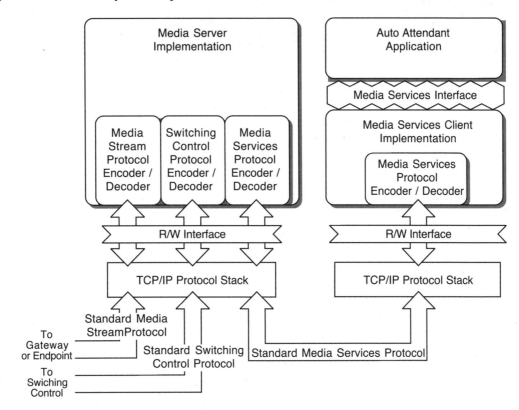

13.10. Ecommerce Business

Jane is the webmistress for a popular online shopping web site. The need to provide a comprehensive customer support solution has transformed the business from being centered around the web server to being centered around a customer relationship management (CRM) solution. Jane's scenario is summarized in Table 13-10. (Refer to Chapter 2, section 2.12 for the complete scenario description.)

Table 13-10. Call center solution scenario

Name:	Jane
Occupation:	Webmistress for online shopping site
Location:	Customer contact center
Switch:	iPBX
Telephone stations:	Ethernet IP telephone sets
Type of computers:	Desktop and server
Type of CT Interface:	Client-server
Applications used:	Web-based CRM software for screen-pop, chat, and e-mail
	Media services applications for caller identification, call queueing, and automated self-service
	Screen-based telephones
Implemented by:	Jane, contract programmers
Customization:	N/A
Future Plans:	Remote workers
Benefits:	Makes the business model possible
	Improves customer satisfaction
	Minimizes business overhead

13.10.1. CT System Configuration

The CT system configuration for Jane's CT solution is shown in Figure 13-23.

Figure 13-23. Ecommerce solution scenario configuration

Jane's configuration involves the following key components:

- CRM Server

    At the heart of the system is an application server that runs a customer relationship database and a number of applications used for accessing this data. The CRM database tracks all customer information and transactions. It can be accessed (subject to access privileges) through a web interface or through an XML query engine. The CRM server includes a web server and a number of web-based applications that support web chat and web callback.

- iPBX

    The new customer contact department's telephone system is an iPBX consisting of a server that runs call processing and switching control and an IP telephony gateway that bridges the company's existing PBX (with its access to the PSTN) and Internet-based telephone calls with the internal iPBX switching fabric. Using an iPBX rather than a conventional PBX allows the customer contact department to grow, to support remote workers, and to deploy new IP-based technologies such as video conferencing more easily in the future. The IP telephony gateway allows the customer support staff to place and accept calls within the company's phone system, across the PSTN, and over the Internet.

- Media Server
  A media server featuring text-to-speech (TTS) and automatic speech recognition (ASR) resources is employed along with media services applications that allow automated customer access to information in the CRM system. One application handles caller identification and routing of unidentified callers, another handles interaction with the CRM database, and a third handles queuing of calls to available customer contact staff.

The customer contact department's IP network is connected to the Internet over a high speed connection with a firewall that passes only web browser access to the CRM server and Internet-based telephone calls.

13.10.2. CT Software Components

Most of the software components that complete Jane's CT solution are found on her central server. These software components include:

- CRM Database and Business Logic
  A multi-user customer contact database that tracks all information about individual customers and all of their transactions and interactions with the company. This database is accessible through both SQL and XML queries. CRM business logic ensures appropriate management and correlation of individual and aggregate customer data regardless of the access method.

- Web Server
  The CRM server includes web server software, the company's ecommerce system, and all associated content. The content includes links that activate back-end software for live chat and for call-backs. The content includes both HTML content

for standard desktop browsers, as well as VoiceXML content for access to CRM information over the telephone and WML content for delivery to mobile devices.

- E-mail Server

  The CRM server includes an e-mail server software that handles all inbound e-mail directed to the customer contact department. It generates acknowledgments, matches e-mail addresses to customer contact information in the CRM database, and queues e-mail requests to available staff members.

The media server includes the following software components, as illustrated in Figure 13-24:

- Standards-based Media Services Platform

  The media server is running a media server implementation that supports standard APIs for media applications as well as programmatic interfaces for add-in media processing resources. The media server is a client of the iPBX's CTI interface.

- TTS and ASR Resources

  In addition to media processing resources for the playback of recorded audio data and for tone detection and generation, there are resources for performing text-to-speech and automatic speech recognition. These are used to support the caller identification and VoiceXML applications.

- Caller Identification Application

  The caller identification application uses call associated information, if available, to form a hypothesis for the identity of the caller. Depending on its hypothesis, it uses a different script for confirming the caller's identity against information stored in the CRM database. This interaction uses recorded speech, text-to-speech and ASR to simplify and humanize the process. If the caller is a new customer, the call will be tagged appropriately and handed off to the call queueing application. Once a caller is identified they will be asked if they wish to talk to a staff member or use the automated system.

Depending on the reply, the application will hand-off to the call queueing application or the voiceXML interpreter respectively.

- VoiceXML Interpreter Application
  The VoiceXML interpreter requests VoiceXML documents from the CRM server that are interpreted and then presented to the caller through playback and TTS. Responses are collected using tones and/or ASR and these responses allow the caller to navigate through the CRM system.

- Call Queueing Application
  The call queueing application implements the customer contact department's ACD functionality. It tracks available staff members and routes queued calls as staff members become available. The application plays music, advertising content, and a periodic indication of the expected wait time.

The software components running on the individual computer systems of customer contact staff include a screen-based telephone application for managing telephone calls and their telephone set along with a web browser. Screen-pops of customer information, customer chat sessions, and queued e-mail are driven by the CRM server through a web browser interface.

Figure 13-24. Ecommerce solution scenario media server software

13.11. Conclusion

Regardless of the type or size of your organization, CT will be an important tool for you in the future. Whether your goals are better service, cost savings, increased revenue, improved morale, or simply the ability to be more productive and effective, CT can help you achieve these goals.

Having read this book, you should have sufficient understanding of the terminology, concepts, and technologies of the telephony and CT industries to begin developing plans for CT solutions of your own. This book should continue to serve as a resource as you plan for the implementations of CT solutions or CT components in your environment, and as you attempt to interpret product and service claims from vendors throughout the CT value chain.

Conclusion

Bibliography

There are four reference documents that developers of CTI products
may wish to consult:

- Apple Computer, Inc. *Telephone Manager Developer's Guide*,
  Mac OS SDK. (Cupertino, Calif.: Apple Computer, Inc., 1991).

- Apple Computer, Inc. *Telephony Apple Event Suite*, Mac OS SDK.
  (Cupertino, Calif.: Apple Computer, Inc., 1991).

- Microsoft Corporation. *Telephony Application Programmng Interface
  (TAPI) Reference*, Microsoft Developer Network Library.
  (Redmond, Wash.: Microsoft Corporation, 2000).

- Versit. *Versit Computer Telephony Integration (CTI) Encyclopedia*,
  Vols. 1–6. (Versit, 1996).

These documents are complete reference documents intended for
different audiences as described below.

Bibliography

Telephony Equipment Vendors and Network System Providers

Anyone developing telephony equipment or hardware products that connect to telephone lines should consult the *Versit CTI Encyclopedia* specifications. In particular, the following volumes should be used as reference material in developing the specifications for any CTI products:

- Volume 3: Telephony Feature Set
- Volume 4: Call Flow Scenarios
- Volume 5: Protocols

Hardware Vendors and Operating System Vendors

Those developing CTI client implementations or CTI server implementations should consult these volumes of the *Versit CTI Encyclopedia*:

- Volume 3: Telephony Feature Set
- Volume 4: Call Flow Scenarios
- Volume 5: Protocols
- Volume 6: Versit TSAPI

If the implementation is Mac OS based, consult the *Telephone Manager Developer's Guide*.

If the implementation is Windows-based, consult the Microsoft *Telephony Application Programmng Interface (TAPI) Reference*.

Mac OS Application Developers

Those developing telephony-specific software should consult the following volumes of the *Versit CTI Encyclopedia*:

- Volume 3: Telephony Feature Set
- Volume 4: Call Flow Scenarios

Developers planning to use the Versit TSAPI programmatic interface should refer to:

- Volume 6: Versit TSAPI of the *Versit CTI Encyclopedia*.

Developers planning to use the Telephone Manager programmatic interface should refer to:
- the Apple *Telephone Manager Developer's Guide.*

Those developing mainstream applications that wish to be telephony-aware should consult:
- the *Telephony Apple Events Suite*

Windows Application Developers

Those developing telephony-specific software should consult the following volumes of the *Versit CTI Encyclopedia*:
- Volume 3: Telephony Feature Set
- Volume 4: Call Flow Scenarios

Developers planning to use the Versit TSAPI programmatic interface should refer to:
- Volume 6: Versit TSAPI of the *Versit CTI Encyclopedia.*

Developers planning to use the TAPI programmatic interface should refer to:
- the Microsoft *Telephony Application Programmng Interface (TAPI) Reference*

Multiplatform Application Developers

Those developing telephony-specific software to run on multiple platforms or platforms other than Mac OS and Windows should consult the following volumes of the *Versit CTI Encyclopedia*:
- Volume 3: Telephony Feature Set
- Volume 4: Call Flow Scenarios
- Volume 6: Versit TSAPI

Index

Index

Index

Index

Index

About the Author

Michael Bayer is the president of Computer Telephony Solutions, a consulting firm specializing in the design and implementation of CTI technologies and products. His work in this field dates back over a decade—before the term "CTI" even emerged—and he is recognized as one of the pioneers of the CTI industry. Many important CTI initiatives can be traced to his groundbreaking technical work, extensive promotion for his vision of CTI, and participation in various standards-setting efforts Mr. Bayer has a technical background in computer-mediated communications and human interface design, and his technical credits include developing the Macintosh Telephony Architecture (MTA) and, more recently, playing a leading role in the development of the *Versit CTI Encyclopedia*.

Since founding Computer Telephony Solutions, he has continued to pursue his mission to promote the widespread adoption of CT Plug & Play standards. Computer Telephony Solutions is a member of the ECTF and Mr. Bayer is an active volunteer in this and other organizations dedicated to interoperability and widespread adoption of CT technology.

SOFTWARE AND INFORMATION LICENSE

The software and information on this diskette (collectively referred to as the "Product") are the property of The McGraw-Hill Companies, Inc. ("McGraw-Hill") and are protected by both United States copyright law and international copyright treaty provision. You must treat this Product just like a book, except that you may copy it into a computer to be used and you may make archival copies of the Products for the sole purpose of backing up our software and protecting your investment from loss.

By saying "just like a book," McGraw-Hill means, for example, that the Product may be used by any number of people and may be freely moved from one computer location to another, so long as there is no possibility of the Product (or any part of the Product) being used at one location or on one computer while it is being used at another. Just as a book cannot be read by two different people in two different places at the same time, neither can the Product be used by two different people in two different places at the same time (unless, of course, McGraw-Hill's rights are being violated).

McGraw-Hill reserves the right to alter or modify the contents of the Product at any time.

This agreement is effective until terminated. The Agreement will terminate automatically without notice if you fail to comply with any provisions of this Agreement. In the event of termination by reason of your breach, you will destroy or erase all copies of the Product installed on any computer system or made for backup purposes and shall expunge the Product from your data storage facilities.

LIMITED WARRANTY

McGraw-Hill warrants the physical diskette(s) enclosed herein to be free of defects in materials and workmanship for a period of sixty days from the purchase date. If McGraw-Hill receives written notification within the warranty period of defects in materials or workmanship, and such notification is determined by McGraw-Hill to be correct, McGraw-Hill will replace the defective diskette(s). Send request to:

Customer Service
McGraw-Hill
Gahanna Industrial Park
860 Taylor Station Road
Blacklick, OH 43004-9615

The entire and exclusive liability and remedy for breach of this Limited Warranty shall be limited to replacement of defective diskette(s) and shall not include or extend to any claim for or right to cover any other damages, including but not limited to, loss of profit, data, or use of the software, or special, incidental, or consequential damages or other similar claims, even if McGraw-Hill has been specifically advised as to the possibility of such damages. In no event will McGraw-Hill's liability for any damages to you or any other person ever exceed the lower of suggested list price or actual price paid for the license to use the Product, regardless of any form of the claim.

THE McGRAW-HILL COMPANIES, INC. SPECIFICALLY DISCLAIMS ALL OTHER WARRANTIES, EXPRESS OR IMPLIED, INCLUDING BUT NOT LIMITED TO, ANY IMPLIED WARRANTY OF MERCHANTABILITY OR FITNESS FOR A PARTICULAR PURPOSE. Specifically, McGraw-Hill makes no representation or warranty that the Product is fit for any particular purpose and any implied warranty of merchantability is limited to the sixty day duration of the Limited Warranty covering the physical diskette(s) only (and not the software or in-formation) and is otherwise expressly and specifically disclaimed.

This Limited Warranty gives you specific legal rights; you may have others which may vary from state to state. Some states do not allow the exclusion of incidental or consequential damages, or the limitation on how long an implied warranty lasts, so some of the above may not apply to you.

This Agreement constitutes the entire agreement between the parties relating to use of the Product. The terms of any purchase order shall have no effect on the terms of this Agreement. Failure of McGraw-Hill to insist at any time on strict compliance with this Agreement shall not constitute a waiver of any rights under this Agreement. This Agreement shall be construed and governed in accordance with the laws of New York. If any provision of this Agreement is held to be contrary to law, that provision will be enforced to the maximum extent permissible and the remaining provisions will remain in force and effect.